RESCUED BY

ASP

ACTIVE SERVER PAGES AND ASP.NET

ROB FRANCIS

Kris Jamsa, PhD, MBA,
Technical Editor

04-0581

THOMSON
DELMAR LEARNING

Australia • Canada • Mexico • Singapore • Spain • United Kingdom • United States

THOMSON
DELMAR LEARNING

OnWord Press

RESCUED BY ACTIVE SERVER PAGES AND ASP.NET
Rob Francis

Business Unit Director: Alar Elken	**Channel Manager:** Fair Huntoon	**Senior Project Editor:** Christopher Chien
Executive Editor: Sandy Clark	**Marketing Coordinator:** Karen Smith	**Editorial Assistant:** Jennifer M. Luck
Senior Acquisitions Editor: Gregory L. Clayton	**Executive Production Manager:** Mary Ellen Black	**Art/Design Coordinator:** David Arsenault
Senior Development Editor: Michelle Ruelos Cannistraci	**Production Manager:** Larry Main	**Full Production Services:** Liz Kingslien
Executive Marketing Manager: Maura Theriault		

NOTICE TO THE READER

Publisher does not warrant or guarantee any of the products described herein or perform any independent analysis in connection with any of the product information contained herein. Publisher does not assume, and expressly disclaims, any obligation to obtain and include information other than that provided to it by the manufacturer.

The reader is expressly warned to consider and adopt all safety precautions that might be indicated by the activities herein and to avoid all potential hazards. By following the instructions contained herein, the reader willingly assumes all risks in connection with such instructions.

The Publisher makes no representation or warranties of any kind, including but not limited to, the warranties of fitness for particular purpose or merchantability, nor are any such representations implied with respect to the material set forth herein, and the publisher takes no responsibility with respect to such material. The publisher shall not be liable for any special, consequential, or exemplary damages resulting, in whole or part, from the readers' use of, or reliance upon, this material.

myValue = 7

CONTENTS

Dedication

To my grandparents—one and all—residing in the next world.

Acknowledgments

First, I would like to thank my wife, Shahla, who supported me throughout the writing of this book. I always feel better after talking things through with her. I would also like to thank her parents, Paul and Kaye Vessey, for not only giving us a place to stay when we arrived in New Zealand, but also for allowing me to take over their computer until we could get a place of our own.

Special thanks to my good friends Andy Spence and Mark Dingwall, who moved all of our furniture and belongings not once but twice in as many months!

I'd like to thank the folks at Delmar for being so understanding with deadlines, as I moved from one country to another at the beginning of this project. Especially, I would like to thank Kris Jamsa and Liz Kingslien for their support and efforts throughout this project.

Thanks to Homayoun Yazdani for long distance support and encouragement all the way from Canada. Finally, I'd like to add James Ghaeni, who really wanted to be in the acknowledgments of my last book.

n this lesson you will learn all about Active Server Pages from an overview perspective. You will learn not only what Active Server Pages are but how Web pages often use them, as well as how they came into being. You will also learn about what requirements Active Server Pages need for both development and production environments, including browsers and their differences. Finally you will learn what is the best method for using the remainder of this book, depending on your particular background and skills. By the time you finish this lesson, you will understand the following key concepts:

- The acronym ASP refers to Active Server Pages, although it is often and easily confused with the same acronym referring to other technical terminology.

- Active Server Pages let Web pages be dynamic in nature rather than simply being static HTML pages.

- Programmers create Active Server Pages using a combination of HTML and some scripting language that supports ActiveX. The most common scripting languages are VBScript and JavaScript.

- You can run Active Server Pages on a large number of platforms, including Microsoft Windows, Linux, Sun Unix, Novell Netware, Apple Macintosh and others, using software from Microsoft or third parties.

- Dynamic HyperText Markup Language (DHTML) is reliant on the browser, and makes dynamic changes to Web pages after the browser has loaded the page. Active Server Pages, on the other hand, are browser-independent. With an

Active Server Page, the server makes dynamic changes to Web pages before it sends the pages to the browser.

- Some of the many examples of what people often use Active Server Pages for are regular updating of their pages with regard to shopping carts and e-commerce, date and time, personal greetings or options, and also database front ends.
- ASP+, the next version of Active Server Pages after version 3.0, became ASP.NET. Despite being a complete rebuild, rather than a version 4.0, ASP.NET will have complete compatibility with previous Active Server Pages code.
- Active Server Pages are browser-independent, which makes browser compatibility now nothing more than a past issue.
- Beginners should read this book's lessons in order. Those who already know HTML or VBScript can skim over those lessons, but continue reading the book in order. Those who already know Active Server Pages can simply jump to the lessons that interest you.

What Active Server Pages Are—in a Nutshell

Many people, because of similarities with other terms or computer-related technologies, often misunderstand Active Server Pages and what they actually are. The acronym ASP, which refers to Active Server Pages, is often confused with the same acronym referring to Application Service Provider, which is also common computer jargon. Another use of the acronym would have you believe that it stands for Association of Shareware Professionals—more computer jargon. Active Server Pages are Web pages that include code, as you will read later in this lesson. This code is not necessarily always VBScript, but more often than not it is. This code allows the Web pages to be more dynamic in nature or to accomplish feats that would be difficult or impractical using HTML alone. Quite simply, that is all Active Server Pages are.

Understanding the History of Active Server Pages

In the early to mid-'90s, the majority of Web pages were static and unchanging. If 200 people went to a particular page on a particular site, they would all see the same data. If you had to provide frequently changing data on your Web pages, you had your work cut out for you. However, people's interest in the Internet was growing rapidly, and early forms of e-commerce and electronic advertising caught the eye of the giant Microsoft corporation.

Microsoft released Internet Information Server 1.0 around late 1995 but not much notice was taken of it. Those who were using Windows NT® were happy about having a proper Web server that ran natively on their chosen platform, however, there were obvious oversights and missing features in this first version. The most glaring of these was the omission of any real scripting support.

With the release of Internet Information Server 2.0 in February of 1996, Microsoft first unleashed Active Server Pages to an Internet-hungry audience. At this time the word Internet was in almost every conversation and there was a tremendous awareness growth amongst the general population. Microsoft began in earnest its browser war against Netscape. Active Server Pages made it possible to create dynamic Web pages. The release of Internet Information Server 2.0 also brought some native scripting support. You could do all sorts of mathematical calculations with ease. You could now handle queries and interact with users. There was a learning curve associated with Active Server Pages, however, and it was clear that writing good Active Server Pages was not as simple and straight-forward as HTML.

Active Server Pages took off and became hugely popular in a relatively short span of time. Microsoft increased the amount of documentation and support for Active Server Pages, as well as releasing new versions to cater to growing needs and demands.

Using Different Languages and Platforms

You can write Active Server Pages in a variety of languages—but you will have to know the fundamentals of HTML (HyperText Markup Language) in combination with another language that Active Server Pages supports. Probably the most popular language that meets this criteria is VBScript, although you can use JScript (JavaScript), which also has a sizeable following. This book will concentrate on using VBScript to create Active Server Pages.

Although they are not widely used, there is some support for creating Active Server Pages in Perl, Rexx, or Python as well. Actually, if you are good with Perl and comfortable on Linux, then you may want to consider writing Active Server Pages for Apache Web servers. Linux has definitely made a name for itself as a solid Web server and it is not uncommon to have to support both platforms, Linux and Windows, these days. At the time of publication, you can get free software (GNU) and support for writing Active Server Pages on Apache at the following URL:

http://www.nodeworks.com/asp/

Another alternative approach to writing Active Server Pages is to try Sun Chili!Soft ASP, a commercial product that allows you to write Active Server Pages on a variety of different platforms. Some of the platforms that it supports are Windows, Linux, Solaris, HP-UX and AIX. At the time of publication, you can get Sun Chili!Soft ASP information at the following URL:

http://www.chilisoft.com/

Similar to Sun Chili!Soft, another product you might want to look at for hosting your Active Server Pages on other platforms is Instant ASP from Halcyon Software. Instant ASP will run on any Java-enabled Web server or application server, from Novell Netware to Oracle, from Apple Macintosh to Lotus Domino. The list of platforms that Instant ASP supports is large indeed. At the time of publication, you can obtain more information about Instant ASP from the following URL:

http://www.halcyonsoft.com/products/iasp.asp

The fact that Active Server Pages has been brought to so many platforms shows clearly how popular and widely used it has become. This book concentrates on writing Active Server Pages for Windows platforms.

Differentiating Between DHTML and Active Server Pages

The acronym DHTML stands for Dynamic HyperText Markup Language and is sometimes simply spoken of as dynamic HTML. It is important to understand the difference between DHTML and Active Server Pages, as both claim to provide dynamic Web pages but are in fact two different sets of technologies.

DHTML was introduced by Microsoft with the release of Internet Explorer 4 and has been improved in subsequent releases. As this may already seem to imply, DHTML is reliant on the browser. DHTML allows you, as the programmer, to create Web pages that change dynamically after they have been loaded by the user's browser.

Active Server Pages, on the other hand, are browser-independent. An Active Server Pages server translates the pages into HTML right there at the server before being sent to the user's browser. The dynamic Web page changes occur before the user's browser gets the page.

So the fundamental difference between the two technologies is that Active Server Pages puts the workload on the server and supports any browser, while DHTML requires a specific version of browser, not to mention support files, on the remote user's computer and puts the processing onus squarely on the remote computer. Both technologies are good and have their place in the grand scheme of things.

Some Examples of Active Server Pages

At this early stage, it is good to have an appreciation of what Active Server Pages can do. For this reason it is helpful for you to see some examples of what people often use Active Server Pages for without concerning yourself with the details of how you accomplish such feats. As you progress through this book you will soon grasp the finer points of how to write Active Server Pages.

One of the popular reasons to use Active Server Pages is for shopping carts and online e-commerce sites, such as the one shown in Figure 1.1. Notice the URL has a default .asp after the domain name, which indicates to you the use of Active Server Pages.

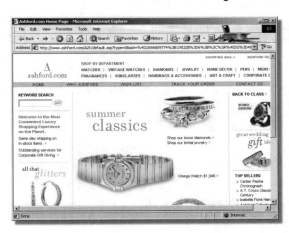

Figure 1.1. A real world example of Active Server Pages powering a shopping cart

Microsoft uses Active Server Pages within its own massive site, as you can see in Figure 1.2. You can easily imagine the amount of constantly changing information and data that they have to deal with, not to mention advertisement rotations. Also, you can use Active Server Pages for providing a personal experience at a site, such as when listing applicable software updates for the user's system or listing items in areas of particular interest to the user.

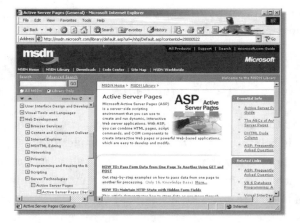

Figure 1.2. Microsoft's own site, showing ever-changing data with the help of Active Server Pages

Although you would not usually publish an employee database on the Web, nevertheless it is a highly popular practice to use Active Server Pages technology within company intranets to support many databases. Active Server Pages provide a convenient front end for databases, giving them not only a Web interface but also allowing such features as only being changed by authorized people and only providing database links to those who should see them. There are many different uses for databases, of course, so a lot of Active Server Pages deal with database interaction both within intranets and on the Internet itself. Figure 1.3 shows an example employee database managed by Active Server Pages.

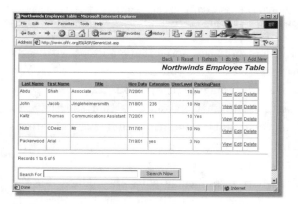

Figure 1.3. An example employee database managed by Active Server Pages

Another very common use for Active Server Pages is its ability to handle frequently changing data with ease. Frequently changing data is such data as the time or the date. Imagine having to update the home page of every site you manage first thing each morning to set the current date. No thanks! In Figure 1.4 you can see another real world example of Active Server Pages, with lots of frequently changing data including the date.

Figure 1.4. Another real world example of handling frequently changing data, such as the date

Conquering ASP+ and ASP.NET

As you will no doubt be aware, the common abbreviation for Active Server Pages is ASP. You will see many references to ASP 2.0 and ASP 3.0, which is what you will be covering in this book. You will also see lots of references to ASP+ and ASP.NET and may be wondering what they are and how they fit into the whole Active Server Pages picture.

Microsoft first started to call their new version of Active Server Pages, ASP+. As the marketing ball began to roll, the name ASP+ became ASP.NET. ASP.NET will become available with the release of Visual Studio.NET. This was not going to be simply version 4.0 of what was already in popular use, but rather would be a complete rebuild from the ground up. Despite being written completely fresh, ASP.NET will have compatibility with previous Active Server Pages code. This means that all the skills that you learn from today's Active Server Pages, such as what you will learn in this book, you can continue to use and build on in tomorrow's Active Server Pages. In fact, you will be in a better position than if you only knew ASP.NET because otherwise you would not understand the existing code that exists in thousands of Active Server Pages sites.

So as not to confuse the issue of compatibility, it is important to realize that there are differences between the current Active Server Pages and ASP.NET. ASP.NET will coexist happily with Active Server Pages so the two technologies can work side by side and both function perfectly well. You can tell the difference between an Active Server Page and one of the new ASP.NET pages without having to examine the underlying code simply by the file extension. Active Server Pages end in .asp, while the new ASP.NET pages end in .aspx.

Most differences between ASP.NET and classic Active Server Pages have to do with changes in the Visual Basic language itself, the Request object, the method of declaring subroutines and functions, the replacing of the Variant type with the Object type, and similar technical nuances. Probably the biggest change from classic Active Server Pages to ASP.NET will be the fact that ASP.NET has done away with VBScript in favor of the full-blown development language Visual Basic.NET. You can also use the new language C# or JScript.

Using Browsers with Active Server Pages

There are many different browsers for the Internet and many different versions of each type of browser. It would not be practical for all of these browsers and their multiple versions to be covered for every reference and example in this book. Unless explicitly stated, any reference to browsers or their usage will be to Microsoft Internet Explorer version 5 or later. Unless you are using an extremely old browser that does not support scripting, you should find that everything that you learn will apply to your particular browser with little if any differences from the descriptions in this book.

On the topic of browsers and Active Server Pages, you might be wondering about the dependencies that Active Server Pages have where browsers are concerned. In most cases it is fair to say that Active Server Pages are browser-independent. By this it is meant that the different versions of browsers that users may have will not affect their being able to view your Active Server Pages. So whether they have Netscape Navigator, Opera, or Microsoft Internet Explorer—whether they have version 1 or version 6—it should no longer have to be a concern.

The exception to this general claim of browser independence is if portions of the script outside of the Active Server Pages code makes use of any browser dependent features. As you are already aware, Active Server Pages are made up of HTML and scripting languages, such as VBScript. Although the VBScript may be inside the Active Server Page sections and therefore completely handled by the server, the HTML may include code for frames, DHTML or some other feature that is browser-dependent. This does not mean that Active Server Pages are not browser-independent. It does mean, though, that an Active Server Page might appear to be browser-dependent because of some other element outside of the Active Server Page sections, yet sharing the same page.

Using This Book

Understandably, Active Server Pages attracts many people from various backgrounds. Although this book covers Active Server Pages right from the beginning and assumes no previous knowledge, it may be beneficial for some people to obtain other books or sources of reference to aid them while those who already have some experience may want to quickly skim over some of the early lessons.

For complete beginners, this book should be read in order to allow you to build a foundation upon which you can continue to add new skills. By the time you reach the end of the book, you will be not only knowledgeable in Active

Server Pages but also quite competent at putting that knowledge into practice. Although this book will cover the basics of HTML (HyperText Markup Language) and VBScript, you might feel that a more in-depth knowledge is desirable. If this is true in your particular case, then there are many entire books devoted solely to this purpose, not to mention a large amount of material readily available on the Internet.

For those who know HTML but not VBScript, simply skim over the next lesson to make sure that you are aware of all the main points or treat it as a brief refresher course. Then, start off more intently with Lesson 3, "Covering VBScript Basics Part 1."

For those who know VBScript but not HTML, start off with the next lesson, "Covering HTML Basics," and then simply skim over the lesson on VBScript before continuing through the rest of the book in order.

For those who know both HTML and VBScript, you can simply skim over Lessons 2 and 3 before starting more intently at Lesson 5, "Installing an Active Server Pages Server." Continue through the rest of the book in order.

Finally, for those of you who have some experience at Active Server Pages already, simply make use of the book by referring to the particular lesson that you are interested in learning about or that you want to brush up your skills on. Make reference to both the index at the back of the book, as well as the table of contents at the front to quickly find what you are after.

Table 1.1 summarizes the various options for using the book according to individual ability and existing knowledge.

HTML	VBScript	ASP	Description
-	-	-	Read the entire book in order.
✓	-	-	Skim Lesson 2. Start from Lesson 3.
-	✓	-	Start from Lesson 2. Skim Lessons 3 and 4.
✓	✓	-	Skim Lessons 2, 3, and 4. Start from Lesson 5.
✓	✓	✓	Read the lessons that interest you.

Table 1.1. Summary of book usage for individuals; check marks indicate areas of existing knowledge

WHAT YOU MUST KNOW

Active Server Pages allows for your Web pages to be dynamic in nature rather than simply being static HTML pages. Active Server Pages are written as a combination of HTML and a scripting language, such as VBScript or JavaScript. Active Server Pages are browser-independent, which makes concerns over browser compatibility a thing of the past. In Lesson 2, "Covering HTML Basics," you will look at HTML more closely and learn the fundamental points for using it to give you a good foundation for programming Active Server Pages. Before you continue with Lesson 2, however, make sure you understand the following key concepts:

- The acronym ASP refers to Active Server Pages, although it is often and easily confused with the same acronym referring to other technical terminology.

- Rather than simply being static HTML, Active Server Pages allows for your Web pages to be dynamic in nature .

- Active Server Pages are written as a combination of HTML and a scripting language. The most common scripting languages are VBScript and JavaScript.

- You can run Active Server Pages on a large number of platforms including Microsoft Windows, Linux, Sun Unix, Novell Netware, Apple Macintosh and others using software from Microsoft or third parties.

- Dynamic HyperText Markup Language (DHTML) is reliant on the browser and makes dynamic changes to Web pages after they have been loaded by the browser. Active Server Pages, on the other hand, is browser-independent and makes dynamic changes to Web pages before they are sent to the browser.

- Active Server Pages are often used by people for regular updating of their pages with regard to shopping carts and e-commerce, date and time, personal greetings or options, and also database front ends.

- ASP+, the next version of Active Server Pages after version 3.0, is now known as ASP.NET. Despite being written from the ground up, ASP.NET will have complete compatibility with previous Active Server Pages code by allowing simultaneous side-by-side functionality.

- Active Server Pages are browser-independent, which makes browser compatibility concerns nothing more than a past issue.

COVERING HTML BASICS

n Lesson 1, "Introducing Active Server Pages," you learned all about what Active Server Pages actually are, as well as getting a fair idea of the sort of practical uses for which programmers apply Active Server Pages. You also learned that to create Active Server Pages, programmers mix HTML (HyperText Markup Language) with other code, such as VBScript. In this lesson, you will take a closer look at HTML and learn the fundamental points that will give you a good foundation for programming Active Server Pages. By the time you finish this lesson, you will understand the following key concepts:

- You should have an understanding of the fundamentals of HTML pages before you start on Active Server Pages.
- You can provide a structured format when you create HTML pages by enclosing your code in an <html> section, within which you add a <head> section and a <body> section.
- Each HTML page should contain a title, which you can easily add through use of the <title> tag. The browser will display the page's title within the title bar.
- There are six levels of heading sizes you can use in HTML pages by using the heading tags <h1>, <h2>, and so on, through <h6>.
- Within an HTML document, you create a paragraph by enclosing sections of text within the <p> tag and the </p> tag. Within an HTML page, you can have multiple paragraphs.

- To emphasize certain portions of your HTML page's text, you can use the <em> and </em> tags, which basically italicize the target text. HTML also supports the <b> and </b> tags which turns bold text highlighting on and off.
- Using the <br> tag you can force line breaks in your text, which is useful when you must maintain some formatting style and do not want the text continuing on the same line.
- Within an HTML document, you can use a nonbreaking space ( ) in place of a normal space character when you want to keep words together even though a space character separates them.
- When you enclose text between the <pre> and </pre> tags, the browser will treat the text as "preformatted."
- You can add many special characters, such as acutes, graves, and cedillas, to your HTML pages by using built-in code specifications.
- HTML supports three different types of lists—unordered lists, ordered lists and definition lists. Unordered lists and ordered lists correspond to bulleted and numbered lists respectively.
- To create links to other documents within your HTML pages, you place the Web address (URL) of the target page within the <a> tag, using the href attribute.
- You can also make links to places within the same page by creating a named reference point and a reference that links to it.
- To add graphics and images to your HTML pages, you use the <img> tag. Image files can be *.jpg, .gif* or *.png* files.
- By using the align attribute, you can flow text around your images.
- Within an HTML document, tables can assist you in page layout problems, as well as provide natural grids for tabular data displays. The main tags that you use with tables are <table>, <th>, <tr> and <td>, as well as their corresponding end tags.

Introducing HTML

HTML is the basis of most Web pages, and Active Server Pages are really no exception. You should have a good understanding of the fundamentals of static HTML pages before you try to program dynamic pages. There are many programs available that can create HTML pages for you. Such programs are wonderful if you do not intend to write your own pages. The problem with these programs is that their generic nature causes them to produce "bloated code" (HTML files that use more statements than necessary to achieve a result). Further, you cannot write crisp Active Server Pages by relying on third-party programs to do all the coding for you.

Learning from Others

If you have access to some programs that produce HTML code for you, you can still use these programs to aid in learning to program HTML for yourself. Using the program, for example, you can create a page that contains a feature that you want. Then, you can view the HTML the program creates for the page. In this manner you can see what code is required for certain features. You can use the same technique for nice Web pages that you find on the Internet. To view source code for any Web page in your Internet Explorer browser, you can simply select the View menu, Source option. For other browsers you should find a similar option for viewing the source code.

Providing Structured Format

Although it is not strictly necessary for an HTML page, it is often good practice to provide a structured format when you create HTML pages. Structured format comprises three key sections—<html>, <head> and <body>. So you enclose the entire code in an <html> section, within which you add a <head> section and a <body> section, as follows:

```
<html>
<head>
   'Statements
</head>

<body>
   'Statements
</body>
</html>
```

Within the heading section that you indicate using the <head> and </head> tags, you would place tags such as the <title> or other similar tags, that you will learn more about as you progress through this lesson. Almost everything else goes into the body section, which is usually the larger of the two sections.

Creating Your First HTML Page

It is now time for you to create your first HTML page. You do not require a special compiler or expensive software to create an HTML page. Any simple text editor will suffice, such as Notepad, which comes with all versions of Windows. Open Notepad from menus by clicking on the Start menu, Programs, Accessories menu, and then select the item Notepad. Windows will open Notepad with a blank document. Type the following HTML code into Notepad:

```
<html>
<body>
Hello world
</body>
</html>
```

Click on the File menu, Save option, and save your document as *c:\hello.htm*. You must save your HTML documents with either a *.htm* or *.html* extension to indicate to browsers and operating systems that they are HTML documents. Next, open the Windows Explorer and locate the file *hello.htm* which you just created. When you double-click on the file's icon, Windows will load your HTML page into your default browser, where you will see the words, *Hello world*, in the main window. Congratulations! You have just created your first HTML page.

Adding a Title

There are a few basic parts of HTML that you will see in practically every HTML page, such as a title. An HTML page's title appears in the title bar of a user's browser when she is viewing it. If the user decides to make a favorite of your page, or if she decides to save your page to disk, the default prompt for the name of the file will be the page title. For this reason, many companies make the title of their Web pages meaningful, such as the company name.

A title is extremely simple to add to your page. It makes use of the <title> tag to mark the start of the title text and then marks the end of the title text with the </title> tag, as follows:

```
<title>The Greatest Page Ever</title>
```

In the example code above, you could replace the text that comes between the start and end tags with any text you want. The only sections that are mandatory are the <title> tag and the </title> tag. To show the <title> tag in proper context, you would place it in the heading section of your html document, as follows:

```
<html>
<head><title>Another Great Page</title></head>
</html>
```

Adding Different Headings

In a textbook you will see headings, some large to indicate a new chapter or lesson, others smaller to indicate sections within the chapter, yet all guiding you to specific information and making it easier to find topics. In the same manner as you find in a textbook, you should provide headings in your HTML documents to guide users to specific points of information. The various headings that you can have should be set simply by importance rather than a specific font and size. To create a heading of greatest importance you simply use the start tag <h1> and an end tag </h1> with the text of your heading in between the two tags, as shown in the following example:

```
<h1>My Home Page</h1>
```

There are six variations in heading importance that you can use, each of which has its own set of matching start and end tags. You will notice that, for all such tag pairs in HTML, there is a pattern of the end tag appearing almost identically to its corresponding start tag except that it has a forward slash as its second character. Table 2.1 shows the six various headings and the start and end tags associated with each.

Importance	Start Tag	End Tag
Heading 1	<h1>	</h1>
Heading 2	<h2>	</h2>
Heading 3	<h3>	</h3>
Heading 4	<h4>	</h4>
Heading 5	<h5>	</h5>
Heading 6	<h6>	</h6>

Table 2.1. The various headings and their associated tags.

So if you want to create a heading for a section within your page, you might use something similar to the following:

```
<h2>A Smaller Section</h2>
```

Figure 2.1 shows an example of the different sizes that result for each of the various headings. The following statements implement that HTML file *headings.htm*:

```
<head><title>Heading importance</title></head>
<body>
<h1>Heading 1</h1>
<h2>Heading 2</h2>
<h3>Heading 3</h3>
<h4>Heading 4</h4>
<h5>Heading 5</h5>
<h6>Heading 6</h6>
</body>
```

Figure 2.1.
The various levels of heading importance

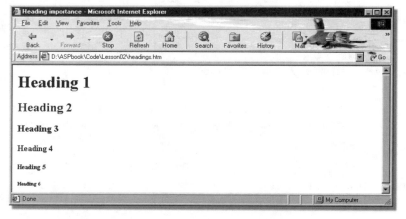

Adding Paragraphs and Emphasis

Just as books organize text on a page within paragraphs, Web pages also often organize text within paragraphs. Using paragraphs in your HTML pages can add clarity for the user by dividing up text into obvious sections. Paragraphs are very simple to add to your pages, too. Like the title and headings, they use a start and end tag pair, namely <p> and </p>. So you would simply use the <p> tag and then follow it with the text of your paragraph before closing it with the </p> tag, as follows:

```
<p>My first paragraph is not very important.</p>
```

You can have multiple paragraphs as well by simply using the paragraph tags in series, one after another, as follows:

```
<p>My first paragraph is not very important.</p>

<p>However, my second paragraph is much more interesting. It
even has more than one sentence!</p>
```

If you want to emphasize certain words in your paragraphs, simply enclose them between an <em> tag and an </em> tag, as follows:

```
<p>However, my second paragraph is <em>much</em> more
interesting. It even has more than one sentence!</p>
```

Basically, you have simply made the emphasized word appear in italics when your Web page is shown. Figure 2.2 shows the result of adding emphasis to a word. The following HTML statements implement the *emphasis.htm* file shown in Figure 2.2:

```
<html>
<head>
<title>Showing Emphasis</title>
</head>

<body>
<h2>Showing Emphasis</h2>
<p>My first paragraph is not very important.</p>
<p>However, my second paragraph is <em>much</em> more
interesting. It even has more than one sentence!</p>
</body>
</html>
```

Figure 2.2.
The result of adding emphasis to a word

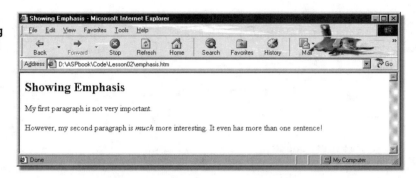

You might also like to add emphasis to the text in your HTML pages by making some of it bold. By surrounding text with a <b> tag and a </b> tag, the browser will display that text in bold. You can also use the <strong> and </strong> tags for a similar effect, or just be aware of them if you see these tags in someone else's pages. Like the <em> tag, surrounding text with an <i> tag and an </i> tag causes the browser to display that text in italics. The following code shows an example of using more of the text style tags:

```
<html>
<head><title>More Text Styles</title></head>

<body>
<p>This <em>word</em> has emphasis.
This <i>word</i> is in italics.
This <b>word</b> is in bold.
This <strong>word</strong> is strong.</p>
</body>
</html>
```

Forcing Line Breaks

One of the original ideas behind HTML was that you did not have to worry about typesetting, text positioning or any other layout issues, as the client's browser would handle all of that. However, there were many specific looks for which more control of the final layout was required by Web page designers. Forcing line breaks is not something that you will probably need often, but on those occasions where you do not want the text to continue on the same line, such as for addresses, you can use the <br> tag to do so. The following example shows an address in a single paragraph, using forced line breaks:

```
<p>Mr. Mickey Mouse<br>
1 Cartoon Drive<br>
Disneyland</p>
```

Note the lack of an end tag when forcing line breaks. This is because the *br* element is an *empty* element, unlike the other elements that you have seen so far. So you will never see </br> in an HTML document, as it is not necessary.

Using NonBreaking Spaces

When you have long sentences or multiple sentences in a paragraph, the browser will automatically wrap the text in the viewable area so the user can see everything. However, there are certain times when you want to keep words together, even though a space character separates them. In these cases you can use a nonbreaking space in place of a normal space character. The users will see a space character when they view the page in a browser. You indicate a nonbreaking space character by placing * * in your HTML page.

For instance, if the company that you were creating Web pages for is *Catchy Chicken*, they may not want their name split on separate lines. The following example shows how you might use a nonbreaking space character:

```
<p>Another incredibly useful service provided by
Catchy Chicken is five levels of <em>free</em> parking
right beside our mega toy store, saving you time and hassle
when you shop with us.</p>
```

Preformatting Text

Rather than abusing the nonbreaking space character by placing it between every single word in a line when you do not want that line to automatically wrap, you should specify that you want to use preformatted text instead. The *pre* element has a start tag <pre> and an end tag </pre>, which you can use to enclose the section that you do not want the browser to wrap. The following code shows an example of using preformatted text:

```
<pre>
<p>This could be a long line of source code.<br>
And this might be another one.<br>
And this could be the final line.</p>
</pre>
```

The preformatted text is usually in a monospace font, such as Courier, and it will not wrap the lines of code regardless of how small the users make their browser.

Adding Special Characters

There are many special characters that you may want to add when creating Web pages using HTML. These may include characteristics such as an acute, grave, tilde or others. Table 2.2 shows you a list of some of the more common special character requirements, although for the entire list you would need to refer to the HTML 4 specification. Where an asterisk is shown in code in Table 2.2, you can either omit the asterisk completely or replace it with the character to which you want to apply the characteristic.

Characteristic	Code	Example
Acute	&*acute;	Á
Grave	&*grave;	À
Circumflex	&*circ;	Â
Tilde	&*tilde;	Ã
Umlaut	&*uml;	Ä
Ring	&*ring;	Å
Cedilla	&*cedil;	Ç
Divide	&divide;	÷
Times	&times;	×
Less than	&lt;	<
Greater than	&gt;	>
Ampersand	&amp;	&
Em dash	&#8212;	—
Quotation mark	&quote;	"
Copyright	&copy;	©
Registered trademark	&reg;	®
Trademark	&#8482;	™

Table 2.2 Some commonly used special characters.

The following code example shows you how to apply two of the special characters in your code, both in a heading and in a paragraph, with the output from the code shown in Figure 2.3:

```
<html>
<title>Special Characters</title>
<h1>The Bah&aacute;'&iacute; Faith</h1>
<p>The Bah&aacute;'&iacute; Faith started in Persia.</p>
</html>
```

**Figure 2.3.
Displaying two
special characters**

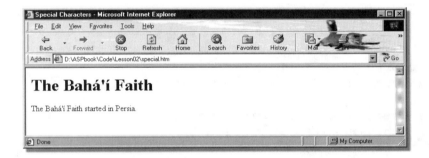

Creating Lists

In HTML there are three different types of lists—unordered lists, ordered lists and definition lists.

Unordered lists are simply a list of points or items that start with a bullet. You would use an unordered list when you want to show various options or points that have no order requirement. The *ul* element is used for unordered lists and it has a start tag <ul> as well as an end tag </ul>. In addition to the *ul* element, you must make use of the *li* element to indicate each list item. List items have both a start tag <li> and also an end tag </li>, however the end tag is optional. The following example shows how you might use an unordered list:

```
<h2>Unordered Lists</h2>
<p>There are <em>three</em> kinds of lists in HTML.</p>
<ul>
  <li>Unordered lists</li>
  <li>Ordered lists</li>
  <li>Definition lists</li>
</ul>
```

Ordered lists have a start tag <ol> and an end tag </ol>, and also use the <li> tag to indicate list items. Ordered lists are very similar to unordered lists, except that they produce numbers, not bullets. This means that an ordered list would be suitable for a set of instructions, as shown in the following example:

```
<h2>Ordered Lists</h2>
<p>Instructions for <em>two minute noodles</em>.</p>
<ol>
  <li>Boil water and pour into cup</li>
  <li>Empty noodles into cup</li>
  <li>Wait for two minutes</li>
</ol>
```

Definition lists allow you to provide a term along with a corresponding definition. A definition list has a start tag <dl> and also an end tag </dl>. Unlike the other two lists, a definition list does not use the *li* element to indicate its list items. The definition term has a start tag <dt> and an end tag </dt>, and the definition itself has a start tag <dd> and an end tag </dd>. The end tags for both the definition term </dt> and also the definition </dd> are both optional and can be safely omitted, as shown in the following example:

```
<h2>Definition Lists</h2>
<p>Definitions for list <em>elements</em> in HTML.</p>
<dl>
  <dt>ul<dd>Unordered list
  <dt>ol<dd>Ordered list
  <dt>li<dd>List item
  <dt>dl<dd>Definition list
  <dt>dt<dd>Definition term
  <dt>dd<dd>Definition
</dl>
```

Figure 2.4 shows what each of the three lists look like when you look at them in a browser.

**Figure 2.4.
The three types
of lists when viewed
in a browser**

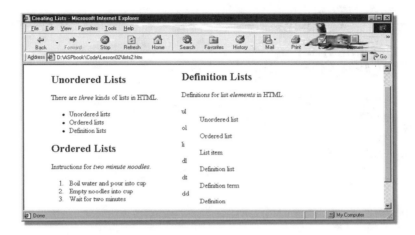

The following HTML statements implement the *lists2.htm* file shown in Figure 2.4:

```
<html>
<head><title>Creating Lists</title></head>

<body>
<table border="0" cellspacing="50">
<tr><td>
<h2>Unordered Lists</h2>
<p>There are <em>three</em> kinds of lists in HTML.</p>
<ul>
   <li>Unordered lists</li>
   <li>Ordered lists</li>
   <li>Definition lists</li>
</ul>

<h2>Ordered Lists</h2>
<p>Instructions for <em>two minute noodles</em>.</p>
<ol>
   <li>Boil water and pour into cup</li>
   <li>Empty noodles into cup</li>
   <li>Wait for two minutes</li>
</ol>
</td>
```

```
<td>
<h2>Definition Lists</h2>
<p>Definitions for list <em>elements</em> in HTML.</p>
<dl>
  <dt>ul<dd>Unordered list
  <dt>ol<dd>Ordered list
  <dt>li<dd>List item
  <dt>dl<dd>Definition list
  <dt>dt<dd>Definition term
  <dt>dd<dd>Definition
</dl>
</td></tr>
</table>
</body>
</html>
```

Adding Links to Other Pages

Of course, one of the most wonderful elements of HTML pages is the ability to click a link that will take you off to another page. Imagine how much more annoying it would be if you had to actually type the URL (Uniform Resource Locator) or address of each and every page that you want to view. The Internet would not be nearly as popular as it is today.

To add links in your own Web pages, you must use the <a> tag. To help remember this element, you might like to think of *a* for *a*ddress. The link also requires an end tag </a>. Between the two tags you type the wording for the link, which usually appears in the browser as underlined blue text. Inside the start tag <a> you must add the reference to the page you want to link to. You add this reference by using the *href* attribute, and then provide the path of the page you want to link to. For example, if you have a page that resides in the same directory that you want to make a link to, you might do something similar to the following:

```
<a href="contacts.htm">Company contacts</a>
```

If the page that you are linking to is at another site, you must provide the full address or URL, as shown:

```
<a href="http://www.abc.com/contacts.htm">ABC Company contacts</a>
```

Note: *Many books and magazines refer to the reference to another page as the* ***link target.***

Adding Links to Areas in the Same Page

Besides linking to other pages, you can also make links to places within the same page. You might want to do this if you have a long page with a list or table of the various sections at the top of the page, such as a FAQ (Frequently Asked Questions) page. To create links to areas in the same page, you basically need two parts—a named reference point and a reference that links to it.

To create a named reference point or target name, as it sometimes known, you still use the <a> tag and the </a> tag. However, you do not use the *href* attribute but the *name* attribute instead. For instance, you might have a section with the heading, *Getting More Information*, and you want to add a link to it at the top of your page. Within the heading tags you would add a named reference point, as follows:

```
<h3><a name="more-info">Getting More Information</a></h3>
```

Now, back at the top of the page where you want to make the actual link, you must specify this target name with a # character directly preceding it, as follows:

```
<a href="#more-info">Getting More Information</a>
```

Working with Images

You can easily add graphics and images to your HTML pages using the <img> tag, which is an *empty* element. When using an <img> tag you must first specify the image source using the *src* attribute, and follow this with the path and name of the image file. Image files can be *.jpg, .gif* or *.png* files. To add an image by the name of *fancy_title.png* to a page, for example, you would use an <img> tag similar to the following:

```
<img src="fancy_title.png">
```

You should also use the width and height attributes to specify the width and height of the image in pixels. If you were to add the width and height of the image to the previous example, it would like something similar to the following:

```
<img src="fancy_title.png" width="400" height="60">
```

For those browsers that are unable to display your image, you can provide an alternative description by using the *alt* attribute, as follows:

```
<img src="fancy_title.png" width="400" height="60" alt="Fox
Hunting Mouse">
```

In this case, if the browser is unable to display the image, it would refer to the alt attribute and display the text, Fox Hunting Mouse. The text that you assign to the alt attribute is the text that will appear to the user in place of the image that he cannot see.

Knowing Which Image Type to Use

*Sometimes it is difficult to know which image format you should choose when preparing images for Web pages. Generally, you should try and use **.jpg** files for quality graphics with color gradation, such as a photograph of rolling green hills. However, lower-resolution graphics or images that use a limited number of colors, such as fancy text on a white background, should be **.gif** or **.png** format. There has been a general shift away from using **.gif** files because of liability issues, but you can use **.jpg** and **.png** image formats without such concerns.*

Flowing Text Around Images

You can flow text around your images by using the *align* attribute. You can use the *align* attribute with left or right bias for any image, and the text will flow around on the opposite side. Using the *align* attribute with a center bias results in non-flowing text that is awkwardly spaced; you will probably want to avoid this in your pages. The following code shows an example of how you might set the *align* attribute in your pages. Figure 2.5 shows a good example of an image with left alignment, and how the text flows around it:

```
<title>Using this Book</title>
<h1>Using this Book</h1>
<p>Understandably, Active Server Pages attracts...
<img src="books.jpg" align="left">
...to quickly find what you are after.</p>
```

**Figure 2.5.
Text flowing
around an image**

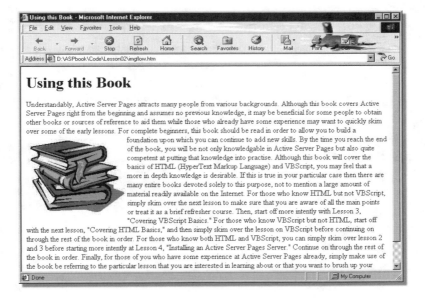

Using Tables

You can use tables in a wide variety of ways in your HTML pages, although you will only learn the basics for now. Tables can assist you with page layout problems, as well as provide natural grids for tabular data displays. You use the start tag <table> and then end tag </table>. Inside your table you assign multiple table rows using the <tr> and </tr> tags. Within each table row you can make use of table headings and table data. Table headings use the start tag <th> and the end tag </th> while table data use the start tag <td> and the end tag </td>. So a simple table might look like the following:

```
<title>Simple RGB Table</title>
<h2>Simple RGB Table</h2>
<table>
<tr><th>Color</th><th>R</th><th>G</th><th>B</th></tr>
<tr><td>Black</td><td>0</td><td>0</td><td>0</td></tr>
<tr><td>Red</td><td>255</td><td>0</td><td>0</td></tr>
<tr><td>Green</td><td>0</td><td>255</td><td>0</td></tr>
<tr><td>Blue</td><td>0</td><td>0</td><td>255</td></tr>
<tr><td>White</td><td>255</td><td>255</td><td>255</td></tr>
</table>
```

Creating Empty Cells in Tables

When you need to create an empty cell in a table, do not leave the table data tags with nothing between them, as follows:

```
<td></td>
```

but rather, use a nonbreaking space character, as follows:

```
<td> </td>
```

You can add borders to your tables by using the *border* attribute, as follows:

```
<table border="10">
```

The value that you assign to your *border* attribute affects the thickness of the surrounding table border. Make your *border* attribute equal to *0* if you want a borderless table. A borderless table is useful for positioning graphics or text in an HTML page without giving the appearance of a table to users.

You can change the amount of padding inside each table cell using the *cellpadding* attribute. You can also change the amount of space between adjacent cells using the *cellspacing* attribute. The two attributes are set as follows, although you might assign different values:

```
<table border="5" cellpadding="10" cellspacing="20">
```

Finally, you can also set the background color of individual cells by using the *bgcolor* attribute. The *bgcolor* attribute takes a hexadecimal number to represent an RGB color value. In a hexadecimal number, the range of values per character goes from *0* to *f*, where f is 15. For instance, the hexadecimal number C7 in decimal terms would be 199. In an RGB value, the first two characters represent the red component (C7), the second two represent the green (C7), and the third two represent the blue (C7), as follows:

```
<th bgcolor="#C7C7C7">Color</th>
```

Figure 2.6 shows a variety of tables and table attributes, and generally depicts graphically the main areas that you now know concerning tables in HTML.

**Figure 2.6.
A variety of tables
and table attributes**

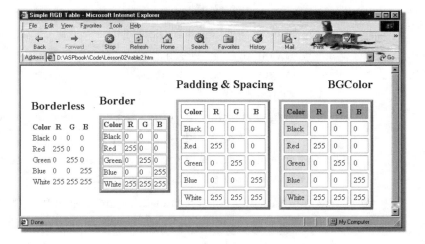

WHAT YOU MUST KNOW

After this lesson, you should have an understanding of the fundamentals of HTML pages which will aid you greatly before you start on Active Server Pages. In Lesson 3, "Covering VBScript Basics Part 1," you will learn the fundamentals of programming in VBScript which, together with HTML, you can then use to create Active Server Pages. Before you continue with Lesson 3, however, make sure you understand the following key concepts:

- Rather than relying on programs that create HTML code for you, understand the fundamentals of HTML pages before you start on Active Server Pages.

- By enclosing your code in an <html> section, within which you add a <head> section and a <body> section, you can provide a structured format when you create HTML pages.

- Through use of the <title> tag and </title> tag, you can easily add a title to each HTML page.

- Simply by using the appropriate heading tags, you can use six various levels of heading sizes in your HTML pages.

- By enclosing various sections within a <p> tag and a </p> tag, you can create multiple paragraphs.

- To emphasize certain portions of your HTML page's text, you can use the <em> and </em> tags, which basically italicize the target text.

- When you do not want the text continuing on the same line, you can use the <br> tag to force line breaks in your text.

- When you want to keep words together even though a space character separates them, you can use a nonbreaking space ( ) in place of a normal space character.

- Preformatting text using the <pre> tag and the </pre> tag stops the browser from automatically wrapping the lines of target text.

- By using built-in code specifications, you can add many special characters, such as acutes, graves, and cedillas, to your HTML pages.

- There are three different types of lists—unordered lists (bulleted), ordered lists (numbered), and definition lists (term and definition).

- Using the <a> tag and the </a> tag together with the *href* attribute, you can add address links in your HTML pages.

- To make links to places within the same page, you can create a named reference point and a reference that links to it.

- Add graphics and images to your HTML pages using the <img> tag. Image files can be *.jpg*, *.gif* or *.png* files.

- You can flow text around your images by using the *align* attribute.

- Tables can assist you in page layout problems, as well as provide natural grids for tabular data displays. The main tags that you use with tables are <table>, <th>, <tr> and <td>, as well as their corresponding end tags.

COVERING VBSCRIPT BASICS, PART 1

 n Lesson 2, "Covering HTML Basics," you learned that programmers combine HTML (HyperText Markup Language) with other code to create Active Server Pages. VBScript and JavaScript are the most common scripting languages that programmers combine with HTML to build Active Server Pages. In this lesson, you will examine the basics of VBScript and cover the fundamentals you will need to be ready to start programming Active Server Pages. By the time you finish this lesson, you will understand the following key concepts:

- VBScript is short for Visual Basic Scripting. It is a subset of the popular Visual Basic language, also from Microsoft.
- When a script executes, the script can store information within a variable that corresponds to a named location in the computer's memory.
- A variable's type specifies the set of values a variable can store (such as integer or floating-point values), and the set of operations the code can perform on the variable (such as addition and multiplication).
- In VBScript, variables are always type Variant. To declare a variable within VBScript, you use the Dim statement to specify the variable's name.
- To prevent errors with variable naming, you should place the statement Option Explicit at the start of your script.
- You assign a value to a variable by using the assignment operator (=).
- When you create a script, you should place comments (remarks) within your script that explain the script's processing. That way, should you or another

programmer later have to change the script, you or they can read the comments to understand the processing the script performs.

- To place a comment within VBScript code, you place an apostrophe (') directly before the comment text. Alternatively, you can type in Rem (short for remark) at the start of a line to indicate a comment. You can also place a Rem comment at the end of a line directly after a colon (:).

- You can perform many different operations with values and variables in VBScript using operators from four main categories—arithmetic, comparison, logical, and concatenation.

- Operator precedence specifies the order in which VBScript performs operations. VBScript evaluates arithmetic operators first, then the concatenation operator, next the comparison operators, and finally the logical operators.

- The scope of a variable refers to the range of code from which it is accessible, while the lifetime of a variable refers to how long the variable's value exists.

Introducing VBScript

VBScript is a scripting language from Microsoft that is a subset of the popular Visual Basic language. VBScript is actually short for Visual Basic Scripting, although even the Microsoft documentation and Web site refer to it simply as VBScript. At the time of publication, the following is Microsoft's URL for scripting languages:

http://msdn.microsoft.com/scripting/

You can click on the link for VBScript from here to access Microsoft's user guide and language reference.

Creating Your First VBScript Program

Before you get into some of the finer details of VBScript, you should first simply create a script so that you can see how easy creating scripts actually is. You can then add greater levels of complexity to your scripts as you gain further knowledge of what VBScript is capable of.

For your first program in VBScript you will simply display a dialog box that contains the text message, *Hello, world*. Your first VBScript application is not an Active Server Page. You will not combine the VBScript statements with HTML. Instead, you will simply build a VBScript application that you can run from within Windows.

When programming any VBScript code, you must use some form of text editor to enter the script's statements. You do not require a fancy program to create scripts. All the Windows family of operating systems come with a simple text editor known as Notepad, which you can use to create your scripts. To open Notepad, click on the Start menu, Programs, Accessories item and then select the Notepad option, as shown in Figure 3.1.

**Figure 3.1.
Opening Notepad
from the Start menu**

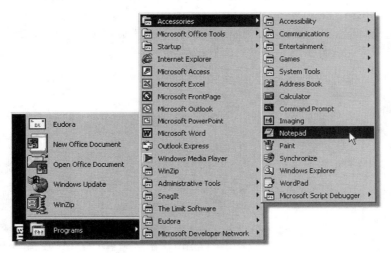

After you select Notepad from the Start menu, Windows will open the Notepad program, which appears as a blank window where you can type your script. For your first script, simply type in the following text:

```
msgbox "Hello, world"
```

Save your script by selecting the File menu, Save As option. Windows, in turn, will open the Save As dialog box in which you can type the name of the file. Within the Save As dialog box, type in *hello.vbs* for the name of the file and then click on the Save button. Your Notepad window should look similar to Figure 3.2.

**Figure 3.2.
Notepad displaying
the VBScript for
hello.vbs**

After saving your VBScript program, use the Windows Explorer to navigate to the directory where you saved *hello.vbs,* and double-click on the *hello.vbs* icon. Windows, in turn, will run your *hello.vbs* script, and a dialog box similar to that shown in Figure 3.3 appears.

**Figure 3.3.
The result of
your first VBScript
program**

Next, within Notepad, modify your script's message to display other text, such as your name, and then save your script and run it again.

Understanding Variables in VBScript

As your scripts become more complex, they must often store information as they execute. To store information, scripts use variables. A variable is a named place in the computer's memory where the script can store and retrieve values as it executes. As its name suggests, a variable's content is subject to variation or change. As your script is running, you use the same variable name even though the program information that the variable contains may change.

A variable's type specifies a set of values a variable can store, and a set of operations a program can perform on a variable. In VBScript, variables are always of type Variant, which at first may sound limiting. However, the Variant can actually represent a number of different data types, as shown in Table 3.1.

Data Type	Description
Boolean	Contains either True or False
Byte	Contains an integer in the range 0 to 255
Currency	Contains a number in the range -922,337,203,685,477.5808 to 922,337,203,685,477.5807
Date (Time)	Contains a number that represents a date between January 1, 100 to December 31, 9999
Double	Contains a double-precision, floating-point number in the range -1.79769313486232E308 to -4.94065645841247E-324 for negative values; 4.94065645841247E-324 to 1.79769313486232E308 for positive values
Empty	Variant is uninitialized. Value is 0 for numeric variables or a zero-length string ("") for string variables
Error	Contains an error number
Integer	Contains integer in the range -32,768 to 32,767
Long	Contains integer in the range -2,147,483,648 to 2,147,483,647
Null	Variant intentionally contains no valid data
Object	Contains an object
Single	Contains a single-precision, floating-point number in the range -3.402823E38 to -1.401298E-45 for negative values; 1.401298E-45 to 3.402823E38 for positive values
String	Contains a variable-length string that can be up to approximately 2 billion characters in length

Table 3.1. The various data types that a Variant can contain

For now, do not worry about all the different data types that a Variant can contain, but rather try to simply be aware that a Variant can, in fact, be many types. Refer back to Table 3.1 whenever you want to check the various types in more detail. But otherwise, you simply declare the Variants in your code and let the computer do all the work for you.

Declaring Variables in VBScript

To declare a variable—which is another way of saying to create a variable—you must use the Dim statement to specify the name that you want to give to your variable. Dim is short for dimension array and has been used in many variations of Basic languages. The following statements use the Dim statement to create (declare) several variables:

```
Dim Value1

Dim apple

Dim RedHerring
```

Within VBScript, you can also declare multiple variables on one line by placing a comma between each successive variable, as follows:

```
Dim orange, apple, banana
```

A variable name in VBScript has some naming restrictions that you must comply with. Namely, it:

- must begin with an alphabetic character
- cannot contain an embedded period
- must not exceed 255 characters
- must be unique within the scope that you declare it

To prevent a mistyped variable name from causing your script to produce strange and possibly difficult to locate errors (bugs), place the Option Explicit statement at the top of your scripts. The Option Explicit statement will make sure that your program does not try to use any variables whose names you have not explicitly declared using a Dim statement.

When you name variables, it is a good idea to use meaningful names that correspond to the actual type of variable or function that it performs. For example, if you are storing the name of the user into a variable you might call that variable username, UserName, usrName, or something similar. You can immediately tell, as can anyone else reading your code, what the variable is supposed to hold.

As another example, you might have to count the total number of entries in a list or some other object. You could name the variable that holds this total totalcount, TotalCount, myTotal or something similar. You will develop your own manner of creating meaningful names. If you already have a set style that you use, such as Hungarian notation (a naming standard that many Microsoft programmers follow), then by all means use that style.

Assigning Values to Variables

After you declare a variable, you can assign a value to it using the assignment operator, which is represented by the equals sign (=) character. The following statements, for example, declare a variable named myValue and then use the assignment operator to assign the value 7 to the variable:

```
Dim myValue

myValue = 7
```

Note that the variable is on the left of the assignment operator and the value that you are assigning to it is on the right. If you have not done any programming before, you might like to think of the assignment operator as meaning either "gets" or "becomes" rather than equals. For instance, if you read the following lines of VBScript and try to work out the value of the variable myValue at the end, you will understand this more clearly:

```
Dim myValue
myValue = 7
myValue = 5
myValue = myValue - 2
```

The value of the variable myValue at the end of this example is 3. If you were to read the lines of code in the same way as a mathematical equation, you would run into problems of inconsistency. The variable myValue cannot equal 7 at the same time as being equal to 5! And there is no possible way that myValue can be equal to itself minus 2! However, by reading the assignment operator as "gets" and remembering that the variable on the left becomes the value on the right, you would see the following pattern of results:

- After line 2 has run, the variable myValue would be 7.
- After line 3 has run, the variable myValue would be 5.
- After line 4 has run, the variable myValue would be its current value (5) minus 2, which is 3.

Adding Comments to Your Scripts

When you start writing more complex scripts, you will soon discover that they become more difficult to read than scripts that consist of only a few lines. Also, you might forget what you were trying to do in your script six months down the road, or you may have to write scripts that others can work with or modify in the future. By adding comments to your scripts, you can prevent a lot of misunderstanding and frustration for others, while making your scripts far more elegant and clear.

To add a comment in VBScript, simply use an apostrophe (') before any text that you want to make into a comment. You can also type in *Rem* (short for remark) instead of an apostrophe. The two comment indicators are interchangeable—except when the comment follows an existing expression. In this case the *Rem* must follow a colon, as you can see in the following few examples:

```
'This whole line is a comment

Rem This whole line is also a comment

CustName = "Martha Vessey" 'input customer name

CustName = "Martha Vessey" : Rem input customer name
```

Using Operators in VBScript

You can perform many different operations with values and variables in VBScript using operators. There are operators for arithmetic, as shown in Table 3.2.

Operator	Meaning	Example	Example Result
+	Addition	x = 3 + 5	x = 8
-	Subtraction (and negation)	x = 10 – 7	x = 3
*	Multiplication	x = 7 * 2	x = 14
/	Division	x = 50 / 2	x = 25
\	Integer division	x = 17 \ 5	x = 3
Mod	Remainder	x = 80 Mod 3	x = 2
^	Exponential	x = 8 ^ 2	x = 64

Table 3.2. Arithmetic operators in VBScript

Although arithmetic operators are probably the most commonly used operators, there are other operators such as comparison operators and logical operators. As your scripts begin to make decisions (using the IF statement), you will use comparison operators on a regular basis. Comparison operators are useful when you want to know whether a variable is greater, less than, or equal to another variable or value. Table 3.3 lists the comparison operators in VBScript.

Operator	Meaning
=	Equal to
<>	Not equal to
<	Less than
>	Greater than
<=	Less than or equal to
>=	Greater than or equal to
Is	Object reference comparison

Table 3.3. Comparison operators in VBScript

Logical operators are useful when you want to check for more than one condition, either of two conditions occurring, or perhaps even a condition not occurring. Using the logical operators, for example, you might test to see whether the user is visiting the site for the first time and the user wants to join the site's mailing list. Table 3.4 lists the logical operators in VBScript.

Operator	Meaning
And	Logical conjunction
Or	Logical disjunction
Not	Logical negation
Xor	Logical exclusion
Eqv	Logical equivalence
Imp	Logical implication

Table 3.4. Logical operators in VBScript

Finally, there is even a concatenation operator (&) that you use in VBScript to join two strings together. The following examples show how to use the concatenation operator with the result of the operation as a comment:

```
myString = "nut" & "shell"    'myString = "nutshell"

Filename = "paper" & ".txt"    'Filename = "paper.txt"
```

You can even use the concatenation operator on variables, as well as multiple concatenation operators in a single line of code, as shown in the following example:

```
Dim a, b, c

a = "apple"

b = "blue"

c = b & " " & a      'c = "blue apple"
```

To quickly recap this section, you have seen that VBScript offers four main categories of operators—arithmetic, concatenation, comparison, and logical.

Operator Precedence

When a VBScript expression contains two or more operators, VBScript operator precedence specifies the order in which VBScript evaluates operators. For example, the following statement uses the addition and multiplication operators:

```
Result = 3 + 7 * 4
```

If VBScript were to perform the addition first, it would calculate the result 40, as shown here:

```
Result = 3 + 7 * 4
Result = 10 * 4
Result = 40
```

Likewise, if VBScript were to perform the multiplication first (which is what VBScript's operator precedence tells it to do), VBScript would calculate the correct result of 31:

```
Result = 3 + 7 * 4
Result = 3 * 28
Result = 31
```

There are occasions when expressions contain operators from different operator categories. In these circumstances, VBScript will evaluate the arithmetic operators first, then the concatenation operators, next the comparison operators, and finally the logical operators.

Overriding Order Precedence

Although understanding order precedence is important, it is also useful to be able to override order precedence in particular situations. You can easily override order precedence in VBScript by using parentheses around certain parts of an expression. VBScript will evaluate the parts of an expression within parentheses before evaluating anything else. Have a look at the following example to see the difference in the same expression when parentheses are used to override the operator precedence:

```
Dim x
x = 7 + 1 * 6 - 2      'x = 11
x = (7 + 1) * (6 - 2)  'x = 32
```

There are quite a few occasions when operators have equal precedence. In these situations, VBScript simply evaluates the operators from left to right. Some examples of equally precedent operators are multiplication and division, addition and subtraction, and finally, all of the comparison operators. Table 3.5 shows the operator precedence for arithmetic and logical operators.

Arithmetic	Logical
Exponentiation (^)	Not
Negation (-)	And
Multiplication and division (*, /)	Or
Integer division (\)	Xor
Remainder (Mod)	Eqv
Addition and subtraction (+, -)	Imp

Table 3.5. Highest to lowest order of precedence for arithmetic and logical operators

Understanding the Scope and Lifetime of Variables

The scope of a variable refers to the range of code from which it is accessible, and it is determined by the place of declaration. The lifetime of a variable refers to how long it exists. You can declare a variable in two places in VBScript—inside a procedure or outside a procedure. In Lesson 4, "Covering VBScript Basics, Part 2," you will examine VBScript procedures in detail.

When you declare a variable inside a procedure, the variable is only available to code within that procedure. Code that is outside of that particular procedure or in other procedures cannot make use of the variable. These variables are known as procedure-level or local variables. The lifetime of a procedure-level variable is from the moment it is declared within the procedure until the moment that the procedure is exited.

When you declare a variable outside a procedure, the variable is available to code anywhere within the entire script. These variables are known as script-level or global variables. The lifetime of a script-level variable is from the moment it is declared until the script itself ends.

Figure 3.4 shows you some code snippets from an earlier example, with the left side depicting in black the scope of the local variables s1, s2, and s3, while the right side depicts in black the scope of the global variable, myValue.

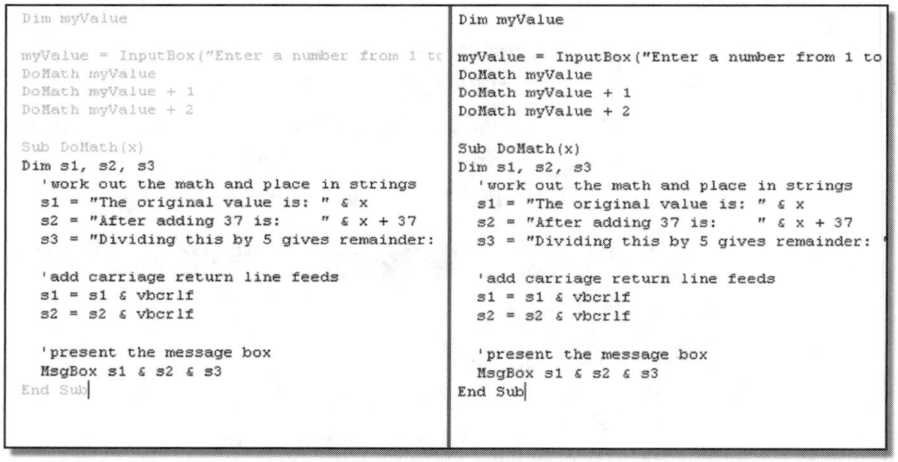

Figure 3.4.
Local variable scope on the left vs. global variable scope on the right

```
Dim myValue

myValue = InputBox("Enter a number from 1 to
DoMath myValue
DoMath myValue + 1
DoMath myValue + 2

Sub DoMath(x)
Dim s1, s2, s3
    'work out the math and place in strings
    s1 = "The original value is: " & x
    s2 = "After adding 37 is:     " & x + 37
    s3 = "Dividing this by 5 gives remainder:

    'add carriage return line feeds
    s1 = s1 & vbcrlf
    s2 = s2 & vbcrlf

    'present the message box
    MsgBox s1 & s2 & s3
End Sub
```

```
Dim myValue

myValue = InputBox("Enter a number from 1 to
DoMath myValue
DoMath myValue + 1
DoMath myValue + 2

Sub DoMath(x)
Dim s1, s2, s3
    'work out the math and place in strings
    s1 = "The original value is: " & x
    s2 = "After adding 37 is:     " & x + 37
    s3 = "Dividing this by 5 gives remainder:

    'add carriage return line feeds
    s1 = s1 & vbcrlf
    s2 = s2 & vbcrlf

    'present the message box
    MsgBox s1 & s2 & s3
End Sub
```

You may think that you will simply make all your variables global or script-level variables rather than local or procedure-level variables—but there are greater memory demands that you should take into consideration. You should aim to make all variables local or procedure-level variables if you can, and only then, when you cannot do what you have to, resort to global or script-level variables.

Aside from the hidden memory demands that global variables have over local variables, global variables can lead to sloppy programming habits that become not only difficult to break, but more importantly, they can be the cause of hard-to-find bugs. When you can use a global variable from anywhere in your script, it can make it difficult to trace a problem.

WHAT YOU MUST KNOW

In this lesson you have seen most of the fundamental concepts of programming VBScript, although there are lots of other features that you will discover as you progress through this book. You understand the importance of variables and also that they can have a limit as far as their scope and lifetime are concerned. In Lesson 4, "Covering VBScript Basics, Part 2," you will further your understanding of VBScript by learning about subroutines and functions, as well as events and conditional statements. Although you will be consolidating your VBScript skills through further examples in Lesson 4, be sure that you understand clearly and feel confident with the following points before moving on:

- A subset of the popular Visual Basic language, VBScript is short for Visual Basic Scripting and is a Microsoft product.

- A variable in VBScript is always of type Variant, and to declare them you use the Dim statement.

- You can use the Option Explicit statement at the start of your script to prevent VBScript from letting your code use any variables you did not explicitly declare using a Dim statement.

- You assign a value to a variable by using the assignment operator.

- Comments can add clarity to your scripts, and you can add them in VBScript by using an apostrophe (') before the comment text. You can also use Rem (short for remark) instead of an apostrophe.

- Using operators from four main categories—arithmetic, comparison, logical, and concatenation—you can perform many different operations with values and variables in VBScript.

- Operator precedence is the order in which VBScript evaluates operators. It evaluates arithmetic operators first, then the concatenation operators, next the comparison operators, and finally the logical operators.

- The lifetime of a variable refers to how long it exists, while the scope of a variable refers to the range of code from which it is accessible.

myValue = 7

Msg ox "The following long sentence needs to be "split" & _
vbcrif across multiple lines because it is SO long." _
& however that " & _
"line of output " & _
"multiple lines of code too!" _
& vbcrlf & vbcrlf & _
"And that's about there is to know about longlines."

COVERING VBSCRIPT BASICS, PART 2

*i*n Lesson 3, "Covering VBScript Basics, Part 1," you learned some of the basics of VBScript, including variables and assigning values, as well as how to perform various operations on variables. In this lesson, you will look at more VBScript fundamentals to complete and enhance your knowledge and to give you the foundation necessary to write Active Server Pages using VBScript. Specifically, this lesson will teach you how to use functions and procedures and how to use conditional statements to make decisions within your scripts. By the time you finish this lesson, you will understand the following key concepts:

- If you have a long line of code, you can wrap the code to the next line, by using the underscore character.
- As programs become larger and more complex, programmers break the programs into smaller, more manageable pieces they call *procedures*. Each procedure should perform a specific task.
- Within VBScript, there are two main types of procedures: subroutines and functions. The main difference between these two types of procedures is that a function will return a value whereas a subroutine will not.
- To make your programs easier for others to read and understand, you should give your subroutines and functions meaningful names that reflect the purpose of the procedure.
- You start a subroutine with a line that contains the Sub keyword followed by the subroutine name. You place the End Sub keywords after the subroutine's last statement.

- You start a function with a line that contains a Function keyword followed by the function name. You place the End Function keywords after the function's last statement.

- To return a value from a function, you must assign the value that you want to return to the name of the function itself within the function code.

- To extend a subroutine's or function's capabilities, your scripts can pass values to procedures, which programmers call *arguments* or *parameters*.

- Conditional statements let your scripts take different courses of action depending on whether specific criteria are met or not. In other words, using conditional statements, your scripts can make decisions.

- An If-Then-Else statement is one method that VBScript supplies for handling multiple conditions. By using it with the ElseIf keyword, you can easily write multiple action statements for different conditions.

- You make use of the Else keyword in all cases where a specific condition has not been met. This can also be a useful error handling feature, such as outputting a message when invalid input has been given.

- The Select Case statement is another method that you can use in VBScript for handling multiple conditions.

Handling Long Lines in VBScript

As you write code, there may be times when you end up with very long statements that make it awkward for someone who is reading the code to see clearly what the script is doing. In such cases, you can split the long lines onto multiple lines by placing an underscore character at the end of a line that you want to continue. You must use a space before the underscore character as well, to prevent VBScript from thinking that the underscore is part of the previous element on the line. Some examples of using an underscore to handle long lines in VBScript are as follows:

```
MsgBox "The following long sentence needs to be split" & _
vbcrlf & "across multiple lines because it is so long." _
& vbcrlf & "Note, however, that " & _
"you can have one line of output " & _
"strung out on multiple lines of code too!" _
& vbcrlf & vbcrlf & _
"And that's about all there is to know about long lines."
```

Figure 4.1 shows the output from the example script, where you can see clearly the result of using the underscore character and how it affects the final output. Even though the argument or parameter for the MsgBox procedure is spread over seven lines of code, with the help of the underscore character, it behaves as if it were all on a single line. Within the message, the symbol *vbcrlf* creates the carriage return linefeed combination that advances the message box text to the next line. If you remove that symbol, the message box will wrap the text at its margins.

Figure 4.1.
The output from the example showing the result of using the underscore character

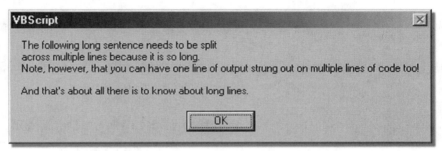

One special point you should note is that you cannot place the underscore character in the middle of a string when handling long lines of code. For instance, placing the underscore within the double quotes, as follows would give you an error:

```
MsgBox "Here is a very, very, long _
sentence that takes up multiple lines."
```

Instead, you must always use the underscore character, for the purpose of handling long lines, outside of strings. So the previous example when written correctly would appear as:

```
MsgBox "Here is a very, very, long " _
& "sentence that takes up multiple lines."
```

Using Procedures in VBScript

As programs become larger and more complex, programmers often break the large programs into smaller, more manageable pieces, each of which performs a specific task. Within VBScript, the programmers group these smaller code fragments into two types of procedures: subroutines or functions.

The main difference between these two types of procedures is that a function will return a value to the code that calls the function, whereas a subroutine will not. In both cases, however, the code becomes not only more manageable, but also aids greatly in code reusability, allowing you to call some complex code or perform a lengthy routine simply by calling the name of the procedure. This saves you from having to repeat lengthy but often used pieces of code throughout your script.

As you will soon learn, VBScript provides a collection of built-in procedures such as the MsgBox and InputBox procedures. However, the ability to create your own functions and subroutines opens up the possibilities a great deal to creating efficient code that achieves your goals.

Creating and Calling a Subroutine

As you already know, a subroutine is a procedure that does not return a value. Subroutines are, however, extremely useful to save you from repeatedly typing repetitive portions of code. When you create a subroutine, you should limit the subroutine to performing a specific task, much like the MsgBox function displays a text message within a dialog box and nothing else. To create a subroutine, you place the subroutine's statements between the Sub and End Sub statements:

```
Sub SubroutineName
    'Statements
End Sub
```

Each subroutine you create within a program should have a unique meaningful name. The name you assign to a subroutine should describe the operation the subroutine performs, such as *ClearTextDisplay* or *InitializeTaxTable*. For instance, perhaps you want to create a subroutine that handles clearing a text display of four different variables. Using all the information given so far, the whole subroutine might look as follows:

```
Sub ClearTextDisplay
    'Clear all the text variables in the display
    ClientNumber = ""
    ClientName = ""
```

```
      ClientPhone = ""
      ClientEmail = ""
  End Sub
```

As you can see, within the subroutine, the statements simply assign empty strings to four different variables.

To call (to use) this subroutine from anywhere within your script, you simply use the name of the subroutine. For example, to call (perform the statements contained in the subroutine) the *ClearTextDisplay* subroutine, you would place the following statement within your code:

```
ClearTextDisplay
```

Just as your scripts may need to store information as they execute, the same is true for functions and subroutines. Within functions and subroutines you can declare *local variables* with names and values that are known only to the subroutine (meaning, the local variables have no scope outside of the subroutine). To declare a local variable you simply use the Dim statement and specify the variable name. The Dim statement must appear after the name of the subroutine but before the End Sub keywords. If the name you use for a local variable is the same as a variable you declared within your program, the subroutine or function code will only affect the local variable.

The following example creates a local variable named *totalcount* within the *MyTest* subroutine:

```
Sub MyTest
Dim totalcount
      'Statements
End Sub
```

Passing Information to Subroutines

When your programs call a subroutine, your programs can pass information to the subroutine using values that programmers refer to as *arguments* or *parameters*. For example, you might pass a subroutine the name of the user and then let the subroutine's code incorporate the user's name into a sentence for output.

Before your code can pass values to a subroutine, you must define the subroutine in a way that specifies the number of parameters it will receive as well as the variable names the subroutine will use to access the values. To begin, you specify the parameter names (the variables the subroutine will use to access the values it receives) within parentheses that follow the name of the subroutine. If the subroutine will support multiple arguments, simply separate the names using commas. The following examples show several different subroutines that support parameters:

```
Sub MyTest(x)
    'Statements
End Sub

Sub AddValues(x, y)
    'Statements
End Sub

Sub MakeMyDay(username)
    'Statements
End Sub

Sub LotsNLots(firstname, lastname, age, gender, hobby)
    'Statements
End Sub
```

To use the subroutines, you again place the subroutine names within your code. In addition, you should specify the values you want to pass to the subroutines within parentheses that follow the subroutine name. You must place the values in the same order you want the subroutine to assign the values to its parameter variables. The following statements illustrate how your code might call the previous subroutines:

```
Sub MyTest(144)

Sub AddValues(5, 3)

Sub MakeMyDay("Clint")

Sub LotsNLots("Billy", "Gates", 11, male, "Computers")
```

The following script creates a subroutine named *DoMath*, which it later calls three times. The example demonstrates how you create a subroutine, call the subroutine, declare and use local variables within the subroutine, and how to use arguments to pass values to a subroutine. The *DoMath* subroutine displays a message that shows the original value the program passes it, along with the value added to 37, and then the remainder of that value is divided by 5. Using Notepad, type in these statements and save the file as *DoMath.vbs*. Then, using Internet Explorer, double click on the program. The program will prompt you for the number you want to use. Enter a value between 1 and 9 (actually, you can enter any number you want—you will later learn how to use conditional processing to check the value the user enters). The program will then display its results within a message. You may want to run the program several times entering different values. Later, you may want to change the *DoMath* subroutine to perform different operations, such as subtracting as opposed to adding the value 37:

```
Dim myValue

myValue = InputBox("Enter a number from 1 to 9...", 1)
DoMath(myValue)
DoMath(myValue + 1)
DoMath(myValue + 2)

' Declare the subroutine
Sub DoMath(x)
Dim s1, s2, s3
  'work out the math and place in strings
  s1 = "The original value is: " & x
  s2 = "After adding 37 is:      " & x + 37
  s3 = "Dividing this by 5 gives remainder: " _
  & (x + 37) Mod 5

  'add carriage return line feeds
  s1 = s1 & vbcrlf
  s2 = s2 & vbcrlf

  'present the message box
  MsgBox s1 & s2 & s3
End Sub
```

Creating and Calling a Function

In the same way that you have seen how to create and call a subroutine, you must also learn how to create and call a function. Functions are incredibly useful when writing VBScript and therefore Active Server Pages, too. A function will not only create an easily reusable procedure for use in your code, as the subroutine does, but it will also return a value. Creating a function is similar to creating a subroutine except that you use the keyword Function instead of Sub. Also when you end the function you use the keywords End Function rather than End Sub. To return a value, you simply assign whatever the value is that you want to return to the name of the function itself. The following basic syntax of a function will make this clearer:

```
Function FunctionName
    'Statements
    FunctionName = return_value
End Function
```

Similarly to naming subroutines, you should name your functions meaningfully so that you and others can tell by the function's name what action it performs or what its purpose is. Also similar to subroutines is the manner in which you handle arguments. In fact, handling arguments is the same in functions, as it is in subroutines. Simply use parentheses after the function name, with the arguments separated by commas within the parentheses. So a simple example of creating a function that takes two numerical arguments, adds them together and returns the result would be as follows:

```
Function AddTwo(x, y)
    AddTwo = x + y
End Function
```

When calling a function, however, there is a difference from calling a subroutine. You must take into account the fact that the function will return a value. So when calling a function, although you use the function's name, you assign it to a variable that is capable of holding the return value. So if you were to call the *AddTwo* function that you just saw, you would have to write something similar to the following:

```
Dim total
total = AddTwo(5, 6)
```

At the end of this code, the variable total would contain the value 11.

In the same way as subroutines, functions can contain local variables that have a lifetime only from when the function is created and the variable declared until the function is exited. The scope of the local variable is only within the function itself. Any script external to the function would not be able to access the local variable. You declare a variable by using the Dim keyword on a new line after the function name and any arguments it might have. So, although not necessary, you could add a local variable to the *AddTwo* function for the purpose of holding the value before returning it, as shown:

```
Function AddTwo(x, y)
Dim result
    result = x + y
    AddTwo = result
End Function
```

Understanding Conditional Statements

As the process your VBScript takes becomes more complex, you will eventually want the code to make decisions based on specific conditions. Within VBScript, conditional statements let your script take a course of action only when it is appropriate to do so. For instance, you might say to yourself, "If it is raining, then I will take an umbrella, else I will leave the umbrella at home." You will only take the umbrella on the condition that it is raining. You can apply the same type of conditions within your scripts. VBScript has an If-Then-Else statement that you can use for conditional programming. In its simplest form, an If-Then statement appears with the following syntax:

```
If (condition_is_true) Then Statement
```

Assuming that you have a variable by the name of *IsRaining*, then an example of this might be:

```
If (IsRaining = True) Then MsgBox "Take your umbrella"
```

You do not have to explicitly state that the condition is equal to True. If you omit the condition equating to True or False then VBScript will use True implicitly. Therefore you do have to explicitly state if you want the condition to be equal to False. You can also omit the parentheses () surrounding the condition if you want to. So the following line of code is exactly the same as the previous line of code:

```
If IsRaining Then MsgBox "Take your umbrella"
```

Be aware, however, that by removing the parentheses and using implicit conditioning you can make your code less readable to both yourself and others. Use your discretion as to whether removing either of these elements in your code could cause problems or confusion later.

You can also expand the If-Then statement to include multiple action statements to take if the condition is met. By supplying all of the action statements on lines after the keyword Then, you can have multiple action statements when a condition is met. You must close your If-Then statement with the keywords End If when you use multiple action statements. You can have a single action statement on a separate line and follow it with the keywords End If, as well. So, you can use the If-Then statement for handling one or more action statements by using the following syntax:

```
If (condition_is_true) Then
    'Statements
End If
```

For example, if you have multiple things that you want to do if it is not raining, then your code might look similar to the following:

```
If (IsRaining = False) Then
    MsgBox "Don't take umbrella"
    WatchTelevision = False
    HangOutWashing = True
    MowLawns = True
End If
```

The conditions that you set do not always have to use Boolean variables (that can only be assigned True or False). You can use conditional statements to handle other variables and values using the operators that you learned about in the previous lesson. For example, the following message will only be shown if the variable *myAge* is equal to 7:

```
Dim myAge
myAge=7
if (myAge = 7) Then MsgBox "You are 7."
```

You will often want to take other actions when a condition is not met. For instance, if you are a security guard and someone has clearance, then you let them through, else you turn them away. By making use of the Else keyword, with the same syntax that you have seen already, the usual form of the If-Then-Else statement is as follows:

```
If (condition_is_true) Then
     'Statements
Else
     'Statements
End If
```

The following statements, for example, show you how you might give a different response depending on the age that the user inputs into the program:

```
'========================================
' Age.vbs - is a VBScript that will
' give a different response depending on
' the user's age.
'========================================

Dim myAge

myAge = InputBox("Enter your age...", "Age")

If myAge>17 then
   MsgBox "You are an adult."
Else
   MsgBox "You are a child."
End If
```

Type the example script above into *Notepad* and save the script as *age.vbs*. Run *age.vbs* a few times by double-clicking on its icon, entering in different ages each time to gauge the program's response. Figure 4.2 shows you what the dialog first looks like when you run *age.vbs* and Figure 4.3 shows you the result of entering 17 as your age.

Figure 4.2.
The input dialog box that *age.vbs* presents

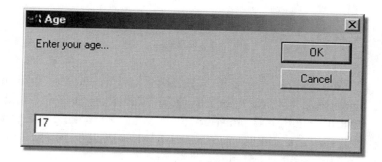

Figure 4.3.
The result of entering 17 as your age

Sometimes you may find that you require more than just a single condition to test against. For instance, using the *age.vbs* example again you might feel that teenagers would take offense at being told they are a child. In this situation you require a further condition. VBScript supplies you with the ElseIf keyword to use in these situations. For those who have done some programming before you may be tempted to separate the ElseIf keyword into two separate words; however, in VBScript you should refrain from doing so and leave them as a single word. You place an ElseIf keyword after the last set of active statements and on a new line. In a similar fashion to the first line of an If-Then-Else statement, after the ElseIf keyword you place your condition, following it with the keyword Then, which in turn is followed by the active statements. You can have multiple ElseIf keywords within a single If-Then-Else statement. Edit your *age.vbs* script so that it now appears as follows:

```
'=========================================
' Age.vbs - is a VBScript that will
' give a different response depending on
' the user's age.
'=========================================

Dim myAge

myAge = InputBox("Enter your age...", "Age")

If myAge>17 then
   MsgBox "You are an adult."
ElseIf myAge>12 then
   MsgBox "You are a teenager."
Else
   MsgBox "You are a child."
End If
```

Now when the user enters 17 as her age, she will see a message box similar to that shown in Figure 4.4, which is much less offensive. Notice how you now have a script that can handle multiple conditions. This is a very powerful option that you can make use of to change the outcome of your Active Server Pages later on.

Figure 4.4.
The result of entering 17 for your age and having multiple conditions

Very briefly, you should also be aware of VBScript's alternative method for handling multiple conditions: the Select Case statement. You create a Select Case statement by beginning with the keywords Select Case, following it with an expression, such as a variable, which you then compare to various cases. When comparing for these various cases, you must express each by stating the Case keyword at the start of a new line and follow it with the condition. Place all of the active statements for each case before the next Case keyword. You can use Case Else if the expression that you are comparing does not match any other cases. You close a Select Case statement with the keywords End Select. The following shows the syntax of a Select case statement:

```
Select Case expression
Case condition_1 'statements
Case condition_2 'statements
Case condition_3 'statements
Case condition_n 'statements
Case Else 'statements
End Select
```

As an example of using the Select Case statement in VBScript, have a look at the following script, which takes input from the user and gives different responses depending on the input. Also note the use of the Case Else keywords to handle situations where incorrect input is given. Using Notepad, type the example script below and save the script as *abc.vbs*.

```
Dim myABC

myABC = InputBox("Please select A, B or C...", "ABC")

Select Case UCase(myABC)
Case "A" MsgBox "A is for apple."
Case "B" MsgBox "B is for banana."
Case "C" MsgBox "C is for cake."
Case else MsgBox "You did not enter A, B or C."
End Select
```

Run *abc.vbs* a few times trying different entries. You will notice that the script handles upper or lower case correctly. This is because of the function UCase, which changes the arguments given to it into uppercase. There is also an LCase function that changes strings to lowercase. Both UCase and LCase are built-in functions within VBScript that you can use anywhere in your scripts.

WHAT YOU MUST KNOW

In this lesson you have seen most of the fundamental concepts of programming VBScript, although there are lots of other features that you will discover as you progress through this book. You can create your own procedures and understand the difference between subroutines and functions. You have seen the powerful art of conditional statements through such concepts as the If-Then-Else statement and the Select Case statement. In Lesson 5, "Installing an Active

Server Pages Server," you will learn how to set up a server and in doing so lay the path for creating Active Server Pages where you will continue to learn about programming VBScript. Go over the following points and be sure to understand them fully before moving on to Lesson 5:

- By using the underscore to break up long lines, you can make your code more legible without having to concern yourself with lengthy strings or variable names causing the line to become too long.

- A function returns a value. A subroutine does not return a value. Besides this key point, both procedure types, functions and subroutines, are fairly similar.

- When you want to start a subroutine in VBScript, you use the Sub keyword. After your statements, close a subroutine with the keywords End Sub.

- Meaningful names for your subroutines and functions help you to see at a glance the purpose of various procedures.

- You can use the Function keyword in VBScript to indicate the beginning of a function. To indicate the end of the function, the keywords End Function are used.

- The main purpose of any function is to return a value, which you accomplish by assigning the value to be returned to the name of the function itself.

- You can pass information to both subroutines and functions through arguments. Both procedure types can then use the information held in these arguments within the procedure's code.

- By using conditional statements in your scripts, you can execute specific sections of code only when it is appropriate to do so.

- You can handle writing multiple action statements for different conditions by using an If-Then-Else statement. These statements range from very simple one line affairs to lengthy blocks of code containing multiple conditions.

- Whenever a specific condition is not met, you can make use of the Else keyword to handle the particular case. The Else keyword can also be a useful error handling feature of sorts.

- Another method that VBScript offers for handling multiple conditions is the Select Case statement. The Select Case uses one expression and then matches it against multiple cases, each indicated clearly by the Case keyword.

INSTALLING AN ACTIVE SERVER PAGES SERVER

*i*n Lesson 3, "Covering VBScript Basics Part 1," you learned the fundamentals of how to create functional code using VBScript. You are now well on your way to acquiring all the basics for writing Active Server Pages. In this lesson, you will learn how to set up an Active Server Pages server so that you can actually work on some Active Server Pages and have the ability to view them locally on your computer. By the time you finish this lesson, you will understand the following key concepts:

- Active Server Pages are not the same as HTML pages. You cannot simply double-click an Active Server Page to view it the same way you can with an HTML page; you must have some form of Active Server Pages server set up first.
- Depending on the operating system you are using, the different names and methods of installing Active Server Pages servers can be rather confusing at first. However, the variations are very similar in practical use.
- The Add/Remove Programs tool within the Control Panel lets you add Personal Web Manager or its equivalent for almost all Windows operating systems.
- You must use the Network tool within the Control Panel to add Peer Web Services on Windows NT 4.
- Personal Web Manager shows you where your home directory is and what your Web server's name is, which is enough information to let you publish new pages and view them from a browser.

- Personal Web Manager lets you change the location of your Web server's home directory, which lets you set up multiple home directories easily, although you can only have a single home directory in use at any particular time.
- You can also use Personal Web Manager to start and stop, as well as pause and continue your Web server by using the Properties menu.
- The *ipconfig* utility is one method you can use to obtain your IP address. By substituting your IP address for your Active Server Pages server name in a browser's Address field, it is possible to access your Active Server Pages from the Internet.

Understanding the Need for an Active Server Pages Server

Active Server Pages are not the same as HTML pages. You cannot simply double-click an Active Server Page to view it in the same way you can with an HTML page; you must have some form of Active Server Pages server set up first. The reason for this is that the server must translate Active Server Pages into raw HTML before passing the statements on to the browser. When you simply double-click an Active Server Page and have it open in a browser, the browser (which does not execute the code within the Active Server Page) does not handle the page as you would expect.

The following Active Server Pages code, which you do not have to understand at this point, creates the file *test.asp*. When you open the file from a server that can process the Active Server Pages code, the file will display a brief sentence several times and in varying font sizes. The code for the Active Server Page, *test.asp*, is as follows:

```
<html>
<body>
<% For i = 1 To 7 %>

   <font size=<%= i %>>
   The cat in the hat came back.<br>
   </font>

<% Next %>
</body>
</html>
```

Using the same Active Server Page, *test.asp*, you can clearly see the difference between opening the page with and without using an Active Server Pages server. At this moment, the code that makes up the page does not matter, as the focus is on the differing results and applies to any Active Server Pages. Figure 5.1 shows the result of opening the file *test.asp* without using an Active Server Pages server, such as entering the *test.asp* file's path directly in a browser Address (URL) field, while Figure 5.2 shows the result of opening the same page properly through an Active Server Pages server.

Figure 5.1. Opening *test.asp* without using a server

Figure 5.2. Opening *test.asp* using a server

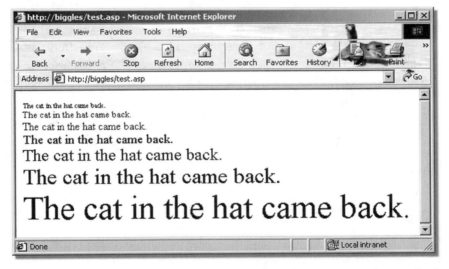

Notice that in Figure 5.2, the browser was able to display the result that you would expect from the Active Server Page *test.asp*. It did not have to understand the code, because the Active Server Pages server had translated the page entirely into HTML.

Learning the Variations in Active Server Pages Servers

There are many different variations of Active Server Pages servers available, and it can be confusing when you are trying to learn what you must do for your system. On Windows 95, 98 and Me, you can find an Active Server Pages server known as Personal Web Server. On Windows NT 4, the server is known as Peer Web Services. And then to make things interesting for you, on Windows 2000 the server is known as Personal Web Manager. Do not worry, though, as you will soon see that there are many similarities between all the different versions.

One last point worth mentioning is Microsoft's full-blown product for hosting Web sites, Internet Information Server or IIS, as it is more commonly known. Although Personal Web Server or its equivalent is capable of hosting Active Server Pages, if you are setting up a company intranet for hundreds or thousands of users, you should really use Internet Information Server for hosting your Active Server Pages.

Installing Personal Web Server on Windows 98

To install the Personal Web Server on a Windows 98 computer, you must first click on the Start menu Settings option and then the Control Panel option. Windows will open up the Control Panel window from which you must open the Add/Remove Programs item by double-clicking on its icon. Windows will display the Add/Remove Programs Properties window, in which you should select the Windows Setup tab. From the list of components, select Internet Tools so that it is highlighted and then click on the Details button, as shown in Figure 5.3.

Figure 5.3. Selecting the Details button for the Internet Tools option

After you click on the Details button, Windows presents you with the Internet Tools window. Inside this window is a list of components that comprise the set of Internet tools that come with each version of Windows. Select the Personal Web Server component by clicking in the check box on the left side so that there is a check mark visible. The Internet Tools window should now look similar to Figure 5.4. Click on the OK button to close the Internet Tools window.

Figure 5.4. Selecting the Personal Web Server check box

Within the Add/Remove Programs Properties window, click on the OK button to close the window and actually make the software changes. Windows will request the installation CD if it cannot automatically locate the necessary drivers.

Installing Personal Web Server on Windows 95/Me

Windows 95 and Windows Me do not have Personal Web Server, but you can easily add it by downloading the NT 4 Option Pack from Microsoft. The NT 4 Option Pack is freely downloadable, as of publication, at the following address:

http://www.microsoft.com/msdownload/ntoptionpack/askwiz.asp

After following the instructions on the Microsoft site, being sure to choose the option for Windows 95 (this also works for Windows Me) and not any of the NT 4 options, run the wizard that you download and select the Install option from the Download Options dialog box. Click on the Next button. The wizard will then present you with installation options from which you should select the full installation, as shown in Figure 5.5.

Figure 5.5. Select the full installation of Personal Web Server

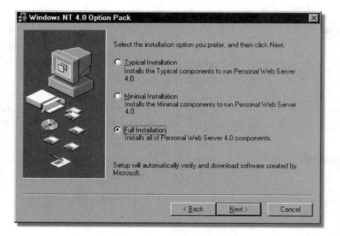

Click on the Next button to continue. Windows will then download all of the necessary files to run Personal Web Server on your computer. When it has completed, Windows opens up a window with the title Microsoft Personal Web Server Setup. Click on the Next button to continue with setup. The setup program presents you with a smaller dialog box from which you can select specific components. Within this list of components, scroll down using the scroll bar to the right of the list until you come to Personal Web Server (PWS), and highlight it by clicking on it once, as shown in Figure 5.6.

**Figure 5.6.
Highlighting the
Personal Web Server
(PWS) component**

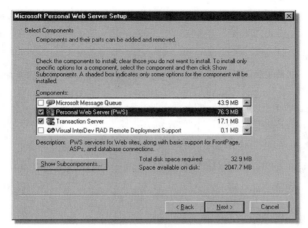

After you highlight the Personal Web Server (PWS) component, click on the Show Subcomponents button. Setup will display another dialog box with three subcomponents. Highlight the Documents subcomponent by clicking on it once, and then click on the Show Subcomponents button. Setup will present yet another dialog box to you, this time with the title bar name Documentation. From within the Documentation dialog box, make sure that the Active Server Pages subcomponent has a check mark next to it, as shown in Figure 5.7, before clicking on the OK button.

**Figure 5.7.
Making sure the
Active Server Pages
subcomponent has
a check mark next
to it**

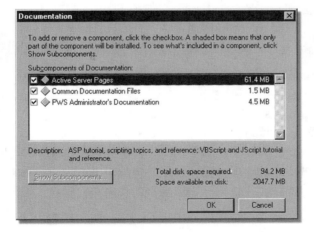

Within the Personal Web Server (PWS) dialog box, click on the OK button to close it. Within the Microsoft Personal Web Server Setup dialog box, click on the Next button to continue. Simply click on the Next button to accept the default

path for your Web publishing home directory. The setup program will now install all of the chosen components and will ask you to restart Windows.

Installing Peer Web Services on Windows NT 4

Under Windows NT 4, the Active Server Pages server is known as Peer Web Services instead of Personal Web Server or Manager. To install Peer Web Services you must do the following:

1. Click on the Start menu Settings option and then select Control Panel.
2. This will cause NT 4 to open up the Control Panel window, from which you must double-click on the Network icon.
3. Windows NT 4 will display a dialog box from which you must select the Services tab and then click on the Add button.
4. Windows NT 4 will then display another dialog box and from the list of network services available double-click on Peer Web Services.
5. Windows NT 4 will then run through the installation process, after which you must click on the OK button.
6. In the Microsoft Peer Web Services dialog box, you must select the services you want to install before clicking on the OK button.
7. In the Publishing Directories dialog box, simply accept the defaults unless you want to particularly specify the directories and then click on the OK button.

Installing Personal Web Manager on Windows 2000

To install an Active Server Pages server on Windows 2000, is similar to installing for Windows 98 but with a few differences. Click on the Start menu, Settings option and then select Control Panel. As a result, Windows will open the Control Panel window within which you must double-click on the Add/Remove Programs icon. Windows 2000 will display the Add/Remove Programs dialog box, which contains a list of sections down the left hand side. Click on the Add/Remove Windows Components section, as shown in Figure 5.8.

**Figure 5.8.
Selecting the
Add/Remove
Windows
Components
section**

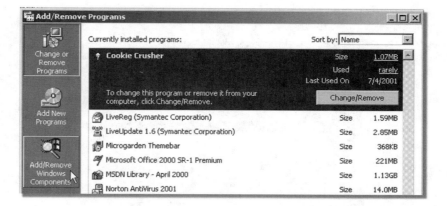

Windows 2000 will then open up the Windows Components Wizard dialog box that displays a list of possible components. From this component list, select the Internet Information Services (IIS) component so that it has a highlight across it and then click on the Details button, as shown in Figure 5.9.

**Figure 5.9.
Click on the
Details button**

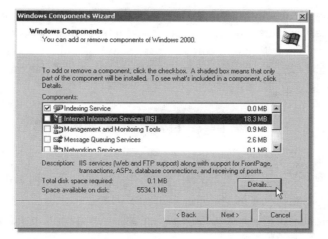

Windows 2000 will now open the Internet Information Services (IIS) dialog box. This dialog box lists all the subcomponents of Internet Information Services that you can install. Select Personal Web Manager so that it has a check in the check box to its left, as shown in Figure 5.10. Windows 2000 will automatically check other dependent components, such as the Internet Information Services Snap-In, if they are not already installed.

**Figure 5.10.
Selecting Personal
Web Manager**

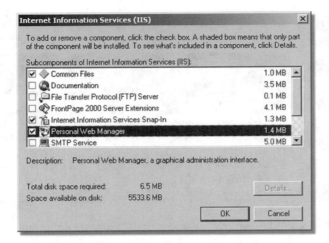

Click on the OK button within the Internet Information Services (IIS) dialog box. When the Internet Information Services (IIS) dialog box closes, click on the Next button within the Windows Components Wizard dialog box. Windows will request the installation CD if it cannot automatically locate the necessary drivers. You will get a dialog box stating that you have successfully completed the Windows Components Wizard. Clicking on the Finish button will close the wizard. Click on the Close button within the Add/Remove Programs dialog box to close it.

Using Personal Web Manager

After successfully installing an Active Server Pages server, you now have to work out how to manage it, such as stopping and starting the server or discovering the location of the home directory. The tool for this purpose is Personal Web Manager, which you can open in different ways depending on your particular operating system.

Accessing Personal Web Manager

For Windows 98, click on the Start menu, Programs, Accessories, Internet Tools and Personal Web Server item before finally selecting Personal Web Manager, as shown in Figure 5.11. Similarly, for Windows 95 or Me, click on the Start menu, Programs, Personal Web Server item before finally selecting Personal Web Manager.

**Figure 5.11.
Starting the
Personal Web
Manager from menus**

For Windows 2000, you must go to the Start menu, Settings option and select Control Panel. When the Control Panel window opens, double-click on the Administrative Tools icon, as shown in Figure 5.12.

**Figure 5.12.
Double-click the
Administrative
Tools icon**

Within the Administrative Tools window, double-click the Personal Web Manager icon, as shown in Figure 5.13. You can copy the Personal Web Manager shortcut to the Desktop or some other more convenient location to make it more easily accessible.

**Figure 5.13.
Double-click the
Personal Web
Manager icon**

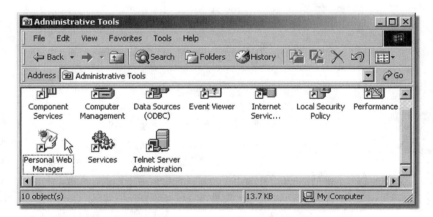

When you open Personal Web Manager, regardless of the particular Windows operating system that you are running it on, you will see a window similar to that shown in Figure 5.14.

Figure 5.14.
The Personal Web
Manager program

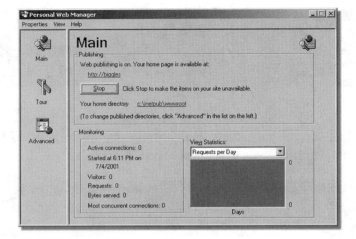

There are a few important pieces of information that you can get from Personal Web Manager, as well as being able to customize certain settings. Personal Web Manager is split in different sections, each of which has an icon in the left bar. To switch between different sections within Personal Web Manager, simply click on the various icons. When you open Personal Web Manager, the default section is the Main section, which contains the actual name of your Web server, as well as the home directory.

Knowing Your Web Server Name and Home Directory

The name of your Web server is very important. You use it to view your Active Server Pages within your browser. Personal Web Manager gives you this name in blue text directly after the words, "Your home page is available at:"

Changing Your Web Server's Name

*Although Personal Web Manager shows you what your Web server's name is, it may not always be what you want. For instance, it is more difficult to type in and remember **http://GRXZ_675F3B** than perhaps **http://biggles**. However, you can change the name of your Web server simply by changing the name of your PC.*

For instance, in Windows 2000 you would have to double-click on the System icon within the Control Panel window. Then you would have to select the Network Identification tab and click on the Properties button. In the Identification Changes dialog box, you can simply type in a new computer name before you click on the OK button. You will see a prompt from Windows to restart your computer before the new settings can take effect.

You can also see clearly where your Web server files are kept. This is known as your home directory. The Personal Web Manager writes this path in blue text directly after the words, "Your home directory." The default path is usually *c:\inetpub\wwwroot,* but Personal Web Manager on your computer will tell you exactly where your particular home directory is.

Armed with the knowledge of both the name of your Web server and the home directory of your Web server, you can now work out how to host Active Server Pages. As an example, assume that your Web server's name is *http://biggles* and that its home directory is *c:\inetpub\wwwroot.* If you place an Active Server Page, say *test.asp*, inside your home directory you could view it from a browser by typing in the URL *http://biggles/test.asp* and then pressing the Enter key.

To further cement this very important concept, if you create a folder beneath your home directory, say green, and then place inside an Active Server Page, say *test2.asp*, you would have a path to *test2.asp* of *c:\inetpub\wwwroot* \green\test2.asp. From a browser, you would access this file by typing in the URL *http://biggles/green/test2.asp* and then pressing the Enter key.

Changing Home Directory Paths

You can use Personal Web Manager to change your home directory. To do this you must complete the following steps:

Open Personal Web Manager and then click on the Advanced section. Personal Web Manager will change the panel on the right to reflect your selection, with the Advanced Options, as shown in Figure 5.15.

**Figure 5.15.
The Advanced
Options of Personal
Web Manager**

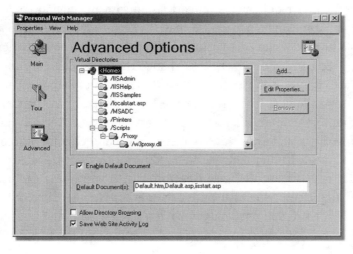

Within the Advanced Options section, click on the Edit Properties button. Personal Web Manager will then display the Edit Directory dialog box, as shown in Figure 5.16.

Figure 5.16.
Closing the Edit
Directory dialog box

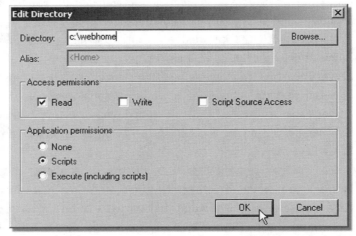

In the Directory field, type in the path to the new home directory, then click on the OK button to close the Edit Directory dialog box.

Having Multiple Home Directories

Imagine a scenario where you have fifteen clients, all of whom require you to test thoroughly offline before uploading any changes to their actual Web sites. Having a separate computer for each client would work, but it would be expensive, not to mention impractical when you take into consideration all the storage space and power requirements for setting up so many computers. If you just have one computer, you do not want to have to keep copying files into and out of your single home directory. The solution is simply to have multiple home directories.

You already know how to change the home directory for your Active Server Pages server using Personal Web Manager. You simply set up all of your client sites as folders, then change the home directory to the top-level folder of the client you are working on at that particular time. There is no need to move files anywhere, and everything remains simple and easy for you to manage. This does not mean having multiple home directories that are all simultaneously active, but rather multiple home directories with which you can set the focus so that any particular one of them is active at a particular time.

For example, you might have a client, ABC, and another client, XYZ. The files for ABC you could store in *c:\sites\abc* and the files for XYZ you could store in *c:\sites\xyz*. If you are currently working on the ABC site and you must switch over and work on the XYZ site, all you have to do is change the home directory in Personal Web Manager from *c:\sites\abc* to *c:\sites\xyz*.

Stopping and Starting Your Web Server

When you are using Personal Web Manager to manage a live site, possibly on a local intranet, you can very easily start and stop the Web server by using the toggle button in the Main section of Personal Web Manager. You can also use the options in the Properties menu to achieve the same result. You can start or stop the Web server, as well as pause and continue the Web Server.

Simply by looking at the Main section of the Personal Web Manager, you can see what state your Web server is in.

Accessing Your Server from the Internet

You might want to have others access your Active Server Pages server and the pages that you have in it from wherever they are on the Internet. To do this you will have to give them your IP address. One way to find out your IP address is to use the *ipconfig* utility. The *ipconfig* utility uses a command shell and is a command-line program. To open up a command shell from Windows NT or Windows 2000, simply click on the Start menu and select Run. Windows will display the Run dialog box, within which you must type in the characters *command* or *cmd* before pressing the Enter key. The Run dialog box will close, launching a command shell with a prompt. In this command shell, simply type in *ipconfig* and press the Enter key. The *ipconfig* utility will list your IP address directly above your subnet mask, as shown in Figure 5.17.

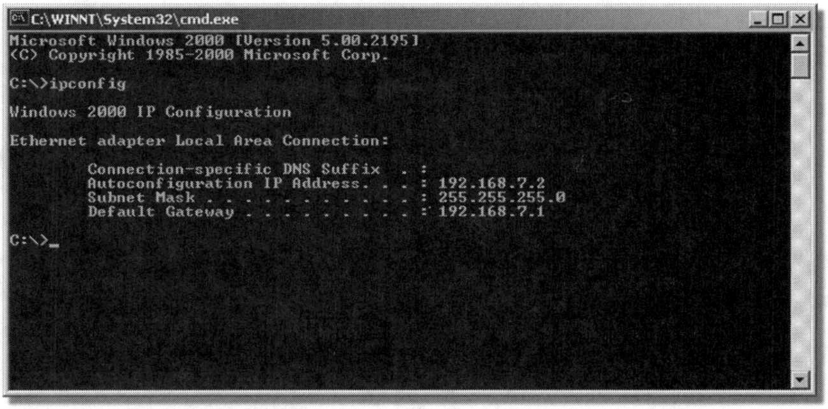

Figure 5.17.
The *ipconfig* utility, listing your IP address in a command shell

On Windows 95, 98, and Me, you can use a graphical version of ipconfig. To open the GUI version of *ipconfig*, simply click on the Start menu and select Run. Windows will display the Run dialog box, within which you must type in the characters *winipcfg* before pressing the Enter key. The Run dialog box will close, launching a window with the title IP Configuration, which lists your IP address clearly, as shown in Figure 5.18.

Figure 5.18. The graphical version of *ipconfig*, clearly showing your IP address

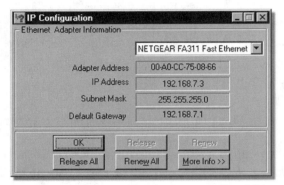

Once you have found out your IP address, whichever method you use, you can simply use that in place of your Active Server Pages server name when typing in the URL to your Active Server Pages. For example, if your server name is biggles and your IP address is 192.168.7.2, you would simply replace biggles with your IP address in the browser's URL or Address field, as shown:

```
http://biggles/test.asp
```

becomes

```
http://192.168.7.2/test.asp
```

WHAT YOU MUST KNOW

You cannot simply double-click an Active Server Page to view it in a browser; you must have some form of Active Server Pages server set up first. The Add/Remove Programs tool in the Control Panel allows you to add Personal Web Manager or its equivalent for almost all of the Windows operating systems. Personal Web Manager shows you where your home directory is and what your Web server's name is, which is enough information to allow you to publish new pages and view them from a browser. In Lesson 6, "Writing Your First Active

Server Page," you will begin to put together all the elements of HTML and VBScript, along with your newly set up Active Server Pages server to produce Active Server Pages that you can view from a browser. Before you continue with Lesson 5, however, make sure you understand the following key concepts:

- ❌ Active Server Pages are not the same as HTML pages. You must have some form of Active Server Pages server set up first before you can view an Active Server Page properly in a browser.

- ❌ There are different names and methods of installing Active Server Pages servers, depending on the operating system that you are using. In practical use, however, the different variations are very similar.

- ❌ For almost all the Windows operating systems, the Add/Remove Programs tool in the Control Panel allows you to add Personal Web Manager or its equivalent.

- ❌ To add Peer Web Services on Windows NT 4, you must use the Network tool within the Control Panel.

- ❌ Personal Web Manager shows you what your Web server's name is and the location of your home directory, which is enough information to let you publish new pages and view them from a browser.

- ❌ You can change the location of your Web server's home directory using Personal Web Manager. This enables you to set up multiple home directories easily, although you can only have a single home directory in use at any particular time.

- ❌ You can start and stop, as well as pause and continue your Web server by using the Properties menu within Personal Web Manager.

- ❌ Using the *ipconfig* utility, you can discover your IP address and use it in place of your Active Server Pages server name so that your Active Server Pages can be accessible from anywhere on the Internet.

WRITING YOUR FIRST ACTIVE SERVER PAGE

n Lesson 5, "Installing an Active Server Pages Server," you learned how to set up an Active Server Pages server on your computer to let you work with Active Server Pages locally, as well as to view them using your browser. In this lesson, you will begin putting all the pieces together and, within the environment that you created in the last lesson, start programming Active Server Pages, using all the skills that you have learned so far in this book. By the time you finish this lesson, you will understand the following key concepts:

- An Active Server Page consists of HTML statements and code. Within the page, you indicate Active Server Page code to the server by placing the code statements between a <% start tag and a corresponding %> end tag.
- HTML pages have drawbacks when trying to handle dynamic data or information, such as the current time or day of the week.
- There are many date and time functions available to Active Server Pages that you can easily use within the pages you create.
- By using loops (iterative constructs such as the For and While statements) within Active Server Pages, you can reduce a lot of repetitive work and handle situations where the exact number of iterations you require is unknown.
- By using forms, you can add buttons, text entry fields, and other objects to your Active Server Pages to collection information from the user. To create such forms, you use the HTML <form> tag and the <input> tag.
- Within an HTML page, you can have the browser notify your code when a special action occurs (such as a mouse click) through what are known as *events*. Your code, in turn, can perform specific processing when an *event* occurs.

Using Tag Syntax with Active Server Pages

At this point you know how to create simple HTML pages and simple VBScript applications, but have not combined the two. To start, you can write an Active Server Page as if it were simply HTML and then later use the <% and %> tags to indicate to the server which parts of the page contain code that it must execute before handing the page to the browser.

Within an Active Server Page, the server considers any statements it finds between the <% and its corresponding %> tags as something to execute before handing the page to the browser.

Creating the Time Page Statically

Imagine that you want to display the current time in an HTML page. Using the time of 8:03 PM, you quickly put together a static HTML page by typing the code that follows into a text editor, such as Notepad, and you then save the file as *time_static.htm*:

```
<html>
<head>
  <title>Static time page</title>
</head>

<body>
  <h2>Static time page</h2>
  <p>The time is currently 8:03 PM.</p>
</body>
</html>
```

Next, when you double-click on the *time_static.htm* file, the browser loads your file with the HTML statements, and displays your message that the time is 8:03 PM, as shown in Figure 6.1. If you close the browser and again double-click on the *time_static.htm* file, the result is exactly the same. In fact, each time you open the page within a browser, the time is static, meaning it does not change. The HTML file is only correct for that particular minute. When the time is different from what has been written in the code, (8:03 PM in the example) this static page will display the wrong time.

Figure 6.1. The result of running statically coded time in a browser

Creating the Time Page Dynamically with Active Server Pages

By using Active Server Pages, you can overcome the problem of rapidly changing data, such as the time or days of the week. You do not have to rewrite all your existing HTML code from scratch to make use of the power of Active Server Pages. You only need to make some modifications to the areas that you want to be dynamic.

For instance, in the previous section you saw some static HTML code that could only display the same time no matter when the page was seen. Using that same code, all you have to do is replace the text, *8:03 PM,* with some VBScript statements within tags. You will have a single page that can dynamically update the value of the time it displays. The file itself must use the *.asp* extension rather than *.htm* or *.html,* and you must save it within your home directory. Again using Notepad or another text editor, enter the following code to the Active Server Page. Figure 6.2 shows the result of using Active Server Pages to display the current time:

```
<html>
<head>
  <title>ASP time page</title>
</head>

<body>
  <h2>ASP time page</h2>
  <p>The time is currently <% = time %>.</p>
</body>
</html>
```

Figure 6.2.
The result of using
Active Server Pages
to handle time

You can refresh this page or have someone else view the same page, and you will get a different result as the time changes. Unlike the static HTML page, Active Server Pages lets you have dynamic content with little effort.

When the server processes the browser's request for the Active Server Page, the server scans the file for the <% tag. In this case, when the server encounters the tag, it executes the following VBScript statement that directs VBScript to display the current time:

```
= time
```

When the server encounters the %> tag, it knows it has completed that section of VBScript code. In this case, the file contained only one VBScript statement. As your Active Server Pages become more complex, you may place VBScript code (or JavaScript code) throughout the file.

Introduction to Date and Time Functions

You can actually do quite a lot of useful work with the time and dates, using Active Server Pages. You have already seen how to display the current time dynamically with your first Active Server Page. There is also a Date function that returns the current date. Further, you can simply use the Now function to return both the date and the time. Just to quickly name some of the many date and time functions before you see briefly how they can be used in your Active Server Pages, some of the following may be of interest to you:

CDate, Date, DateAdd, DateDiff, DatePart, DateSerial, DateValue, Day, FormatDateTime, Hour, IsDate, Minute, Month, Now, Second, Time, Timer, TimeSerial, TimeValue, Weekday, Year.

You might also like to look at the Microsoft documentation for these functions or learn more about them from a VBScript book, although many of them are quite straightforward. The following Active Server Page script shows an example of some of these date and time functions, and Figure 6.3 shows the result of running the script:

```
<html>
<head>
  <title>ASP Date and Time Functions</title>
</head>

<body>
  <h2>ASP Date and Time Functions</h2>
<table cellpadding="30">
<tr><td><p>
Date: <% = Date %><br>
Time: <% = Time %><br>
Now: <% = Now %>
</p></td>

<td><p>
Minute: <% = Minute(Time)%><br>
Weekday: <% = Weekday(Date)%><br>
Year: <% = Year(Date)%>
</p></td>

<td><p>
IsDate: <% = IsDate("January 14, 1974")%><br>
FormatDateTime(2): <% = FormatDateTime(Now,2)%><br>
FormatDateTime(4): <% = FormatDateTime(Now,4)%>
</p></td></tr>
</table>
</body>
</html>
```

Figure 6.3. The result of some of the date and time functions

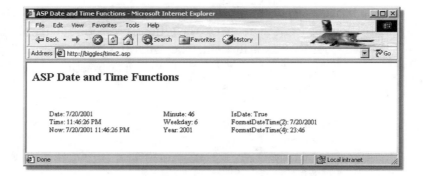

Creating Loops in Active Server Pages

In Lesson 4, "Covering VBScript Basics Part 2," you learned how to use the If-Then-Else statement within your scripts to perform conditional processing. Just as there may be times when a script must perform specific processing when a condition is true, there may also be times when a script must repeat a series of statements, either a specific number of times or while a condition is true.

There are four main loops in VBScript, each of which you can use to save yourself work. You will come across these loops as you work your way through this book and also in other VBScript material and documentation. Table 6.1 briefly describes the four VBScript loops.

Loop	Purpose	Example
For Next	Repeats statements a specific number of times	For i = 1 to 10 'Statements Next
For Each	Repeats statements once only for each item in an object	For Each i in Object 'Statements Next
Do While	Performs a loop's statements at least one time and then possibly repeats the statements based on a condition	i = 0 Do While i < 10 'Statements i = i + 1 Loop
While Wend	Performs a loop's statements repeatedly only while the condition is met	i = 1 While i < 5 'Statements i = i + 2 Wend

Table 6.1. The VBScript looping control structures

The following example, which resides in the file *loop.asp*, uses a For-Next loop that displays the same line of text seven times, with the font size progressively increasing. The result of programming this loop is shown in Figure 6.4. Notice again the use of acutes, slight marks above certain vowels, in the code and their final appearance:

```
<%@ LANGUAGE = "VBScript" %>
<HTML>
<BODY>
<% For i = 1 To 7 %>
  <FONT SIZE=<% = i %>>
  The Bah&aacute;'&iacute; Faith started in Persia.<BR>
  </FONT>
<% Next %>
</BODY>
</HTML>
```

**Figure 6.4.
Showing the result
of programming
a loop**

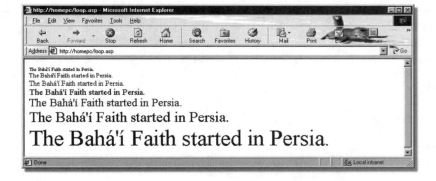

To understand the significance of how much work a loop can save you, consider the HTML source code that creates the same output. You can see the difference between the number of lines the same result takes in Active Server Pages in comparison to the amount in HTML. This example merely deals with seven iterations; imagine the number of lines of work you can save for a loop that has hundreds of iterations! Also, consider the fact that you might not always know how many iterations a loop could require beforehand, which means that it would be impossible to create the long form in HTML code instead of using a loop. The following code is the HTML that the Active Server Pages server returns to the browser, as a result of the earlier code that produced Figure 6.4:

```html
<HTML>
<BODY>
  <FONT SIZE=1>
  The Bah&aacute;'&iacute; Faith started in Persia.<BR>
  </FONT>
  <FONT SIZE=2>
  The Bah&aacute;'&iacute; Faith started in Persia.<BR>
  </FONT>
  <FONT SIZE=3>
  The Bah&aacute;'&iacute; Faith started in Persia.<BR>
  </FONT>
  <FONT SIZE=4>
  The Bah&aacute;'&iacute; Faith started in Persia.<BR>
  </FONT>
  <FONT SIZE=5>
  The Bah&aacute;'&iacute; Faith started in Persia.<BR>
  </FONT>
  <FONT SIZE=6>
  The Bah&aacute;'&iacute; Faith started in Persia.<BR>
  </FONT>
  <FONT SIZE=7>
  The Bah&aacute;'&iacute; Faith started in Persia.<BR>
  </FONT>
</BODY>
</HTML>
```

Viewing HTML Source Code of Active Server Pages

If you ever want to see the HTML source code that results from any of your Active Server Pages, you can easily do so by using a simple feature of your browser. First, open the page you want to view in your browser. Then, click on the View menu and select the Source option. Your browser will open a new window that displays the HTML source code of the current page.

Working with Forms

Active Server Pages utilize forms to collect information from users. Before you can begin to retrieve and process the information that users have put into these forms, you need to learn how to combine HTML and VBScript to create forms. With the basics of form creation in hand, you can later concentrate on the data collection side of Active Server Pages with a clearer focus.

An HTML form might consist of buttons and text entry fields or other similar form attributes with which users are able to make choices and submit textual data. So basically, the form is a container for the visible controls the user sees. In HTML you add a form to your page by placing *form* elements within the <form> tag and the ending </form> tag. So the code for a simple, empty form might look as follows:

```
<html>
<form>
</form>
</html>
```

The <form> tag can itself contain attributes, of which the two most important are the *action* and *method* attributes. Table 6.2 briefly describes the <form> tag attributes.

Attribute	Description
accept	Specifies a comma-separated list of acceptable content types
accept-charset	Specifies a comma-separated list of character encodings for input data that the server accepts
action	Sets the target for the form to be sent to
enctype	Specifies the content type to use when submitting the form to the server
method	Specifies the Get method or Post method for submitting the form
name	Specifies a name by which to access the form from script

Table 6.2. The various attributes of the <form> tag

The *action* attribute specifies the target (the remote address) to which the browser will send the form data and usually takes the form of an HTML or Active Server Page. The *method* attribute can be either "Get" or "Post," and you will learn more about the significance of these two methods in Lesson 8, "Using the Request Object."

To set an attribute within the <form> tag, you write the name of the attribute, followed by an equal sign and the attribute's value. The following statement, for example, assigns values to the *method* and *action* attributes:

```
<form method="POST" action="http://biggles/form2.asp">
```

As you will see, you place other tags between the <form> and </form> tags to make up the body of the form. These other tags include the <input>, <select>, and <textarea> tags. Within a form, you create buttons and single line text boxes (the <textarea> tag is for multiline text areas) using the <input> tag *type* attribute. Table 6.3 lists several different values you can assign to the *type* attribute within an <input> tag.

Form Attribute	*Type* Name
form button	button
submit button	submit
reset button	reset
image button	image
radio button	radio
check box	check box
single-line text entry field	text
password entry field	password
file upload field	file
hidden field	hidden

Table 6.3. The various *form* attributes available for the <input> tag and their actual *type* names

In addition to the *type* attribute, the <input> tag supports several other attributes, some of which are optional. As was the case with the <form> tag, to specify an attribute within the <input> tag, you provide the attribute name, followed by an equal sign (=) and then a value, as shown here:

```
<form>
<input name="nnn" type="ttt" value="vvv">
</form>
```

In practice you would not use "nnn" for the name of your <input> tag, but rather something more meaningful, such as User or City. Likewise, "ttt" would be an actual attribute type, such as "button," and instead of "vvv" you would assign the text that you want to display on the button to the *value* attribute. The following <input> tag illustrates the use of attribute values:

```
<input name="SeaButton" type="button" value="Sea">
```

Adding Buttons to Your Forms

The following example will show you how to create a form with a single button on it. Within the HTML statements, the heading section contains only standard HTML, which you have seen many times already. However, the body section contains a <form> tag in its simplest form, as well as an <input> tag.

The <input> tag has a *name* attribute that you will not use right now, a *type* attribute that selects the use of a button, and finally a *value* attribute that specifies the text to display on the button. Using Notepad, type in the following statements and save your file as *button.asp* within your home directory:

```
<html>
<head>
<title>Working with buttons</title>
</head>

<body>
<h3>Working with buttons</h3><hr size="4">
<form>
<input name="SeaButton" type="button" value="Sea">
</form>
</body>
</html>
```

To view the contents of this file, you must type the file name in the address field of your browser. The script, in this case, does not do anything exciting. However, you can see the button that you just set up. Because all the code in *button.asp* is really just HTML (the file does not contain any VBScript), you could also have named the file *button.htm* and it still would have run.

Handling Events in VBScript

An event in VBScript is basically a form of notification from the operating system that some particular action is occurring. For example, you may wish the operating system to notify your script when the user clicks his mouse on a button. By creating special procedures, you can have the code within those procedures execute whenever the specific event occurs.

A simple, but useful example of handling an event in VBScript is the *OnClick* event. You can use the *OnClick* event with certain objects, such as the button on a form. To write a subroutine that will execute whenever the event occurs, you must assign the subroutine the name of the object, followed by an underscore and the name of the event. For example, the following <input> tag contains a button which has been named SeaButton:

```
<input name="SeaButton" type="button" value="Sea">
```

To create a subroutine to handle the *OnClick* event for the button, you would name the subroutine *SeaButton_OnClick*, as shown here:

```
Sub SeaButton_OnClick
   'Statements
End Sub
```

The following example shows you how to create code that responds to the user clicking on the button. Within the SeaButton_Onclick subroutine, the code calls the *MsgBox* function to display a message each time the user clicks the SeaButton. To create the button, the HTML statements use an <input> tag within a form:

```
<html>
<head>
<title>Working with buttons</title>
<script language="VBScript">
<!--
Sub SeaButton_OnClick
   MsgBox "She sells seashells by the sea shore."
End Sub
-->
</script>
</head>
```

```
<body>
<h3>Working with buttons</h3><hr size="4">
<form>
<input name="SeaButton" type="button" value="Sea">
</form>
</body>
</html>
```

If you run the application and then click on the button, the application will display the message box as shown in Figure 6.5.

Figure 6.5. Using a form and button to create a response

Working with More Form Attributes

In a similar manner to adding a button to a form, you can easily add single-line text entry fields, radio buttons, check boxes, reset buttons, and submit buttons to a form. The following example will incorporate several *form* attributes, showing you how you can use these features in your own scripts. The example does not use scripting but instead relies completely on HTML.

The <form> tag uses the *Post* method and sends the submitted data to a file called *form2.asp*, which you will create later in this lesson. The format of the <form> tag is as follows:

```
<form method="POST" action="http://biggles/form2.asp">
```

To add a single-line text entry field to the form, you must use the <input> tag and set its *type* attribute to *text*. The following <input> tag, for example, creates a single-line text field named myName and assigns the field the initial value Rob:

```
<input type="text" name="myName" value="Rob">
```

To add radio buttons to the form, you also use the <input> tag, but you set the *type* attribute to *radio*. Because radio buttons, by definition let the user select only one of a set of buttons, for each radio button in a set you must use the same name. However, you assign each radio button a different *value*. You can pre-check a specific radio button by using the keyword CHECKED as a *stand-alone* attribute inside the <input> tag. To specify text that the user can see next to the radio button, simply type the desired text outside and immediately after the close of the <input> tag. The following code, for example, shows how you might add radio buttons to a form:

```
<input type="radio" name="myColor" value="blue">blue<br>
<input type="radio" name="myColor" value="red" CHECKED>red
```

Similarly, you can add check boxes to your forms. However, unlike radio buttons, you must assign a different name to each check box. Like radio buttons, you can pre-check any or all check boxes using the keyword CHECKED inside the <input> tag. Once again, type some text immediately after the close of the <input> tag to give the user some indication of what the check box represents. The following <input> tag creates a check box named NewUser:

```
<input type="checkbox" name="NewUser" value="1" CHECKED>1
```

Most of the forms you will encounter on the Web contain a reset button that lets you reset the form's values to their defaults and a submit button that lets you submit your data to the server. To add a reset button and a submit button to a form, you use an <input> tag, without requiring a *name* or *value* attribute. (You can use the *value* attribute if you want to change the button text.) As discussed, the reset button directs the browser to reset the fields in the form back to their default values. The submit button, in turn, directs the browser to submit the data to the target that you specify in the <form> tag's *action* attribute. The following statements illustrate how to add a reset and a submit button to a form:

```
<input type="reset">
<input type="submit">
```

Open up a new document in Notepad and name the file *form.asp*. Then, type in the following statements to create the form shown in Figure 6.6. Make sure you replace the target server in the <form> tag's *action* attribute from *biggles* to your own personal server:

```
<html>
<head>
  <title>Working with Form Attributes</title>
</head>

<body>
  <h3>Working with Form Attributes</h3>
  <hr size="2">
<form method="POST" action="http://biggles/form2.asp">
<p>Enter your name...
<input type="text" name="myName" value="Rob"></p>

<table cellspacing="40">
<tr><td>
<p>Choose a color...<br>
<input type="radio" name="myColor" value="purple">purple
<input type="radio" name="myColor" value="blue">blue<br>
<input type="radio" name="myColor" value="yellow">yellow
<input type="radio" name="myColor" value="green">green<br>
<input type="radio" name="myColor" value="orange">orange
<input type="radio" name="myColor" value="red" CHECKED>red
</p></td>

<td>
<p>Select one or more numbers...<br>
<input type="checkbox" name="chk_1" value="1" CHECKED>1
<input type="checkbox" name="chk_2" value="2">2
<input type="checkbox" name="chk_3" value="3">3
<input type="checkbox" name="chk_4" value="4" CHECKED>4
</p>
</td></tr>
</table>

<hr size="2">
<input type="reset">
<input type="submit">
</form>
</body>
</html>
```

When you load the file from inside your browser, your browser will display the form and its default values, as shown in Figure 6.6. Using the single-line text entry field, you can enter in your name. You can select one of six various colors. You can also check multiple check boxes numbered 1 to 4. The Reset button lets you reset the fields in the form to their defaults, and the Submit Query button will send the data to the target. When you stop to think about it, you have now created a complex form capable of submitting data to a server. However, you do not yet have a target file to receive that data, which brings you to the next section.

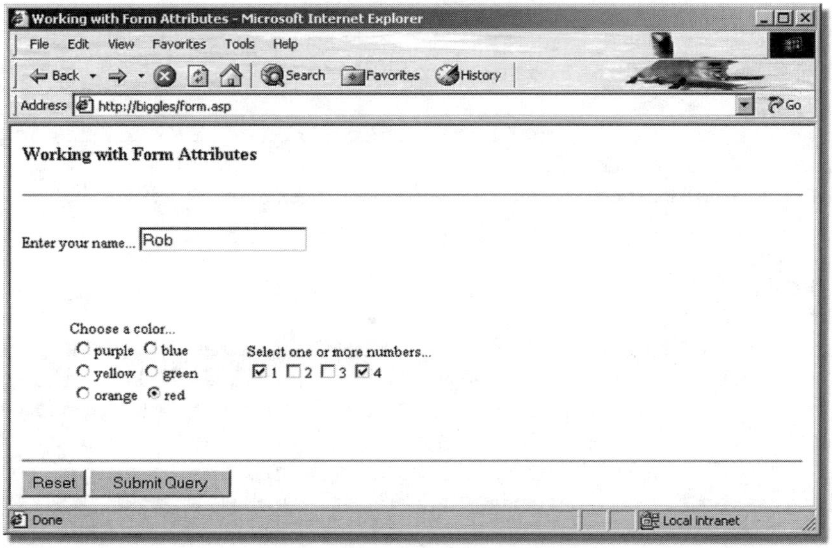

Figure 6.6. The default form layout and selections after running *form.asp*

Sneaking a Peek Ahead

Depending on how you want to use the data that a user submits by filling in a form, the statements that make up the Active Server Page that processes the incoming data will differ. One application, for example, might simply redisplay the data. A second might append the data to a text file, while a third application might store the data within a database. Using the following script, you can extract the relevant information from within the form data. Note that the code makes use of the *Request* object, which you will examine in detail in later lessons. For now, do not concern yourself with the object.

The Active Server Page *form2.asp* will process the data that you submit from the form in the *form.asp* application. The code will first receive the data using the *Request* object. Then the code will output a sentence that contains the date and time, as well as incorporating the form data itself within the sentence. Type the following script into Notepad and save the code as *form2.asp*:

```
<html>
<head>
  <title>Receiving Data in ASP</title>
</head>

<body>
  <h3>Receiving Data in ASP</h3>
  <hr size="4">
<p>On <% = date%> at <% = time%>,
<% = Request.Form("myName")%> bought
<% Dim total
total = cint("0") + Request.Form("chk_1")
total = total + Request.Form("chk_2")
total = total + Request.Form("chk_3")
total = total + Request.Form("chk_4")
%>
<% = total%> large <% = Request.Form("myColor")%>
seashells on the sea shore.</p>
</body>
</html>
```

To retrieve data from the submitted form, the Active Server Page uses the *Request* object. For example, to retrieve the value of the myName field, the code uses the following statement:

```
Request.Form("myName")
```

The code uses a variable named *total* to count the number of check boxes that are checked. If a check box has been selected, the statement Request.Form("checkbox_name") will return the value *1*. If the check box is not selected, the statement will return the value *0*. The function *CInt()* initializes the variable total to be of type *Integer* rather than of type *String*.

Make sure that *form.asp* and *form2.asp* are both in your home directory and then open up *form.asp* so that you can submit the form. Again, within *form.asp*, be sure that the *action* attribute is set for your own server and not set for *biggles*. Try entering different data into *form.asp* to submit, especially checking different check boxes, and then viewing the result. After evaluating the data, *form2.asp* should produce a page similar to that shown in Figure 6.7.

**Figure 6.7.
The result of
form2.asp receiving
the form data**

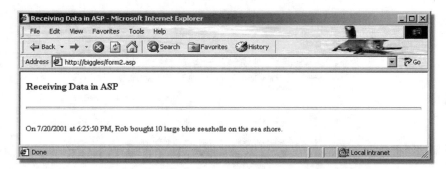

Receiving Data in ASP - Microsoft Internet Explorer

File   Edit   View   Favorites   Tools   Help

⇐ Back ▾ ⇒ ▾ ⊗ ⊡ ⌂ | ⊗Search ⊠Favorites ⊛History

Address 🔲 http://biggles/form2.asp

Receiving Data in ASP

On 7/20/2001 at 6:25:50 PM, Rob bought 10 large blue seashells on the sea shore.

Done   Local intranet

WHAT YOU MUST KNOW

In this lesson, you learned how to create simple Active Server Pages that display the current date and time, as well as pages that use VBScript looping constructs to execute statements one or more times. You also learned how to create HTML forms using the <form> and <input> tags, and how to submit the form's data to an Active Server Page for processing. In Lesson 7, "Understanding the Active Server Pages Model," you will step back and take a more holistic view of Active Server Pages so that you will understand the Active Server Pages model and how the different objects interact and work together. Before you continue with Lesson 7, however, make sure you understand the following key concepts:

- By placing your Active Server Page code between a <% tag and a corresponding %> tag, you indicate to the server what parts to evaluate.

- HTML pages do not handle dynamic data or information, such as time or days of the week, as well as Active Server Pages do.

- Many date and time functions are available that you can make use of in your Active Server Pages including Date, Time, Now, FormatDateTime, Year, Weekday, and others.

- By using loops in your Active Server Pages, you can save a lot of repetitive work and also handle situations where the exact number of iterations you require is unknown.

- The four main loops in VBScript are For-Next loops, For-Each loops, Do-While loops and While-Wend loops.

- By using forms, you can use Active Server Pages for data collection. You use the <form> tag when creating a form, along with its two major attributes: *action* and *method*.

- The <input> tag is extremely useful when working with forms. It allows you to add a large variety of different form attributes including buttons, single-line text entry fields, check boxes, radio buttons, and others.

- Events in VBScript let the operating system (or browser) notify your code when a specific action occurs. By writing a subroutine for a specific event, you can execute code whenever it occurs.

UNDERSTANDING THE ACTIVE SERVER PAGES OBJECT MODEL

n Lesson 6, "Writing Your First Active Server Page," you learned how to start programming simple Active Server Pages, using many of the skills that you have learned so far in this book. In this lesson you will step back and take a more holistic view of Active Server Pages. You will learn in greater detail how Active Server Pages actually work and the technology driving it. You will understand not only the Active Server Pages object model but the overall manner in which the different objects interact and work together. By the time you finish this lesson, you will understand the following key concepts:

- When a user requests an HTML page, the browser sends the request to the server, which in turn looks for the file and then returns it to the user's browser.

- When a user requests an Active Server Page, the request is sent to the server, which in turn passes it to the Active Server Pages server component. The Active Server Pages server component translates the script into HTML and passes this to the server, which in turn passes it to the user's browser.

- A popular standard, the Common Gateway Interface (CGI) basically allows a remote application or script to run from within a Web page.

- Internet Server Application Programming Interface (ISAPI) applications are nothing more than Dynamic Link Library (DLL) files. The Active Server Pages server component is an ISAPI application named ASP.DLL.

- CGI applications require the server to load each instance of the application separately, while ISAPI applications only require a single instance to be loaded by the server. This makes ISAPI applications faster and less resource intensive than CGI applications.

- The Active Server Pages server component, ASP.DLL, contains seven built-in objects that make up the Active Server Pages object model. Each built-in object may contain properties, collections, methods, and events.
- The seven objects that make up the Active Server Pages object model are the Application, ASPError, ObjectContext, Request, Response, Server, and Session objects.
- All scripting languages, when used inside a Web page for client-side scripting, are within the <script> and </script> tags. Client-side scripting is completely browser dependent, and the processing power to run the script is placed squarely on the client's computer.
- Server-side scripting is always contained between the <% and %> tags. With server-side scripting, the handling of the script is done, as the name suggests, on the server and is not dependent on the user's browser.

Understanding How Static HTML Pages Are Loaded

To appreciate how an Active Server Page is loaded by a browser and the entire process that it goes through from the initial page request right through to the final display for the user, you must first take a look at how an HTML page is loaded by a browser, for comparison.

An HTML page is merely a file of text. It is not dynamic but rather the opposite—remaining static and unchanging for long periods of time regardless of who or how many people decide to view it. When a request for an HTML page is made from the user's browser, it sends this request to the server. The server, in turn, searches through its files to locate the page. If the page cannot be found, the server will send back an error, often in the form of a 404 error HTML page. Assuming the page is found the server simply sends this back to the user's browser, where it is displayed on the screen.

Figure 7.1 shows the entire three-step process of loading an HTML page in a graphical form for clarity.

Figure 7.1.
The steps the
browser and server
perform when load-
ing an HTML page

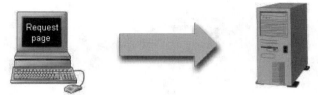

User requests HTML page from server

Server searches for HTML page

Server sends HTML page to user's browser

Understanding How Active Server Pages Are Loaded

Unlike the process with loading HTML pages into your browser, the process of loading Active Server Pages has a lot more steps and requirements. It is important to recall that Active Server Pages are dynamic in nature and that you cannot simply have a ready-made HTML page waiting to be sent to the user. The Active Server Pages server component must create each HTML document after translating all the relevant script sections in each Active Server Page.

First, the user submits a request for the Active Server Page from the server. The server, in turn, passes this request on to the Active Server Pages server component. The Active Server Pages server component then translates any script in the Active Server Page and produces a HTML page, which it passes to the server. The server then passes the HTML page back to the user's browser. Figure 7.2 shows the process of loading an Active Server Page.

Figure 7.2.
The steps the
browser, server,
and Active Server
component perform
to load an Active
Server Page

User requests Active Server Page from server

Server hands request to ASP server component

ASP server component translates script into HTML

ASP server component hands HTML page to server

Server sends HTML page to user's browser

Learning About the Roles of CGI and ISAPI

Now that you know the role that the Active Server Pages server component plays in processing requests for Active Server Pages from the user, you must learn what the Active Server Pages server component actually is.

When the Internet was first adding something more than static pages to its repertoire, the technology that was giving it this ability was known as Common Gateway Interface (CGI). CGI is actually a standard for interfacing external applications with Web servers. CGI applications were very popular because you could do more than simple static pages. You could basically run an application from the Web and depending on your programming preferences, you could use anything from C++ to Perl. It did not matter whether you made use of a compiled program or a script. CGI faced some early criticism over its security holes. The biggest problem with CGI applications, however, was the fact that every time a user made a request that involved the CGI application, the server had to load a separate instance of the application into memory. This was, as you can no doubt imagine, extremely taxing on the server's limited resources.

Microsoft put support into Internet Information Server, not only for CGI applications, but also for a new technology of its own design, known as ISAPI. ISAPI stands for Internet Server Application Programming Interface, although you will hardly ever see any reference to anything other than the shorter acronym. ISAPI actually performs a similar task to CGI and some might question the need for such a technology. But ISAPI was designed to address CGI's biggest flaw—it only requires the server to load a single instance of its applications to cater to multiple requests. As a direct result of this fact, ISAPI applications are not only less taxing on the server to run, but are also faster than their CGI application cousins. The speed comes about because there is no longer any loading time, which includes disk access, as well as a greater availability of memory.

ISAPI applications are nothing more than Dynamic Link Library (DLL) files. Each ISAPI application is actually a single DLL file, of which there are many available both commercially and for free. Programmers can write their own ISAPI applications to cater to their precise requirements.

The Active Server Pages server component is nothing more than an ISAPI application. The name of its actual DLL file is ASP.DLL.

Learning About the Active Server Pages Object Model

The Active Server Pages server component, ASP.DLL, contains built-in objects that are readily available to you when you create your own Active Server Pages applications. As you will shortly see, the term for your collections of HTML and script files is known as an Active Server Pages application. In some books and online references it may often be referred to as an ASP application. There are seven main objects that comprise the Active Server Pages object model, as shown in the following list:

- Application object
- ASPError object
- ObjectContext object
- Request object
- Response object
- Server object
- Session object

These seven objects are known as the intrinsic Active Server Pages objects. They are all built-in and do not require additional files, making up the core of Active Server Pages functionality. Each of the objects can contain properties, collections, events, and methods. In later lessons, you will learn in detail about each of the seven intrinsic objects of the Active Server Pages object model, as well as the various properties, collections, events, and methods, if any, that they support. Right now, however, it would be better to quickly understand what each of these features are and what the difference is between them.

Properties Store an Object-Specific Value

A property is simply an attribute of an object to which you can assign a value. For example, an object might have three attributes: a minimum value, a maximum value, and a current value. The properties of this imaginary object might be Min, Max, and Current. Your code might assign values to the Current property, according to user input, only if it is within the range of the Min and Max properties. The property is really just the tag or label to which a value is attached. Some of the objects in the Active Server Pages object model also have properties that you can make use of.

Collections Are Containers That Hold Objects

A collection is really just an object that contains other objects. Usually the objects that the collection contains all have some relation to each other, which makes them easy to access from your code, as well as being easy to locate as a programmer. For example, the Request object has a Cookies collection that contains a list of cookies and their values, which you can use to present the user with a more personal page. Some of the objects in the Active Server Pages object model also have collections that they expose and that you can make use of.

Methods Are Executable Functions

A method is like a statement or function that you apply to one or more objects. As an example, think in the abstract for a moment and pretend that painting something red is a method known as PaintRed(). But you must indicate what you want to paint red—the wall, the door, or the fence. So you have both a method (PaintRed), and also various objects (wall, door, fence) to which you can apply the method. Some of the objects in the Active Server Pages object model also have methods that they expose and that you can make use of.

Events Let Objects Respond to Specific Actions

An event is basically something that occurs as a direct result of an action. It is like being notified by the operating system that a certain action is occurring. You can use these events to perform some code. A simple example of an event might be the clicking of a button. You might want some code to execute whenever this event occurs. By using the OnClick() event of the button, you can actually execute some particular code whenever this event occurs. If you think of the button in this simple example as the object, then the OnClick() event that it exposes to you is the handle that you can use to get hold of this notification and execute your own code. Some of the objects in the Active Server Pages object model also have events that they expose and that you can make use of.

Putting It All Together—Reviewing the Intrinsic Objects

In summary, the Active Server Pages object model comprises seven intrinsic objects, each of which may contain properties, collections, methods, and events. These make up the core of Active Server Pages. A short explanation of each of the seven intrinsic objects follows, in alphabetical order, showing some of the interaction between objects, as well as their purpose.

*The Application object is basically all of the files contained within a virtual directory and all of its subdirectories. This is why collections of files that make up a particular Active Server Pages site are referred to as Active Server Pages applications or ASP applications. You can access all of the global variables and objects that are in the **global.asa** file, through use of the Application object.*

The ASPError object provides you with greater error handling information than the Err object that comes with VBScript and is a new addition in Active Server Pages 3.0. You use the ASPError object to catch bugs and errors in your scripts and code.

The ObjectContext object, a new addition in Active Server Pages 3.0, replaces the old ScriptingContext object, which existed in the old Active Server Pages object model. Basically, the ObjectContext object handles transactions, where an entire sequence of steps either succeeds or fails as a whole. You can use the ObjectContext object with Microsoft Transaction Server (MTS) and COM+ Component Services.

The Request object handles all of the requests from the user, including data from cookies, forms, and security certificates. You use the Request object when getting information from the user.

The Response object handles sending data back to the user. You have already made use of the Response object's Write method, both explicitly and implicitly. The Response object's IsClientConnected property checks to see if the connection between the user (or client) is still active, which relies on the Session object behind the scenes.

The Server object is basically your Web server. You can use the Server object to incorporate some of the server's own functionality into your Active Server Pages. You can use the Server object to handle script timeouts or execute other Active Server Pages from within an Active Server Page.

The Session object maintains a connection between your Active Server Pages application and each user. The Web is basically a connectionless medium, at least when using HTTP 1.0 standards, and so the Session object prevents users from having to authenticate repeatedly.

You now have an overview of the seven intrinsic objects that make up the Active Server Pages object model. In the following lessons you will look more closely at each of these important objects, as well as the various features that they expose to you as a programmer.

Coming to Grips with Client-Side Scripting

Although, you have already seen the basics of VBScript and probably understand it quite well, it is important for you to understand the difference between server-side scripting and client-side scripting.

Client-side scripting began when Netscape Communications and Sun Microsystems teamed up to create something that would allow interactivity within Web pages. What they came up with was a scripting language that the Netscape Navigator browser would support. They first named this scripting language LiveScript, although with all the marketing hype concerning Java and the Web, LiveScript was renamed to JavaScript. This renaming has led to a lot of confusion, as JavaScript is nothing to do with Java although its syntax has similarities.

JavaScript was limited to its interaction outside of the browser to attempt to make it secure. If JavaScript simply took over your computer and was able to do whatever it wanted, there would have been chaos over the security hole. After some teething problems though, JavaScript became not only popular but also safe and secure to use.

JavaScript awakened the interest of many people, with its ability to make Web pages interactive. Using JavaScript, you could perform mouseovers, changing the content of the page depending on the location of the user's mouse. You could create forms that would react when the user made use of the buttons. In short, the Web became interactive.

Microsoft, in the true spirit of competition, created not just one but two scripting languages to compete with JavaScript for client-side scripting. One was VBScript, which is exactly the same syntax as the extremely popular Visual Basic. More people use and know Visual Basic than any other programming language, so VBScript quickly became popular because programmers realized that they already knew this new language and that there was practically no learning curve for them. Microsoft also created JScript, an adaptation of JavaScript with some changes of their own. Other scripting languages, such as PerlScript, Python, and Awk, also added to the confusion.

With all of these scripting languages and various versions, together with the fact that each language depends on the browser to support it, development became a nightmare. A script that works for one user's browser does not work for another's. And a script that runs on a third user's browser will not run on the first user's. It all became too messy with so many competing scripting languages and browser requirements. In June of 1998, the European Computer Manufacturers Association (ECMA) released the standard for an Internet scripting language called ECMA-262, or ECMAScript. Microsoft, Netscape and others in the industry took on this standard and put support for it into their browsers. Informally, people continue to call ECMAScript by the old name of JavaScript.

Now that you know something of how the scripting languages came into being, it is time to look at some of the details for using them with client-side scripting. Regardless of which scripting language you use in your Web page, you must enclose your scripting code between the tag <script> and the closing tag </script>. The following example shows how to open a message box using VBScript and client-side scripting:

```
<html>
<head>
<script language="vbscript">
<!--
sub hello()
   MsgBox "Hello, world"
end sub
-->
</script>
</head>

<body onload="hello()">
</body>
</html>
```

By modifying just the code within the <script> tags, you could do the same thing using JavaScript, as shown in the following code snippet:

```
<script language="javascript">
<!--
function hello()
{
  alert("Hello, world")
}
-->
</script>
```

The key point for you to note here is that all scripting languages, when used inside a Web page for client-side scripting, are within the two <script> tags. You may also have seen the two tags <!-- and --> which denote a comment in HTML. This is really nothing more than a common safety net in case your client-side script is loaded by a browser that does not support that particular scripting language, or any scripting language as the case may be.

When code is written for client-side scripting, the processing power to run the script and do all the work is placed squarely on the client's computer. This can be a good thing because it relieves the server of this burden. The obvious downside to this is that you are relying on the client's computer to have the right files and versions to support what you are doing. Client-side scripting is completely browser dependent.

Coming to Grips with Server-Side Scripting

You now know all about client-side scripting, or at least the history of it and how it works. In comparison, server-side scripting, which is what Active Server Pages is all about, takes a different approach to creating interactive Web pages. The handling of the script is done, as the name suggests, on the server and is not dependent on the user's browser.

Recall from earlier in this lesson that when a user requests an Active Server Page, the request is given to the Active Server Pages server component, ASP.DLL, before the resulting HTML page is sent back to the user's browser. The Active Server Pages server component will open the Active Server Page and must determine which sections containing scripting languages are for it and which are client-side scripts.

Just as client-side scripting is always contained between the <script> and </script> tags, server-side scripting is always contained between the <% and %> tags. You came across this tag syntax in Lesson 6. So far, you have only seen the short form of the Response object's Write method; however, you can write lengthy scripts between the <% and %> tags, as well. The following two server-side scripts are identical, except one uses the short form of the Response object's Write method:

```
<%=time%>
```

and

```
<%Response.Write time%>
```

If you must use the character sequence %> somewhere within your code, the Active Server Pages server component will treat this as the end of your script. So the following example is the wrong way of doing things:

```
<%

Response.Write "This is an example of setting the width of table data..."
Response.Write "<td width=75%>"
Response.Write "And that's all we have time for."

%>
```

In this example the script would end abruptly as soon as the Active Server Pages server component hit the first %> characters. The third Response.Write line would not have been processed properly and the result would not be what you would expect. Instead, you should use the escape sequence %\> to output the characters and still avoid abruptly ending your script. So the same example written correctly would be:

```
<%
Response.Write "This is an example of setting the width of table data..."

Response.Write "<td width=75%\>"

Response.Write "And that's all we have time for."

%>
```

As with the client-side scripting, you can specify which scripting language you want to use with your server-side scripting. The default is to use VBScript; however, you can explicitly specify which language by using one of the following lines:

```
<%@   language="vbscript" %>
```

or

```
<%@ language="jscript" %>
```

WHAT YOU MUST KNOW

In this lesson you have seen not only a greater overview of Active Server Pages, but also gained an understanding of the technologies driving Active Server Pages, such as ISAPI and server-side scripting. The Active Server Pages object model comprises seven intrinsic objects that are always available to you when developing your Active Server Pages. The seven intrinsic objects make up the core of Active Server Pages, and they include the Application, ASPError, ObjectContext, Request, Response, Server, and Session objects. In Lesson 8, "Using the Request Object," you will learn about the Request object and all the collections it has, as well as its single property and method. You will cover in detail the Get and Post methods and how to use the Request object to retrieve data from the user. Before you continue with Lesson 8, however, make sure you understand the following key concepts:

- There is a different process that is gone through when a user requests an HTML page and an Active Server Page. With an HTML page, the server simply finds the file and sends it to the user's browser, but with an Active Server Page, the server passes the request to the Active Server Pages server component to translate into HTML before receiving it back and being able to send it to the user's browser.

- The Common Gateway Interface (CGI) lets scripts or applications run from a Web page.

- CGI applications and ISAPI applications differ in loading instances of applications into server memory. CGI applications require separate instances, while ISAPI applications require only a single instance, making ISAPI applications faster and less resource intensive than CGI applications.

- Internet Server Application Programming Interface (ISAPI) applications are Dynamic Link Library (DLL) files. ASP.DLL is the name of the Active Server Pages server component, which is an ISAPI application.

- Seven intrinsic objects make up the Active Server Pages object model, with each object containing different properties, collections, methods, and events, if any.

- The Application, ASPError, ObjectContext, Request, Response, Server, and Session objects are the seven intrinsic objects that make up the Active Server Pages object model.

- Browser dependent and using the client computer's processing power to run scripts, client-side scripting places all its script, regardless of scripting languages, within the <script> and </script> tags.

- Server-side scripting, in comparison, is not dependent on the user's browser and uses the server's processing power to run scripts. You always place server-side script between the <% and %> tags.

USING THE REQUEST OBJECT

n Lesson 7, "Understanding the Active Server Pages Object Model," you looked at some of the key technologies driving Active Server Pages, as well as the Active Server Pages object model. In this lesson you will take a close look at one of those intrinsic objects in the Active Server Pages object model—the Request object. By the time you finish this lesson, you will understand the following key concepts:

- There are two main request methods used by Active Server Pages that you use to send data back to a server—Get and Post. Unlike the Post method, the Get method appends the data to the end of the target URL.

- You set the method to be Get or Post using the <form> tag's method attribute.

- To find out the total number of bytes that has been sent, you can use the TotalBytes property, which is a read-only property.

- The BinaryRead method reads in binary data from the user's request and for an argument you can pass it the value of the TotalBytes property.

- When the user submits a form to the server using the Post method, you can use the Form collection to read in the data.

- All collections in Active Server Pages have three properties in common. They are the Count property, the Item property, and the Key property.

- As the Item property is the default property of collections, you can make use of an implicit call rather than an explicit call when putting it to use. You specify the exact name of the field whose value you want to return, as the sole argument of the Item property.

- Using the Key property and the Count property, you can find out the number of elements and what their field names are. Combining this information with the Item property allows you to retrieve the values of all fields.
- The QueryString collection works in the same way as the Form collection, except that you use the QueryString collection with the Get method.
- To create some form of persistence or personalization in Web pages, you can read in data from cookies using the Cookies collection. Cookies can also have a cookie dictionary, which is basically a list of fields and values.
- The Request object is one of the seven intrinsic objects that comprise the Active Server Pages object model. It contains five collections, one method, one property, and no events.

Introducing the Request Object

Basically, the Request object is what you use for handling requests from the user or client. So you use the Request object whenever you want to retrieve data that has come from the user, be that in the shape of a form, a cookie or otherwise. Table 8.1 shows you an overview of the Request object in terms of the collections, events, methods, and properties that it contains.

Collections	Events	Methods	Properties
ClientCertificate	-	BinaryRead	TotalBytes
Cookies			
Form			
QueryString			
ServerVariables			

Table 8.1. An overview of the Request object

Clearly, the Request object is made up mostly of various collections, with only a solitary method and property. There are no events associated with the Request object.

Knowing the Difference Between Get and Post

In Lesson 6, "Writing Your First Active Server Page," you learned the basics of how to create a form that the user could fill in to send data to a server. At that time you were using the Post method, but you were not really made aware of all that this implies. There are two main request methods that you use to send data back to a server—Get and Post.

The Get method can retrieve all types of data but has a limit on the amount of data (about 2,000 characters) that it can retrieve. The reason for this limit is because the Get method appends the data to the end of the target URL in clear text using a lot of question marks and percentage signs, as you will probably have seen when visiting many Web pages. A URL will usually have a limit on its size, depending largely on the browsers and servers involved, which is of course passed on to the Get method.

The Get method also raises questions about security. Due to passing clear text to the end of the URL, the Get method makes a very poor choice for handling sensitive data such as passwords, income, and personal information. You would most commonly use the Get method to ask the server to "get" information, such as a document, and send it to the user's browser.

The Post method, on the other hand, can contain much more information than the Get method and does not place all the information on the URL. Instead, the Post method transfers the data mostly in the body of the request being sent to the server. The Post method is typically used when submitting form data to a server for processing.

You should always use the Post method when you want to make changes, such as when fielding form submissions that change a database. You must never use the Get method in these instances. The Post method is the safest method to use and is often more efficient for servers to handle, making them faster than the Get method.

To help make things clearer, take a look at the following example script for a form in which the user submits their favorite color. The name of the field that they are assigning a value to before submitting is *usercolor*. Notice that the method attribute of the <form> tag is set to the Get method:

```
<html>
<head><title>Binary Form</title></head>
<body>
<form action="binary.asp" method="get">
Type in your favorite color below:<br>
<input type="text" name="usercolor" value="purple">
<input type="submit" value="Submit this color">
</form>
</body>
</html>
```

Figure 8.1 shows the URL with the appended information from the Get method, as a result of running this code and submitting the default color of purple. Notice how clearly you can see both the name of the field and the value that is being assigned to it. Imagine if this was your users' credit card details or passwords when they were using a site that you created.

If you change nothing more than the method of the request to the Post method, as in the following line of script, you will see a noticeable difference in the URL field.

```
<form action="binary.asp" method="post">
```

After making the change in the example script, as shown above, and running the script again, you will notice that when you submit the default color of purple, nothing is appended to the URL, as shown in Figure 8.2.

Reading in Binary Data from the User

There are two elements of the Request object that you use when retrieving binary data from the user. The first of these elements is the TotalBytes property, which is the only property of the Request object. The TotalBytes property is a read-only property that tells you the total number of bytes that has been sent. This means basically that you cannot assign a value to this property, as you can with some other properties. The only action you can take with the TotalBytes property is to read the value that it contains by assigning it to a variable, as shown:

```
Dim myTotalBytes
myTotalBytes = Request.TotalBytes
```

The second element that you use when retrieving data from the user is the BinaryRead method. The BinaryRead method is the sole method of the Request object. There are no other methods to learn with this object. The BinaryRead method reads in binary data from the user's request and only takes one argument and that is the number of bytes to read in. Strangely enough, you know exactly how to retrieve the total number of bytes to read in already. Combining what you already know with the BinaryRead method, you are able to write Active Server Pages script to handle reading binary data. The following code would serve as a simple form where the user inputs his favorite color:

```
<html>
<head><title>Binary Form</title></head>
<body>
<form action="binary.asp" method="post">
Type in your favorite color below:<br>
<input type="text" name="usercolor" value="purple">
<input type="submit" value="Submit this color">
</form>
</body>
</html>
```

You will notice that in the code above, you were using the form's *action* attribute to submit the form to a page called *binary.asp*. You can see the code for *binary.asp*, which does not do anything other than read in the binary data from the form, as shown:

```
<html>
<%
Dim myTotalBytes
Dim myBData

myTotalBytes = Request.TotalBytes
myBData = Request.BinaryRead( myTotalBytes )
%>
</html>
```

Although you can now read in binary data from the user, it would be fair to say that in practice you would probably seldom do so. You would have to parse or sort through a lot of data including header information to get to the information portion that you were after, which is a lot of work unless you purchase commercial software to make the task easier. Rather, you would more typically make use of the collections in the Request object to retrieve data from the user.

Reading Data with the Form Collection

The Request object's Form collection is a really nice and simple way to retrieve information from the user. As its name implies, you use the Form collection when the information being sent to the server comes as the result of a form submission from the user. When you intend to use the Form collection to retrieve data, you must ensure that the form that the user submits is using the Post method. The Form collection will then handle all of the data extraction from within the main body of the request without you having to do anything.

The Form collection has three properties that are all very straightforward. The three properties are the Count property, the Item property, and the Key property. The most important and most often used of these three properties is the Item property, which is the default property of the Form collection.

The Item property allows you to retrieve the value that has been assigned to a particular field. So if you know that the name of a field in the form is *usercolor*, you could retrieve the value that has been assigned to *usercolor* by using the Form collection's Item property to assign it to a variable, as shown in the following code snippet:

```
Dim myColor
myColor = Request.Form.Item("usercolor")
```

As mentioned earlier, the Item property is the default item of the Form collection, which means that if you do not explicitly state which property you are referring to, the server will use the Item property. In practice, you will most likely use the implicit call of the Item property. So the previous example could also be written as follows:

```
Dim myColor
myColor = Request.Form("usercolor")
```

Looking at a more complete example, you will see the entire process from creating fields through to using the Form collection's Item property to retrieve the form data. Create a file with the following script and save it as *icecream_form.asp*:

```
<html>
<head>
<title>Ice Cream Form</title>
</head>

<body>
<h2>The Ice Cream Form!</h2>
<form action="icecream.asp" method="post">
<h3>Please enter your name:</h3>
First name:<input type="text" name="firstname"><br>
Last name:<input type="text" name="lastname">
<h3>Select your favorite ice cream flavor below:</h3>
<input type="radio" name="flavor" value="chocolate">
Chocolate<br>
<input type="radio" name="flavor" value="strawberry">
Strawberry<br>
<input type="radio" name="flavor" value="vanilla">
Vanilla<br><hr>

<input type="reset" value="Reset Information">
<input type="submit" value="Submit Information">
</form>
</body>
</html>
```

Notice that there are lots of fields but when you think about it properly, only three of these fields will be useful information. The *firstname* and *lastname* fields are quite straightforward. The *flavor* field is a single field even though there are multiple occurrences of it. The *flavor* field can be only one of chocolate, strawberry or vanilla but not more than one. If you had it otherwise, and received three *flavor* fields—one with chocolate, one with strawberry and one with vanilla—you would not be able to do anything with them. Also the submit and reset fields are not necessary to be sent to the server, as their entire function and usefulness is at the user's end.

For example, assume that your user, Homayoun Yazdani, loads the page *ice-cream_form.asp*, enters his name in the correct fields, and selects the Vanilla option before submitting the form, the browser might look something similar to that shown in Figure 8.3.

Figure 8.3.
After loading
the page
icecream_form.asp
and entering in data

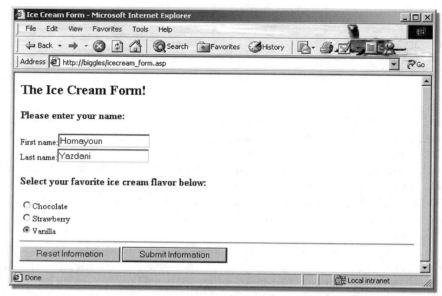

Now that you know what to expect, you must prepare the Active Server Page, *icecream.asp*, which will handle retrieving the data from the form using the Request object's Form collection. All you will do in this example is read in the data, assign it to variables, and then display it back to the user. Use the implicit form of the Form collection's Item property, so that the code should look something similar to the following:

```
<html>
<head>
<title>Ice Cream Submission Successful</title>
</head>

<body>
<h2>Ice Cream Submission Successful</h2>
<%
Dim name1, name2, userflavor
name1 = Request.Form("firstname")
name2 = Request.Form("lastname")
userflavor = Request.Form("flavor")
%>
The user, <%=name1%> <%=name2%>, likes <%=userflavor%>
ice cream.
</body>
</html>
```

If you recall the <form> tag in *icecream_form.asp*, you set the target using the active attribute to be *icecream.asp*. So when your Active Server Page loads as a result of form submission from a user, it will look similar to Figure 8.4.

Figure 8.4.
The result of
icecream.asp
receiving form data

You now know the main use of the Form collection; however, there are still two more properties that you should look at for completeness. The Count property simply tells you how many items there are in the collection. So if you had a form with five fields to pass to the server, and then you submit the form, the Count property would return five. The Count property is a read-only property, as you would never have any need to assign a value to it.

The final property is the Key property. The Key property allows you to retrieve the name of a particular field. By using the Key property together with the Count property you can iterate through all of the fields of data without knowing how many there are or what their names are. These two properties could be useful if you were to set up a generic Active Server Page that could accept data from many different forms, as shown in the following example:

```
<html>
<head>
<title>Generic Form Submission</title>
</head>

<body>
<h2>Generic Form Submission</h2>
<%
Dim name, value, i, totalcount
totalcount = Request.Form.Count
for i=1 to totalcount
  name = Request.Form.Key(i)
  value = Request.Form(name)
```

```
    Response.Write name & ":   " & value & "<br>"
next%>
</body>
</html>
```

To see an example of your generic form, try using your page *icecream_form.asp*, but just change the target to be your new page *generic.asp* in the <form> tag, as follows:

```
<form action="generic.asp" method="post">
```

The result of your page, *generic.asp*, after receiving a submission from *icecream_form.asp* would look similar to that shown in Figure 8.5, but bear in mind that *generic.asp* could handle information from any form submission, which could be very handy to you for form debugging purposes among other uses.

Figure 8.5.
The result of
generic.asp
receiving a form
submission

Reading Data with the QueryString Collection

Just as you use the Forms collection with the Post method, you use the QueryString collection with the Get method. The QueryString collection is basically the twin of the Forms collection except that you use it for request information that is sent using the Get method, rather than the Post method. Like the Forms collection, the QueryString collection has three properties and—surprise, surprise—they are the Count property, the Item property, and the Key property.

To see an example of the QueryString collection and all its properties, you can simply edit the files from the previous example. Firstly, in *icecream_form.asp* you would have to change the method in the <form> tag to use the Get method, as follows:

```
<form action="generic.asp" method="get">
```

After you have made the change to the Get method for the form, you must replace every occurrence of Request.Form in the file *generic.asp* that you made earlier so that the Active Server Pages section looks like the following:

```
<%
Dim name, value, i, totalcount
totalcount = Request.QueryString.Count
for i=1 to totalcount
   name = Request.QueryString.Key(i)
   value = Request.QueryString(name)
   Response.Write name & ":   " & value & "<br>"
next%>
```

Loading the page *icecream_form.asp* and submitting data will result in the same final page as you saw with the Forms collection example, except that the URL will contain all of the fields and their values as well.

Handling Information Regardless of Method

You know that to handle data that is sent from a form using the Post method you use the Request object's Form collection. Similarly, you know that to handle data that is sent using the Get method you use the Request object's QueryString collection. You may not be aware, however, that you can actually handle either method using the same statement. Basically, by not specifying any collection, Active Server Pages will try all of the collections, one at a time, in an attempt to handle the data. As an example, if you submit a form regardless of method and want to retrieve the value of the flavor field, you might use something similar to the following:

```
Dim myValue
myValue = Request("flavor")
```

Reading Cookies from the User

Cookies in general are small files of clear text that contain information, such as user preferences, to create some form of persistence when changing pages within a site or for subsequent visits. You can use the Request object's Cookies collection to retrieve cookies from the HTTP body of the user's requests.

The basic use of the Cookies collection is very straightforward. Like all collections in Active Server Pages, the Cookies collection has an Item property which is set as the default property, as well as a Key property and a Count property. You will most often only use the implicit form of the Item property to retrieve a specific value from the cookie, as shown in the following code snippet:

```
Dim usercolor
usercolor = Request.Cookies("lastcolor")
```

To understand cookies properly, it is important to realize the whole process that you must undertake. This section is dealing with reading in values from cookies, but the other side of the coin, as it were, is writing those cookies in the first place, which you will learn in the next lesson, "Using the Response Object." The two parts are necessary for the whole process to work. In other words, you cannot write the cookie without having the form to submit first, yet when writing the form (what you are doing now), you have no cookie to read! For now, simply understand how to read in data using the Cookies collection and then in the next lesson you will complete your knowledge of cookies when you learn to write them.

Open up Notepad to create a new Active Server Page and call it *cookie_form.asp*. In *cookie_form.asp*, you are going to create a form that not only submits data to the server but also personalizes the form itself, depending on the user's last choice of color. The page will not only specify what the last selection was but will also select that particular radio button for the user. It will do this by using a function within the Active Server Pages portion of the page. You can use this concept for a variety of practical situations, such as shopping carts, layout preferences, remembering a client's last search, and many others. The entire example code for *cookie_form.asp* is as follows:

```
<html>
<head>
<title>Cookie Form</title>
</head>

<body>
<h2>The Cookie Form!</h2>
<%
Dim usercolor
usercolor = Request.Cookies("lastcolor")
Response.Write "On your last visit you selected the color <b>"
& usercolor & "</b>."

Function GC(color)
Dim result
If usercolor=color then
   result = color & chr(34) & " CHECKED>" & color
Else
   result = color & chr(34) & ">" & color
End If
GC = result
End Function
%>
<form action="cookie.asp" method="post">
<h3>Select a color below:</h3>
<input type="radio" name="myColor" value="<%=GC("purple")%>
<input type="radio" name="myColor" value="<%=GC("blue")%>
<br>
<input type="radio" name="myColor" value="<%=GC("yellow")%>
<input type="radio" name="myColor" value="<%=GC("green")%>
<br>
<input type="radio" name="myColor" value="<%=GC("orange")%>
<input type="radio" name="myColor" value="<%=GC("red")%>
<br><hr>
<input type="reset" value="Reset Information">
<input type="submit" value="Submit Information">
</form>
</body>
</html>
```

Although you will have to complete the section on using the Response object with the Cookies collection to see the full effect of this code, you can see the eventual outcome of *cookies_form.asp* in Figure 8.6.

**Figure 8.6.
Using cookies
to create more
personal pages**

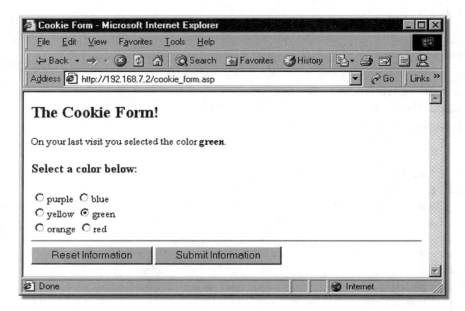

Cookies can also have a cookie dictionary, which is similar to a collection. This is basically where the cookie contains a list of other fields and values. For example, you may write a cookie that is called *UName* and inside that cookie you store three subkeys: *firstname*, *middlename*, and *lastname*. You can check if a cookie has subkeys by using the Cookies collection's HasKeys before you reference them, as follows:

```
Dim anysubkeys, value1, value2, value3
anysubkeys = Request.Cookies("UName")
if anysubkeys then
  value1 = Request.Cookies("UName")("firstname")
  value2 = Request.Cookies("UName")("middlename")
  value3 = Request.Cookies("UName")("lastname")
end if
```

Additional Collections of the Request Object

The ClientCertificate collection and the ServerVariables collection are also part of the Request object; however, both of these collections you will learn about in Lesson 24, "Being Security Conscious with Active Server Pages." At the same time as you learn about these two collections, you will also be learning about other relevant security objects in Active Server Pages.

WHAT YOU MUST KNOW

One of the seven intrinsic objects that comprise the Active Server Pages object model, the Request object contains mostly collections, with only one method and one property. Using the Forms collection with the Post method and the QueryString method with the Get method, you are able to easily retrieve data from user requests and form submissions. By reading in data from cookies using the Cookie collection, you are able to personalize your Active Server Pages and provide a state of persistence for the user. In Lesson 9, "Using the Response Object," you will learn how to send data back to the user, including working with the buffer, splitting responses, and creating cookies. Before you continue with Lesson 9, however, make sure you understand the following key concepts:

- Get and Post are the two main request methods that you use to send data back to a server. The Get method appends the data to the end of the target URL, whereas the Post method sends it in the body of the HTTP request.

- Using the <form> tag's method attribute, you can set the method as Get or Post by assigning "get" or "post" respectively.

- The TotalBytes property allows you to find out the total number of bytes that the user has sent. Together with the BinaryRead method, you can read in binary data from the user's request.

- You can use the Form collection to read in data whenever a user, using the Post method, submits a form to the server.

- There are three properties that all collections in Active Server Pages have in common: the Item property, the Count property, and the Key property.

- By not explicitly specifying a collection property, the server will use the default property, such as the Item property.

- The only argument that you need to supply with the Item property is the name of the field whose value you want to return.

- The Key property will return the actual field name of any given element, and the Count property will return the total number of elements.

- Except for the fact that the QueryString collection works with the Get method and the Form collection works with the Post method, the QueryString collection works in the same way as the Form collection.

- By using cookies and reading in cookie data via the Request object's Cookies collection, you are able to add a personal touch to your Active Server Pages. Basically lists of fields and values, cookie dictionaries are an additional feature of cookies.

USING THE RESPONSE OBJECT

n Lesson 8, "Using the Request Object," you took a close look at the Request object and how you can make use of it to read in data from the user. In this lesson you will take a close look at another one of those intrinsic objects that make up the Active Server Pages object model—the Response object. Just as you learned how to handle user requests and receiving data from the user, now you will learn how to handle responses to those requests and how to send data back to the user. By the time you finish this lesson, you will understand the following key concepts:

- You use the Response object whenever you want to send data to the user, regardless of whether it is text strings, cookies, or otherwise.
- The Response object is made up almost totally of methods and properties. There is one collection—the Cookies collection—and no events.
- Arguably the most often used method in Active Server Pages, the Response object's Write method has an explicit and implicit form, and you use it to send text strings back to the user in the main body of the response.
- Taking only a single argument, the BinaryWrite method can be used to send binary data to the user in the body of the response. The BinaryWrite method does not convert the binary data to characters.
- Setting the Buffer property enables you to send the response to the user in pieces or all at one time, giving you greater control.

- Along with the Buffer property, you can make use of the Clear, Flush, and End methods to aid with buffer control. Nothing will be added to the response after a call to the End method.

- The Redirect method allows you to easily redirect the user's request to another page.

- Using the Response object's Cookies collection, you can write cookies to the HTTP body of the response. There are four main elements to writing a cookie: the Domain, Path, Expires, and Security elements.

- Using the Response object's Expires property and the ExpiresAbsolute property, you can have greater control over the caching of your Active Server Pages.

- Avoid continuing with lengthy or time consuming scripts unnecessarily by checking that the user is still connected first. The Response object's IsClientConnected property will give you the information that you require for this decision.

- By using the Response object's ContentType property, you can specify the type of content to be audio, video, text, or otherwise.

- Through use of the PICS property, you can add a PICS label to your Active Server Pages to allow rating services to rate their contents accordingly.

- If you need to specify the character set to use with your Active Server Pages, you can use the Response object's CharSet property.

- Using the Status property, you can send standard status lines, such as the well-known "404 Not Found" status line that occurs when a user requests a page that cannot be found.

- More of an advanced option, you can use the Response object's AddHeader method to add additional headers to a response.

- The AppendToLog method takes a single argument of a text string which you supply and appends it to the end of the Web server's log file.

Introducing the Response Object

Basically, the Response object is what you use for handling responses to the user or client. So you use the Response object whenever you want to send data to the user, be that in the shape of a line of text, a cookie, or otherwise. Table 9.1 shows you an overview of the Response object in terms of the collections, events, methods, and properties that it contains.

Collections	Events	Methods	Properties
Cookies	-	AddHeader	Buffer
		AppendToLog	CacheControl
		BinaryWrite	CharSet
		Clear	ContentType
		End	Expires
		Flush	ExpiresAbsolute
		Redirect	IsClientConnected
		Write	PICS
			Status

Table 9.1. An overview of the Response object

Clearly, the Response object is made up mostly of various methods and properties, with only a solitary collection. There are no events associated with the Response object.

Using the Write Method

The Response object's Write method is what you use when you want to send text back in the main body of the response to the user. The Write method is arguably the most often used method in Active Server Pages. Already throughout this book you have been using the Write method in one form or another. The explicit form of the Write method simply takes a string for an argument, as in the following few examples:

```
<%
Dim s1, s2
s1 = Request.Form("username")
s2 = "Hello"
Response.Write s2 & " world"
Response.Write "I am feeling <b>bold</b> today."
Response.Write s2 & " " & s1
%>
```

This example code would produce the output shown in Figure 9.1.

Figure 9.1.
The output from
the Write method
example

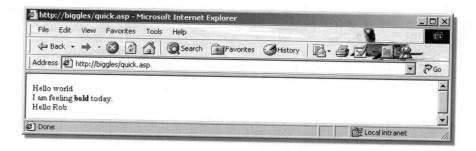

As you will already have seen many times, you can also use an implicit form of the Response object's Write method by simply using an = character, as shown:

```
<%="Hello world"%>
```

Writing Binary Data

Unless your Active Server Pages are dealing with lots of binary data, the BinaryWrite method will probably not be an attractive option for you to use. Syntactically simple, the BinaryWrite method can be deceivingly complex to put into practice. You might use it to access a database of graphical data and send the data for particular images directly back to the user's browser using the BinaryWrite method. The BinaryWrite method takes just one argument, the array of binary data to write, and it does not convert the binary data to characters. The following is a very simple example of using the BinaryWrite method:

```
<%
Dim total, binpic
total = Request.TotalBytes
binpic = Request.BinaryRead(total)
Response.BinaryWrite binpic
%>
```

Working with the Buffer

You can specify whether you want your Active Server Page to be sent back to the user all at once or line by line (the default), as data is written. To do this, you must set the Response object's Buffer property to either True or False.

Setting the Buffer property to False will be the same as the default and results in lines of text being sent back as they are written out. If you use any of the Response object's Clear, Flush or End methods when the Buffer property is set to False, a runtime error will occur.

However, if you set the Buffer property to True, you can control when the response is sent back to the user. If you are handling transactions, rather than outputting each successful portion one line at a time, it would be more prudent to wait until the transaction is complete so that the user gets either a success page or a failure page rather than a mixture of both. Setting the Buffer property to True means that you can make use of the Clear, Flush, and End methods, which can be very useful to you for troubleshooting. By using the Clear method, anything in the buffer is cleared. By using the Flush method, anything in the buffer is flushed or sent to the user. By using the End method, anything in the buffer is flushed and no more data from the script will be sent to the user.

You must set the Buffer property before the <html> tag or you will get a runtime error. Enter the following script into a new file and save it as *buffer.asp*:

```
<%Response.Buffer = True%>
<html>
<%
'Sending the first part of the page
Response.Clear
Response.Write "This will be sent with the first part."
Response.Write "<br>And so will this.<br><br>"
Response.Flush

'Sending the first part of the page
Response.Clear
Response.Write "This will be sent with the second part."
Response.Write "<br>Followed closely by this."
Response.End

'This next section is AFTER Response.End
Response.Clear
Response.Write "This will not be sent at all."
Response.Write "And neither will this."
%>
</html>
```

Figure 9.2 shows the result of loading *buffer.asp*, portraying clearly that no further Response.Write methods were processed after the Response.End method.

Figure 9.2.
The result of
loading the
buffer.asp
example script

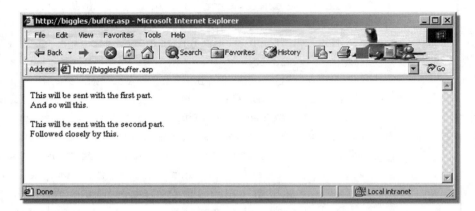

Redirecting the User's Request Elsewhere

Sometimes you may want to redirect a user's request for a particular page to another page. There are a few reasons that you might want to do this, such as for testing purposes, for security reasons, or for temporary site reorganizing, to name a few. Whatever your reasons, you can easily redirect a user's request by using the Response object's Redirect method, as shown in the following examples:

```
Response.Redirect "transaction2.asp"
Response.Redirect "/graphics/3dhome.asp?focus=rendering"
Response.Redirect "http://www.ivywall.com/"
```

Learning Other Redirection Methods

Although the Response object's Redirect method is very simple and effective, there are known problems with some HTTP 1.0 browsers that switch from Post method to Get method, as a result of this method. Errors will start occurring because the Get method cannot handle the amount of data that the Post method can. This can cause frustration on your part, as well as for your site's users, not to mention being difficult to track down because all the script would appear to be correct. You will discover later in this book that the Server object's Transfer and Execute methods can achieve the same result without the client browser problem. This problem is apparently remedied in HTTP 1.1 browsers.

Sending Cookies to the User

In Lesson 8, "Using the Request Object," you learned how to read the cookie information sent by the user to the server. Now you will learn how you can actually send cookies to the user, storing the relevant information that your Active Server Pages require. You can use the Response object's Cookies collection to write cookies to the HTTP body of the response. It is important to realize that you are not simply writing a file to the user's hard drive, but rather you are *offering* the cookie to the user within the HTTP body of the response. Depending on the user's browser settings, the user's computer will perform one of these actions: save the cookie immediately, prompt for possible saving of the cookie, or reject all cookies outright.

The first part of writing a cookie is to decide on the name of the variable you are placing in the cookie. In the case of this example, you will use *lastcolor*. Simply assign the value, such as a variable by the name of *usercolor*, to the element *lastcolor*, as in the following code:

```
Response.Cookies("lastcolor") = usercolor
```

After that there are four main elements to writing a cookie: the Domain, Path, Expires, and Security elements.

The Domain element tells the cookie to be sent to the server if it matches the domain being sent to. If you want to be more specific and only have the cookie sent when the domain and a deeper path match, then you supply the value of the path in the Path element. If you want all pages within a domain to receive the cookie, then you simply leave the Path element as a forward slash character (/).

You can use the Expires element to have the cookie only last for a week, a month, or some other duration of time that you specify. Simply enter in a valid date between hash marks (#) and assign it to the Expires element.

Finally, the Secure element is for use with Secure Sockets Layer (SSL) connections. You simply assign the Boolean value of True if you want to use Secure Sockets Layer or False otherwise. Not that you should really be storing any sensitive data in a cookie, as it is simply kept in clear text, but the option of using Secure Sockets Layer makes it difficult for hackers to intercept.

Now that you understand how to handle cookies using the Response object, you can continue on with the other half of the cookie pages that you began in the previous lesson. You can save the following code into a file and call it *cookie.asp*. Be sure to change the Domain element to your server name or IP address.

```
<html>
<head>
<title>Cookie Form Response</title>
</head>

<body>
<h3>Cookie Form Response</h3>
<% '=====================================
    'First, retrieve the data from the user
    '=====================================
Dim usercolor
usercolor = Request.Form("myColor")
%>
You selected the color <b><%=usercolor%></b>.<br><br>
<a href="cookie_form.asp">Back to color selection</a>
<% '=====================================
    'Actually write out the cookie
    '=====================================
Response.Cookies("lastcolor") = usercolor
Response.Cookies("lastcolor").Domain = "192.169.7.2"
Response.Cookies("lastcolor").Path = "/"
Response.Cookies("lastcolor").Expires = #August 1, 2002#
Response.Cookies("lastcolor").Secure = False
%>
</body>
</html>
```

The result of this code does not show a lot in the way of personalization; however, if you click on the link *Back to color selection*, you will discover that the cookie input form, *cookie_form.asp*, is personalized and remembers not only your last choice of color but selects it for you as well. This was seen in the cookie section of the previous lesson. Figure 9.3 shows the output from the cookie response code.

Figure 9.3.
The result of
the cookie
response code

Setting Page Expiration

You can set whether the user's browser will cache the page and, if it is to be cached, for how long. Using the Response object's Expires property and assigning 0 to it will mean that the user's browser will not cache the page, as shown:

```
Response.Expires = 0
```

Otherwise if you do want to cache the page, simply assign the value in minutes. It is important to be aware that if you assign more than one value to the Expires property, the lowest value will be taken, regardless of order. So in the following example of setting the Expires property, the Expires property will be set to 10 in the end because it is the lowest value:

```
<%
Response.Expires = 80
Response.Expires = 10
Response.Expires = 35
Response.Expires = 20
%>
```

Note: Assigning no expiry means that the default 24-hour cache will apply.

You can also set an expiry that is based on date and time using the ExpiresAbsolute property. The user's browser will use the cached page if the date and time are not later than what you specify. Simply assign a standard date in month, day, year format and then follow it with the time. If no time is set then the server uses midnight. Remember to place the entire date and time between hash (#) characters, as shown in the following examples:

```
Response.ExpiresAbsolute = #January 7, 2002#
Response.ExpiresAbsolute = #November 25, 2002#
```

On the topic of keeping a cache of your pages, you have an option of allowing proxy servers to cache your page in order to speed up the response for users. You use the Response object's CacheControl property for this, and simply remain with the default value of private or assign public if you do want to allow a cache, as shown:

```
Response.CacheControl = "public"
```

If your pages change extremely frequently this may not be a good idea.

Checking if the User is Still Connected

Imagine having a very lengthy or time consuming script or a script that could be taxing on the server's resources, such as a long loop. Rather than simply running through the entire length of these scripts, you may prefer to check to see if the user (or client) is still connected to your server. Otherwise, you might be wasting time and precious server resources, not to mention the possibility of causing unnecessary delays or lags in responding to other requests. By using the Response object's IsClientConnected property, you can easily tell if the user is still connected.

The IsClientConnected property is a read-only property that returns True if the user is still connected or False otherwise. The following is a brief example of using the IsClientConnected property:

```
<%
'This is a long loop so check for user first
If Response.IsClientConnected Then
  'perform long loop here
End If
%>
```

Setting the ContentType Property

The HTTP header contains information on the type of content that is held within the body of the page. You can set the type of content by using the Response object's ContentType property. You simply assign a text string to the ContentProperty that is based on a type and a subtype, with a forward slash separating the two, as shown in the following example:

```
<%
If ctype="jpg" Then
   Response.ContentType = "image/jpeg"
ElseIf ctype="mpg" Then
   Response.ContentType = "video/mpeg"
Else
   Response.ContentType = "text/plain"
End If
%>
```

You can see a more complete list of the types and subtypes that the ContentType property supports, although this list is subject to change with approval of additions, in Table 9.2.

Type	Subtype	Description
Application	Active	This is mostly binary data
	ODA	
	PostScript	
Audio	Basic	This is audio data
Image	GIF	This is image or picture data
	JPEG	
Message	External-body	This is encapsulated message data
	Partial	
Multipart	Alternative	This is multiple parts of independent data
	Digest	
	Mixed	
	Parallel	
Text	Plain	This is text data
	RichText	
Video	MPEG	This is video data

Table 9.2. The various types and subtypes that the ContentType property supports

Adding a Platform for Internet Content Selection Label

You are probably aware that there are some Web sites on the Internet with explicit sexual content that we do not want our children to see. Many different organizations, such as SafeSurf (set up by parents), the Recreational Software Advisory Council (RSAC), and the Internet Content Rating Association (ICRA) are working to try to protect people from being subjected to Web sites that hold dubious or potentially harmful content. These organizations use a rating system to screen content, and Web sites can add a Platform for Internet Content Selection (PICS) label to participate in this rating scheme. To add a PICS label to your site, you use the Response object's PICS property which takes the label in text string format as its only argument. The following is only an example of using the PICS property to add a PICS label, and you should not enter in the exact code for your own site but rather obtain your own PICS label (they are free):

```
<%
Dim pl
pl = "(pics-1.1 " & chr(34)
pl = pl & "http://www.icra.org/ratingsv02.html" & chr(34)
pl = pl & " l gen true for " & chr(34)
pl = pl & "http://www.ivywall.com/" & chr(34)
pl = pl & " r (cz 1 lz 1 nz 1 oz 1 vz 1) " & chr(34)
pl = pl & "http://www.rsac.org/ratingsv01.html" & chr(34)
pl = pl & " l gen true for " & chr(34)
pl = pl & "http://www.ivywall.com/" & chr(34)
pl = pl & " r (n 0 s 0 v 0 l 0))"
Response.PICS(pl)
%>
<html>
```

Rating for www.ivywall.com

The above PICS label in the example shows that the site *http://www.ivywall.com/* contains no nudity, no sexual content, no violence, and no offensive language. ICRA holds a database of label holders so client browsers will ignore forgeries. Note, the PICS property must be set prior to the <html> tag or you will get a runtime error.

Specifying the Character Set to Use

You can use the Response object's CharSet property to specify the character set that you want to use with your Active Server Pages. Usually, you will not have to set the CharSet property but will instead use the default character set of ISO-LATIN-1. The following is an example of changing the character set to ISO-LATIN-2:

```
<%Response.Charset("ISO-LATIN-2")%>
<html>
```

You must set the character set prior to the <html> tag or you will get a run-time error.

Note: *Although the Microsoft documentation calls it a property, the CharSet property is really a method, and you cannot assign a value to it in the usual manner that you use with properties.*

Setting a Status Line

You can return a standard status line as a part of your response to a user request to handle a variety of situations that might occur. Most likely you will have come across a "404 Not Found" page before. This page is very common, and is used when a user attempts to go to a page that does not exist or cannot be found by the server. Although it is classed as a client error, take note of it because it could also occur if you have a bad link in your pages. The following example code shows you how to use the Response object's Status property, by assigning a "404 Not Found" status to it:

```
<%
Dim st
st = "404 Not Found — Unfortunately, the page that you "
st = st & "requested cannot be found. Please check that"
st = st & " the URL has the correct address."
Response.Status = st
%>
```

The following list shows you the general range for standard status lines according to current HTTP specifications:

- 1xx: Informational—Not used, but reserved for future use.
- 2xx: Success—The action was successfully received, understood, and accepted.
- 3xx: Redirection—Further action must be taken in order to complete the request.
- 4xx: Client Error—The request contains bad syntax or cannot be fulfilled.
- 5xx: Server Error—The server failed to fulfill an apparently valid request.

You can also see some of the more common status values and the short description that accompanies them in Table 9.3.

Status Value	Short Description
200	OK
201	Created
202	Accepted
204	No Content
301	Moved Permanently
302	Moved Temporarily
304	Not Modified
400	Bad Request
401	Unauthorized
403	Forbidden
404	Not Found
500	Internal Server Error
501	Not Implemented
502	Bad Gateway
503	Service Unavailable

Table 9.3. Common status values and their accompanying short description

Sending Additional Headers in Your Response

As its name implies, the Response object's AddHeader method will add additional headers to a response. The AddHeader method does not replace previous headers, and you cannot remove a header once you add it to the response. Sending additional headers is not a common method that you would implement, and it is here only for completeness. However, you can usually accomplish everything you require in Active Server Pages without resorting to this advanced method. The following is an example of sending an additional header:

```
<%Response.AddHeader "CustomError", "User not listed"%>
<html>
```

You must call the AddHeader method before the <html> tag or you will get a runtime error. Also, you must not put an underscore in the arguments to avoid ambiguity with built-in definitions.

Logging Information with the Response Object

Sometimes it is useful to be able to go through the logs of a Web server to find out the cause of a problem. Rather than rely purely on logging that you did not implement, you can append a string of up to 80 characters to the Web server's log file using the Response object's AppendToLog method. The AppendToLog method takes a single argument of a text string, and you can apply this method numerous times throughout your scripts.

```
<%
Dim errtext
errtext = "Error loading database at " & time
Response.AppendToLog errtext
%>
```

Depending on the platform that you are running your Active Server Pages server on and what software you are using, your results may differ. The log files are stored in the directory *winnt\system32\logfiles\w3svc1* using a pattern of *exyymmdd.log*, where *yy* is the year, *mm* is the month, and *dd* is the day. As a general guide, you can turn logging on or off in Personal Web Server by going to the advanced section and changing the check box, as shown in Figure 9.4.

**Figure 9.4.
Changing your logging status by using Personal Web Server**

WHAT YOU MUST KNOW

To send data to the user you use the Response object, with its many methods and properties. The Response object's Write method is probably the method that you will use more than any other throughout your writing of Active Server Pages. You can also use the Cookies collection to offer cookies to your users to provide a means for user-state permanence. In Lesson 10, "Using the Server Object," you will learn how to make use of some of the server's built-in functionality, such as calling another Active Server Page from within an Active Server Page, controlling script timeouts, and making use of the CreateObject method to create instances of all sorts of objects. Before you continue with Lesson 10, however, make sure you understand the following key concepts:

- The Response object comprises methods and properties, a single collection, and no events.

- The Write method has an explicit and implicit form, either of which you can use to send text strings back to the user in the main body of the response.

- To send binary data to the user in the body of the response, you can use the BinaryWrite method, which does not convert the binary data to characters and takes an array of binary data as its sole argument.

- Together with the Buffer property, you can make use of the Clear, Flush, and End methods to send the response to the user in pieces or all at one time. After you call the End method, however, no more data will be written to the body of the response.

- You can redirect a user's request to another page, simply by calling the Response object's Redirect method and passing it the new destination as its argument.

- The Response object's Cookies collection lets you send cookies in the body of the response, leaving it up to the user to accept or reject. The four main elements to writing a cookie are the Domain element, the Path element, the Expires element, and the Security element.

- Together, the Expires, the ExpiresAbsolute, and the CacheControl properties give you better control over the caching of your Active Server Pages, including denying caching and caching on proxy servers.

- The IsClientConnected property lets you end lengthy or time consuming scripts rather than executing them unnecessarily by checking if the user is still connected.

- You can specify the type of content to be application, audio, image, message, multipart, text, or video, by using the ContentType property.

✕ Adding a PICS label to your site with the aid of the PICS property, you can allow rating services to rate the contents of your Active Server Pages accordingly.

✕ Using the CharSet property, you can specify the character set to use with your Active Server Pages, such as ISO-LATIN-2.

✕ You can easily send status messages that conform to HTML protocols in your response by using the Status property.

✕ The Response object's AddHeader method will let you add additional headers to a response, although you cannot remove a header after it has been added, nor can you overwrite existing headers.

✕ By using the AppendToLog method, you can append up to 80 characters of text to the end of the Web server's log file. However, multiple calls to the AppendToLog method are permissible, letting you save as much text as you require.

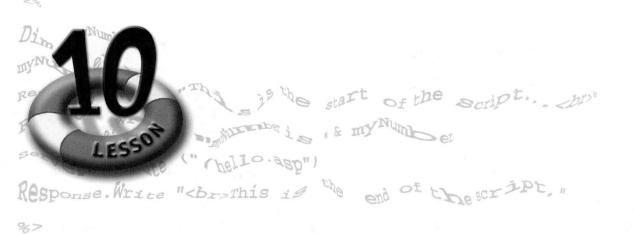

*i*n Lesson 9, "Using the Response Object," you took a close look at the Response object and its many methods and properties, as well as cementing your knowledge of cookies in Active Server Pages. Building on the skills you have already acquired, you will now learn how to make use of some of the server's built-in functionality, such as controlling script timeouts, making calls from within an Active Server Page to other pages, and creating instances of a variety of objects. By the time you finish this lesson, you will understand the following key concepts:

- The Server object is mostly a mixture of miscellaneous methods that give you access to some of the server's built-in functionality.

- You can use the Server object's ScriptTimeout property to set a timeout for your Active Server Pages.

- The CreateObject method lets you create an instance of objects that are not already built into Active Server Pages.

- The Execute method allows you to call other scripts from within an Active Server Page and then return to the original page afterwards. By using the Execute method you can split your long scripts into smaller, more manageable portions.

- You can use the GetLastError method if you want to customize error handling for your Active Server Pages. It is an advanced method despite the simplicity of its syntax.

- To display HTML code in the browser you should use the HTMLEncode method, otherwise the browser will treat it as actual commands.

- The MapPath method lets you find the physical path of a file by passing it the virtual path.
- Similar to the Execute method, the Transfer method lets you transfer control to another script from within an Active Server Page but does not return to the original script afterwards.

Introducing the Server Object

Basically, the Server object is a mixture of miscellaneous methods, some of which are incredibly useful and others that you might never find a practical use for. However, that being said, it is important for you to be aware of all that it offers so that you can make use of any of its methods if the need arises. Table 10.1 shows you an overview of the Server object in terms of the collections, events, methods, and properties that it contains.

Collections	Events	Methods	Properties
-	-	CreateObject	ScriptTimeout
		Execute	
		GetLastError	
		HTMLEncode	
		MapPath	
		Transfer	
		URLEncode	

Table 10.1. An overview of the Server object

The Server object is made up mostly of methods, with a single property, and no collections and no events.

Setting a Timeout for Your Active Server Pages

If you have an Active Server Page that takes a long time to process, you should look at setting a longer timeout. The default timeout for your scripts or Active Server Pages is 90 seconds, therefore anything longer than 90 seconds will return an error to the user unless you alter the timeout explicitly. You can use the Server object's ScriptTimeout property to set a timeout for your Active Server Pages. The ScriptTimeout property takes a single argument, which is simply the number of seconds before the script times out. The following example shows how you might set the timeout of a script to be three minutes:

```
<%Server.ScriptTimeout = 180%>
```

In practice, most users will not sit for 180 seconds, happily waiting for your script. In today's speed oriented world, people are more likely to stop the page from loading if it takes too long and appears to be hung whenever they attempt to access it. Most probably you will decrease the timeout for your scripts to something less than one and a half minutes, rather than increase them. One nicer method of dealing with scripts that take a long time is to have the user connect to an interim page that responds almost immediately with a message and then calls the long script using something such as the Response object's Redirect method. This way you give your user feedback before embarking on a lengthy wait.

Using the CreateObject Method

The Server object's CreateObject method is an extremely useful method that you will use often in your Active Server Pages. Basically, the CreateObject method allows you to create an instance of nonintrinsic objects; that is, objects that are not already built into Active Server Pages. When you think about how much you get out of the seven intrinsic objects, you will see that the ability to load other objects and all the services that they contain is truly a welcome and desirable feature.

To use the Server object's CreateObject method, you must first declare a variable and then use the Set statement to assign a new instance of the object to that variable. The following is an example of using the CreateObject method to load in the Scripting FileObject, a very useful object for handling files, folders, and drives, as you will learn later in the book:

```
<%
Dim fs
set fs = Server.CreateObject("Scripting.FileSystemObject")
%>
```

You cannot simply assign the CreateObject method to a variable but must use the Set statement. So the following example will result in a runtime error:

```
<%
Dim fs
fs = Server.CreateObject("Scripting.FileSystemObject")
%>
```

You can destroy an instance of an object simply by setting it to Nothing, as in the following example:

```
set fs = Nothing
```

Execute

The Server object's Execute method is similar to the Response object's Redirect method. Prior to Active Server Pages 3.0, the Execute method did not exist. The difference between the two methods seems subtle at first but has a big impact on performance for your site. The Redirect method will send information back to the client, whereas the Execute method will handle the calling entirely on the server. This means that you can split your long scripts into smaller, more manageable portions that can be accessed without the performance loss that would occur with the Redirect method. The Execute method will also return to the calling script.

One point to be aware of is that variables that you declare in the original script are not available in the scripts that you call with the Execute method. To actually use the Execute method in your scripts, all you have to do is pass it a single argument that specifies the path of a file, such as the following:

```
Server.Execute("/hello.asp")
```

The following script shows you an example of using the Execute method to call another script and then return to the original script. It will output some words before and after using the Execute method, as well as declare a variable to show scope limitations. The script is as follows:

```
<%
Dim myNumber
myNumber = 10
Response.Write "This is the start of the script...<br>"
```

```
Response.Write "myNumber is " & myNumber
Server.Execute("/hello.asp")
Response.Write "<br>This is the end of the script."
%>
```

Save this script as a new file using Notepad and call it *execute.asp*. You must also create the other file that you will call from *execute.asp* within Notepad. You must name this next file *hello.asp* to match the reference to it in *execute.asp*. The script in *hello.asp* will simply output some text and also display the value that it has for the variable *myNumber*. The script for *hello.asp* is as follows:

```
<%="<br>Hello world<br>"%>
<%="myNumber is now " & myNumber%>
```

After loading *execute.asp* into your browser, you will see something similar to Figure 10.1. You can tell from this output that the script continues after the Execute method because the line stating that it is the end of the script is displayed. You can also see that variables from the original script have no scope in scripts that the Execute method calls from the output showing the value of the variable *myNumber*.

**Figure 10.1.
The result of loading
the file *execute.asp***

GetLastError

The GetLastError method is an advanced yet simple method that returns an instance of the ASPError object, which you will learn about in detail later in this book. The ASPError object, you will recall, is one of the seven intrinsic objects from the Active Server Pages object model. You can only use the GetLastError method before you send any output to the client. So using the GetLastError after a Response.Write will generally cause a runtime error. You assign the GetLastError object to a variable to create an instance of an ASPError object, as shown:

```
set ASPErrObject = Server.GetLastError
```

You would not usually make use of the GetLastError method and certainly not in your everyday scripts and pages. The GetLastError method handles pre-processing errors, scripting errors, and runtime errors. The place that you would use the GetLastError method if you had to is in special error handling scripts, such as the 500-100.asp script that comes with IIS 5.0. This 500-100.asp script is the default script for handling these types of errors, and only if you want to customize the script in any way or write your own error handling scripts would you be making use of the GetLastError method.

An important point to note about the GetLastError method is that it cannot be used within the same script that the error actually occurs. The following code, for example, because of the syntax error, would never reach the line with the GetLastError method because the Active Server Pages server (IIS 5.0) would send control immediately to the 500-100.asp script:

```
<%
On Error Resume Next
  dim count
  for count=1 to 10
  nex                       : Syntax error
Set ASPErrObject = Server.GetLastError
Response.Write ASPErrObject.ASPCode
%>
```

HTMLEncode

When you want to display actual HTML code in the browser and not have the remote browser treat it as commands, you should use the HTMLEncode method. To use the HTMLEncode method all you must do is pass it the string that you want to encode using the following syntax:

```
Server.HTMLEncode(HTML_string)
```

The following example you clearly shows not only how to use the HTMLEncode method in your scripts but also the difference between using it and not using it:

```
<html>
<head><title>Using HTMLEncode</title></head>
<body>
<font size=5>
<%
Dim myString
myString = "<p>Here is some <b>bold</b> text.</p>"
Response.Write myString
Response.Write server.HTMLEncode(mystring)
%>
</font>
</body>
</html>
```

Save the example script in a new file using Notepad and name it *htmlencode.asp*. Load *htmlencode.asp* in your browser to see the result. The output from *htmlencode.asp* is shown in Figure 10.2, where you can see clearly the difference between two identical lines of text, because the second line is using HTMLEncode.

Figure 10.2.
Two identical lines of text—except that the second line is using HTMLEncode

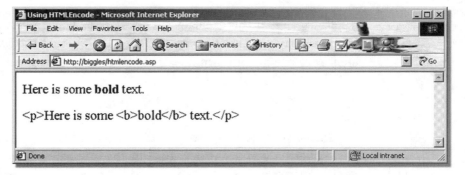

MapPath

Using the Server object's MapPath method, you are able to find out the physical path of a file rather than its virtual path. A physical path deals with a logical drive letter and location on a hard disk, such as "C:\myfile.htm". A virtual path usually begins with a forward slash (/) character and can be thought of as a shortcut to a specific location. You can use Internet Information Server or Personal Web Manager to set up virtual paths for your Active Server Pages server.

The following code shows you how you might simply reference a virtual file and use the MapPath method to discover its physical location:

```
<%
Dim myPath
myPath = Server.MapPath("/webpub/gfxjpg/3dtutorial.asp")
Response.Write "<font size=5>" & myPath & "</font>"
%>
```

Saving this file as *mappath.asp* and loading it into a browser clearly shows the physical location of the file that was referenced using a virtual path. Figure 10.3 shows the output from *mappath.asp*.

**Figure 10.3.
The output from
mappath.asp show-
ing the physical
location clearly**

Be aware that this example will not work on your servers unless you have an identical virtual path setup. Try changing the file that the MapPath method references to one of your own files that you keep in a virtual directory.

Transfer

The Server object's Transfer method is just like the Execute method that you saw earlier only it does not return to the original script. To use the Transfer method, you simply pass it the path of a file as its only argument, as shown:

```
Server.Transfer("/hello.asp")
```

To show the difference between the Transfer method and its cousin, the Execute method, you will create the same scenario that you were using earlier, except that you will, of course, use the Transfer method this time. So the following script will declare a variable and output some text before and after calling the Transfer method:

```
<%
Dim myNumber
myNumber = 10
Response.Write "This is the start of the script...<br>"
Response.Write "myNumber is " & myNumber
Server.Transfer("/hello.asp")
Response.Write "<br>This is the end of the script."
%>
```

Save this script as a new file using Notepad and call it *transfer.asp*. The other file that you will call from *transfer.asp* is *hello.asp*, the same file that you created earlier while testing the Execute method. The script in *hello.asp* will simply output some text and also display the value that it has for the variable *myNumber*. The script for *hello.asp* is as follows:

```
<%="<br>Hello world<br>"%>
<%="myNumber is now " & myNumber%>
```

The result of loading *transfer.asp* into a browser is shown in Figure 10.4. Notice that the only difference between using the Transfer method and using the Execute method is that control did not return to the *transfer.asp* script after using the Transfer method. You can tell this by the clear absence of the line stating that it is the end of the script.

Figure 10.4.
The result of
loading *transfer.asp*

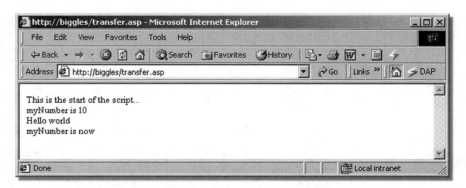

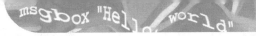

URLEncode

Just as you saw earlier how to encode HTML strings so that they can be displayed within a browser, you can also use the Server object's URLEncode method to encode a string so that it appears as it would when passed to a URL (Uniform Resource Locator). You can use the URLEncode method whenever you are unable to use the Get method but need to send data over the URL. Sometimes this may be useful to you for testing purposes. Using the URLEncode method is extremely straightforward and only takes the string that you want to encode as its sole argument, as the following syntax shows:

```
Server.URLEncode(string_to_encode)
```

So as an example, the following lines of code place the output from the URLEncode method into a variable called *buffer*. To see clearly the result of what the variable *buffer* holds after the URLEncode method, you simply use the Response object's Write method, as shown:

```
<%
Dim buffer
buffer = Server.URLEncode("http://www.ivywall.com")
Response.Write buffer
%>
```

The value of *buffer* after running through this script is the same as what you would see when passed over a URL, as shown:

```
http%3A%2F%2Fwww%2Eivywall%2Ecom
```

Putting It All Together—A Server Object Example

Having covered quite a few rather odd looking methods that the Server object provides, it might be worthwhile for you to try putting together an Active Server Page that incorporates some of these in a more practical sense.

Imagine this scenario: you are a Web site developer and you have a client that is putting together a large HTML resource site. Your client wants to be able to display HTML code within the site's pages to aid in teaching the visitors to the site.

Not only that, but your client does not want the HTML to be encoded within Active Server Pages that he cannot understand. He wants you to simply load in HTML pages that he creates and edits, and then display them within the site's pages.

In the following example, you will produce an Active Server Page that is capable of displaying HTML from a file. It is important to realize that the focus of this exercise is the manner in which you use the Server object and its methods, not how to open and work with a file, which you will learn about in greater detail later in this book. The reason you are seeing any file handling script at all is because of the Server object's CreateObject method. This example shows you a practical use for it.

Working with Constants in VBScript

So far in this book you have been dealing with variables. Variables are great to work with, especially when you want to be able to change the value that a variable holds. However, there are times when you do not want the value in a variable to ever change throughout the course of your script. In these circumstances you should make use of constants.

Think of a constant as a name that represents a value. You declare a constant by using, not the keyword Dim, but the keyword Const instead. After that you simply give the name that you want to use for your constant and follow it with an = character and the value you want to assign. The following examples show how to assign a constant:

```
Const MaxChar = 80
Const logpath = "c:\winnt\logs\"
Const TotalColumns = 9
```

Constants are also very useful when you want to use a static value in many places throughout your script. If you simply use the actual value, you may have not only a lot of work when you have to change that value but also the potential to introduce bugs into your Active Server Pages. By using a constant, there is only one place in your script that you have to change the value which is very good when you want to test different values to see how a script performs best.

```
path = Server.MapPath("/button.htm")
```

The next section that deals with the Server object is the CreateObject method. You will use the CreateObject method to create an instance of the Scripting FileSystemObject so that you can open a file, as follows:

```
set fs = Server.CreateObject("Scripting.FileSystemObject")
```

The last part of the example code that deals with the Server object is the HTMLEncode method. Because you want to display the HTML one line at a time, you place the HTMLEncode method within a loop and work your way through each line of the HTML file. The code for using the HTMLEncode method is as follows:

```
Response.Write Server.HTMLEncode(ts.ReadLine)
```

Remember that the focus here is on the use of the Server object and its methods, not working with files. If you do not understand the part of the script that deals with files, do not worry as you shall learn all about them in due time. So the complete example of using the Server object to load in an entire HTML file would look something like the following:

```
<%Option Explicit%>
<html>
<head>
<title>Teaching HTML</title>
<h2>Teaching HTML</h2><hr>
</head>

<body>
<p>In this page you will review some basic HTML code for
creating a button on a form. The following code shows an
example:<br>
<%
'Declare variables and constants
Dim fs, ts, path
Const ForReading = 1

'Use the MapPath method
path = Server.MapPath("/button.htm")
```

```
'Open the HTML file for reading
set fs = Server.CreateObject("Scripting.FileSystemObject")
set ts = fs.OpenTextFile(path, ForReading)

'Using a loop read in and display each line of the file
While not ts.AtEndOfStream
  Response.Write Server.HTMLEncode(ts.ReadLine) & "<br>"
Wend
%>
</p>
</body>
</html>
```

Save the entire script into a file using Notepad and call it *teachhtml.asp*. Open up *teachhtml.asp* in your browser to see the contents of *button.htm* displayed properly, as shown in Figure 10.5, and as a direct result of using multiple Server object methods.

Figure 10.5. Using a variety of Server object methods to display an entire HTML file

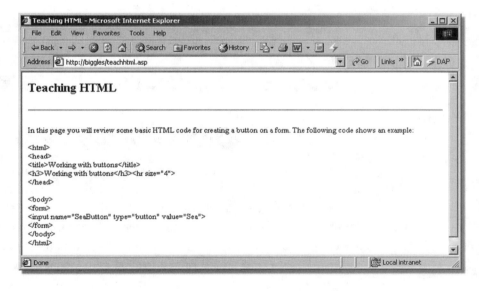

Notice carefully the use of the <br> tag within the While-Wend loop.

```
Response.Write Server.HTMLEncode(ts.ReadLine) & "<br>"
```

If you neglect to place the <br> tag in the loop and after each occurrence of writing a line of HTML code, you would lose all sense of format. Change the script in *teachhtml.asp* by removing the <br> tag from the loop, so that the line of code looks like the following:

```
Response.Write Server.HTMLEncode(ts.ReadLine)
```

Save *teachhtml.asp* and then refresh the page in your browser. Figure 10.6 shows the result of removing the <br> tag from the loop. You can clearly see the impact that it has on the final output and how all formatting of the HTML code that you are trying to display is lost.

**Figure 10.6.
Removing the
tag from the loop
causes an impact on
the final output**

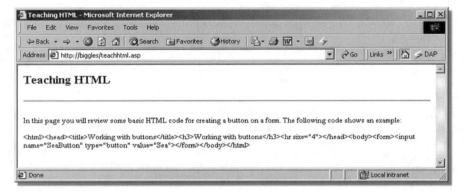

WHAT YOU MUST KNOW

The Server object is mostly a mixture of miscellaneous methods that give access to some of the server's built-in functionality. Some of the more useful elements of the Server object are its CreateObject method, Execute method, and ScriptTimeout property. The CreateObject allows you to create an instance of objects that are not part of the Active Server Pages object model. The Execute method lets you split long scripts into smaller, more manageable portions, allowing you to call other scripts from within your pages. In Lesson 11, "Using the Session Object," you will learn about providing users with a persistent state, saving them from having to continuously authenticate or provide data as they move from page to page within your site. Before you continue with Lesson 11, however, make sure that you are comfortable with the following key concepts concerning the Server object:

- The Server object's ScriptTimeout property lets you set a timeout for your Active Server Pages. The default timeout is 90 seconds.

- There are many objects that are not already built into Active Server Pages, unlike the seven that comprise the Active Server Pages object model. By using the Server object's CreateObject method, you can create an instance of these other objects for use within your scripts.

- The difference between the Transfer method and the Execute method is that the Execute method will return to the original script afterwards. Other than this, the two methods both allow you to run other scripts from within an Active Server Page.

- Because you can return to your calling scripts, using the Execute method means that you can split long scripts into smaller, more manageable portions.

- Despite the simplicity of its syntax, the GetLastError method is really an advanced method that you would use only if you want to customize error handling for your Active Server Pages.

- To prevent the browser from treating a string containing HTML code as actual commands when you want to display HTML code in the browser, you should use the HTMLEncode method.

- By passing the virtual path of a file to the Server object's MapPath method, you can discover the physical path of that file.

USING THE SESSION OBJECT

n Lesson 10, "Using the Server Object," you took a close look at the Server object and the assortment of methods that it provides, including some very useful and powerful functionality. In this lesson you will learn how you can use the Session object to store information that you might require during a particular user session. The Session object provides a persistent state, so variables that you store in the Session object persist for the entire duration of the session and are not discarded when the user jumps between pages in the application. By the time that you finish this lesson you will know the following key concepts:

- The Session object is made up of a mixture of collections, events, methods, and properties, and provides a persistent state that lets you store information that lasts for the entire duration of each session.

- Because HTTP is a stateless protocol (at the application level), to be able to make use of the Session object properly the user's browser must support cookies.

- When a user first connects to an Active Server Pages site, the server assigns some memory for the session, as well as a session identification number. You can access this session identification number using the Session object's SessionID property.

- Using the Session object's Timeout property you can set the amount of time in minutes from the user's most recent request to the server before the session times out and the memory allocated for the session on the server is freed.

- Both the OnStart event and the OnEnd event let you execute script at the beginning and end of each session, which lets you properly initialize variables and close down gracefully.

- By changing the locale through the LCID property, you can present users with either a format of currency, date layout, or other locale influenced elements that they prefer or that your site requires.
- You can store variables and settings that are accessible from any Active Server Pages in a site for the entire duration of a session by using the Session object's Contents collection.
- Remove items from the Contents collection with the Contents.Remove and Contents.RemoveAll methods.
- The StaticObjects collection contains items that are static and unchanging. The items are read-only from your Active Server Pages scripts.
- The Abandon method will clear all the session variables at once; however, the server only clears the variables at the end of the calling script.
- If you are using Web farms for load balancing, be aware of adverse effects that it causes for Active Server Pages that use the Session object.

Introducing the Session Object

The Session object provides you with a persistent state, which means that you can maintain variables or information that will endure for the entire visit, or session. At the application level, HTTP is largely a stateless protocol; that is, HTTP does not hold status information that applications can use from one request to the next. Due to this fact, it is difficult to create Web applications that can retain information for the duration of a user's session. Imagine trying to order ten items from an online catalogue and having to resubmit all the items in your virtual shopping cart every time you change pages.

The Session object relies on cookies, which you have already seen how to work with in Active Server Pages. If a browser does not support cookies, you cannot really make use of the Session object or the functionality that it provides. This is an important point considering the fact that you do not always know or have control over the user's browser environment.

Unlike the other intrinsic objects that you have seen so far, the Session object is a mixture of collections, events, methods, and properties. Table 11.1 shows you an overview of the Session object in terms of the collections, events, methods, and properties that it contains.

Collections	Events	Methods	Properties
Contents	OnEnd	Abandon	CodePage
StaticObjects	OnStart	Contents.Remove	LCID
		Contents.RemoveAll	SessionID
			Timeout

Table 11.1. An overview of the Session object

The Session object is made up of a mixture of collections, events, methods, and properties.

Understanding the SessionID Property

When a user first connects to the Active Server Pages server, two main tasks occur:

- The server assigns memory on the server for the user session and marks it with a unique session identification number.
- The server writes a cookie on the remote user's computer to create session persistence and store information.

To see the unique session identification number that a server has assigned to a session, you use the Session object's SessionID property, as follows:

```
Dim userSID
userSID = Session.SessionID
```

The session identification number is a long variable type, which you can see clearly for yourself by creating an Active Server Page that outputs this value. To output the security identification number simply use the SessionID property, as shown:

```
<%
Response.Write "The current session identification " _
& "number is: " & Session.SessionID
%>
```

The output from this code is shown in Figure 11.1, where you can see the session identification number.

**Figure 11.1.
Viewing the session
identification num-
ber in the browser**

The session identification number is unique for the period since the Active Server Pages server was last started. This is a subtle point that you do not want to misunderstand. The session identification number is not unique at all times. If you were to stop and start your Active Server Pages server, such as IIS or Personal Web Manager, then there is nothing to prevent the server from handing out the same session identification number that it may have handed out before being restarted. So you should never use the Session object's SessionID property to signify a primary key in a database because you cannot guarantee that it is unique.

Setting a Timeout for User Sessions

Just as you learned how to set a timeout for your Active Server Pages scripts, you can also set a timeout for your user sessions. The length of time that the server will maintain a session for a user can be important. The default timeout is usually twenty minutes, which may be perfectly fine for most cases. However, there are times and situations where you may want to either increase the timeout period or decrease it.

Recall that the server allocates memory on the server for each session. Therefore, if you are writing scripts for a client who has a high volume of users visiting that site, every single visitor has memory set aside for their session. This can be extremely taxing on the server's resources unnecessarily, especially if those users do not stay for very long. In these types of cases, you may want to consider setting a lower timeout.

To actually set the timeout for each session, you use the Session object's Timeout property and simply assign the number of minutes that you want to use for a timeout, as shown in the following examples:

```
Session.Timeout = 5    'Set the timeout to 5 minutes
Session.Timeout = 30   'Set the timeout to 30 minutes
```

The timeout does not last from the first instance of connection by a user but rather will begin with a user's last request. So, for instance, if you are using a timeout of 10 minutes and the last request you made was 9 minutes ago, you would only have 1 minute remaining before the server closes the session and frees up the allocated memory. If, however, you make another page request at this 9-minute mark, you would now have 10 minutes remaining again.

Making Use of the Session_OnStart Event

When a user first connects to a site that is using Active Server Pages, the Active Server Pages server automatically creates a session. The Session object provides the OnStart event, which you can use to execute some script of your own at the time of the session creation. The OnStart event is similar to the OnClick event that you saw earlier in the book when dealing with buttons in Web forms. Instead of using the name of a button followed by an underscore and the text OnClick, you use a similar approach but substitute *Session* for the button name and *OnStart* for the text OnClick, as follows:

```
Sub Session_OnStart
    'Statements
End Sub
```

Making Use of the Session_OnEnd Event

In the same manner as the OnStart event where you set up variables and prepare for the session, the OnEnd event gives you a place where you can tidy up and gracefully close a session. The OnEnd event will trigger whenever a session is ended. To use the Session object's OnEnd event, simply create a subroutine by the name of Session_OnEnd, as shown:

```
Sub Session_OnEnd
    'Statements
End Sub
```

Using the Location Identifier

You can set your Active Server Pages to use a specific system locale by setting the location identifier. Setting the locale for an Active Server Page means that you can present information in a manner more suitable for the content or perhaps for the user. For instance, you may have asked a user previously about their preferences for currency so all subsequent requests to pages during the session will show product prices using a currency format that the user prefers. To set the location identifier, you must assign a value to the Session object's LCID property, as shown:

```
Session.LCID = 1041
```

To get a better idea of using the LCID property, you should take a look at an example that demonstrates both reading in the current location identifier and setting it to a new value. You will first read in the current location identifier and display it, as well as display the date and time using the Now function. After changing locales to something different, you can once again output the date and time using the Now function and compare the results. Type the following example of using the LCID property into a new document using Notepad and save it as *lcid.asp*:

```
<%
'look at the current locale
Response.Write "The current LCID is: " & Session.LCID
Response.Write "<br>Now: " & now & "<br><br>"

'change the locale
Session.LCID = 2057
Response.Write "The new LCID is: " & Session.LCID
Response.Write "<br>Now: " & now & "<br>"
%>
```

Load the *lcid.asp* script into your browser. You can see from the results of this script, as shown in Figure 11.2, that changing the location identifier using the Session object's LCID property, made the difference between the U.S. settings of month/day/year and the U.K. settings of day/month/year. Depending on your everyday locale settings, your results may not be quite the same as what is shown here, but you can see the point being made about how the locale can affect your Active Server Pages.

Figure 11.2.
The difference
between locales
changing the output
of the Now function

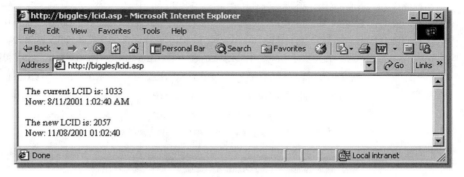

Of course, to be able to change the location identifier effectively, you must know what value to assign to the LCID property. Table 11.2 shows you a handful of system locales and their corresponding location identifier.

Locale	LCID Value
Afrikaans	1078
Albanian	1052
Chinese (Hong Kong)	3076
Danish	1030
Faeroese	1080
French (standard)	1036
German (standard)	1031
Italian (standard)	1040
Japanese	1041
New Zealand	5129
Spanish (standard)	1034
United Kingdom	2057
United States	1033

Table 11.2. A variety of locales and location identifiers

The list of locales in Table 11.2 is obviously not the entire list of locales available. As of the time of writing, you can see a more comprehensive list on the Microsoft site at the following page:

http://msdn.microsoft.com/scripting/vbscript/doc/vsmscLCID.htm

Using the CodePage Property

You have seen how the locale used by your Active Server Pages can alter the final appearance of the response sent to the user. However, there is also the matter of code pages. Many English speaking countries and European countries make use of code page 1252, an ANSI code page, to display the characters that they require. However, if you were to display Japanese you would use code page 932 instead, or if you wanted to use Greek characters you would use code page 1253.

You can see some of the major code page values and their corresponding, self-descriptive character sets in Table 11.3.

Character Set	Code Page Value
Arabic	1256
Baltic	1257
Chinese	950
Cyrillic	1251
Greek	1253
Latin 1	1252
Latin 2	1250
Latin 5	1254
Hebrew	1255
Korean	949
Japanese	932
PRC GBK	936
Thai	874
Vietnamese	1258

Table 11.3. The major code page values and their character sets

To set the code page in your Active Server Pages, you can use the Session object's CodePage property. You assign the particular code page value to the CodePage property, making sure that it is a valid code page value; do not simply make up a value. A simple example would simply be:

```
Session.CodePage = 1252
```

To use the CodePage property in an example where you can see some result of changing the code page value, you will now create an Active Server Page that first uses the Latin 1 character set (code page 1252) and write America, and then change to the Japanese character set (code page 932) and write America in

Japanese katakana. Open up a new document in Notepad and type in the following script, saving the file as *codepage.asp* when you finish:

```
<html>
<body>
<font size=5>
<%
'Write something in English
Session.CodePage = 1252
Response.Write "America<br><br>"

'Write something in Japanese
Session.CodePage = 932
Response.Write "&#65383;&#65426;&#65432;&#65398;"
%>
</font>
</body>
</html>
```

Open up *codepage.asp* in your browser to see the result. The output from *codepage.asp* is shown in Figure 11.3.

Figure 11.3. The result of *codepage.asp* using both Latin and Japanese character sets

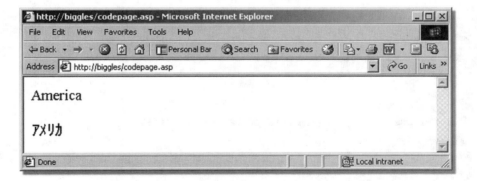

Working with the Contents Collection

The Contents Collection is where you can store many variables for use throughout the session using scripting commands. Session variables are accessible from all Active Server Pages in a site, making it easy to access information. To add a variable to the Contents collection, you must use the following syntax:

```
Session("keyname") = value
```

You can use your own keynames although you should try to make them descriptive. To see the number of items, you use the Count method of the Contents collection, as follows:

```
Session.Contents.Count
```

Reading the value that an item in the Contents collection holds can be either from a variable that holds the keyname or from the keyname itself, as shown:

```
Session.Contents(item)        'Using a variable
Session.Contents("keyname")   'Using a keyname string
```

To make things clearer, try using the Contents collection in a more complete script. The following example will show how you can assign some session variables and will also show how to see the number of items currently stored in the Contents collection, as well as how to iterate through the collection to read back all the items. Open up Notepad and save the following script as *contents.asp*:

```
<html>
<head>
<title>Using the Contents Collection</title>
</head>

<body>
<h2>Using the Contents Collection</h2>
<hr size="4">
<%
'Declare some variables
Dim item
Dim a, b, c
a = "Red"
b = "White"
c = "Blue"

'Add some variables to the Contents collection
Session("first") = a
Session("second") = b
Session("third") = c
```

```
'Find out the number of items in Contents collection
Response.Write "The number of items in Contents " _
& "collection is: " & Session.Contents.Count

'Loop through the Contents collection and read in each item
Response.Write "<br><br>The items in the collection " _
& "are:<br>"
for each item in Session.Contents
  Response.Write "This is item " & item & ": " &
  Session.Contents(item) & "<br>"
next
%>
</body>
</html>
```

Load in the file *contents.asp* to your browser to see the results, which should appear as shown in Figure 11.4.

Figure 11.4.
The result of
loading *contents.asp*
in the browser

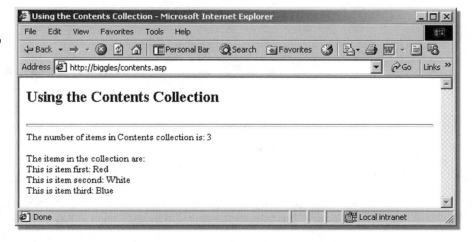

Removing Items from the Contents Collection

There are two ways that you can remove items from the Contents collection using the Session object. One method, the Contents.Remove method, removes a single item and the other, the Contents.RemoveAll method, removes all items. Both of these methods can be very useful when working with the Contents collection.

To remove a single item from the Contents collection, simply call the Contents.Remove method and pass it the keyname of the item you want to remove. For example, if you were to remove the item *second* from the previous example of the Contents collection, you would simply write the following:

```
Session.Contents.Remove("second")
```

Similarly, to remove all the items in the Contents collection, you make use of the Contents.RemoveAll method, as shown:

```
Session.Contents.RemoveAll
```

Note that you can also use numbers to remove individual items but be careful because as items come and go, the place number of individual items within the Contents collection changes. So the following would remove the fourth item in the Contents collection:

```
Session.Contents.Remove(4)
```

Working with the StaticObjects Collection

The StaticObjects collection holds variables that are available to all Active Server Pages in a site and lasts the entire duration of a session. While this may at first sound exactly the same as the Contents collection, there are some subtle differences.

The StaticObjects collection is read-only from within your usual Active Server Pages scripts. The items within the StaticObjects collection are added only in a file known as the global.asa. You add these items using the HTML <Object> tag. You will learn about the *global.asa* file and the <Object> tag in Lesson 15, "Using the Global.asa, Includes and Directives."

Although you will learn how to add the items that the StaticObjects collection contains at a later stage, you can see clearly from the following code that reading the items in is almost exactly the same as with the Contents collection. The following script shows you how to read in all the items in the StaticObjects collection and display them in the browser:

```
<%
Dim item
for each item In Session.StaticObjects
  Response.Write item & "<br>"
next
%>
```

Clearing All the Session Variables

You can also clear all the session variables in one step by using the Session object's Abandon method. Using the Abandon method saves you having to go through each session variable individually to clear it, and you can also feel safe in the knowledge that even if you or someone else adds more session variables at a later date, the Abandon method will still handle them.

To use the Abandon method, you do not need to pass any arguments, as shown:

```
Session.Abandon
```

There is one important point to be aware of with the Abandon method though—the Active Server Pages server will not process the Abandon method until the end of the script that it was called from. So any references to session variables after the Abandon method will still work if those references are in the same script. References in subsequent pages will not work, however.

To understand this more clearly, take a look at the following simple script where you firstly create a session variable with the value of 95 and then use the Abandon method. After using the Abandon method, you output the value of the session variable.

```
<%
Dim a
a = 95
Session("a") = a
Session.Abandon
Response.Write "Same script: " & Session.Contents("a")
%>
```

The result of this script would still output the 95 for the value of the session variable. However, now that the script that the Abandon method was called from has ended, calling a subsequent Active Server Page will show that the session variable has indeed been cleared. A simple script is all that is necessary to show this, as shown:

```
<%= "New script: " & Session.Contents("a")%>
```

Nothing will be returned by the server for this last call to the session variable.

One final point worth noting is that if you do not explicitly call the Abandon method, the server will automatically free memory and clear all the variables when the session ends anyway. So do not feel that you have to call the Abandon method to close your variables in every Active Server Pages script that you write. You should really only use it in practice for giving the user an option to clear their current settings and preferences.

Working with Web Farms

In practice, you may have to implement Active Server Pages in a Web farm environment, where load balancing between many servers takes place. In this situation you have to be aware of complications that arise from using the Session object.

Imagine the following scenario to get a clear picture of the problem you face. A user connects to a site using server A, which stores valid information about the session on its hard disk. The user then goes to another page and happens to get server A again. All is well for the user for the time being, with the server keeping track of the current session. When connecting to yet another page on the site, server A is busy and the user's request is taken care of by server B. Server B has no notion of the user or the current session and requires information to be input again. The user is annoyed at this but continues. After a further page request, the user's request is this time looked after by server C. Once again, there is no notion of the current session. You can imagine the ire of the user at this point.

This is a large issue with the current implementation of available Active Server Pages servers, such as IIS 5.0, and can at present only be properly addressed by using third-party add-ins. There are articles that you can find on the Microsoft site and elsewhere on the Internet that address this problem and in some cases even offer COM+ object solutions. The most important point is that you are aware of the issue of losing sessions in a Web farm environment.

WHAT YOU MUST KNOW

The Session object is very useful for providing users with a persistent state throughout their visit to a site, with the ability to carry preferences, variables, and settings from page to page without requiring intervention on the part of the user. The server assigns a session identification number and default timeout for each session, although you can change the timeout using the Timeout property. You can use the Contents collection to store and retrieve items that can persist for the entire duration of the session. In Lesson 12, "Using the Application Object," you will learn how to share resources among multiple users of the same Active Server Pages application, including how to lock and unlock those resources. Before you move on to Lesson 12, make sure that you understand and feel confident with all of the following key concepts concerning the Session object:

- The key point to the Session object is to provide a persistent state that allows for storing information that lasts for the entire duration of each session.

- Support for cookies is necessary if the Session object is to be made use of widely in your scripts because HTTP is a stateless protocol and there is no other simple means of storing persistent state information on the user's computer.

- The server assigns a session identification number and a block of memory when a user first connects to an Active Server Pages site. The Session object's SessionID property returns this number.

- The Session object's Timeout property basically represents a countdown in minutes before the session is closed. It starts this countdown from the time of the user's most recent request to the server.

- You can execute script in a subroutine whenever a session starts by using a subroutine called Session_OnStart, which utilizes the OnStart event. You can do practically the same thing for the end of a session using the OnEnd event and naming a subroutine Session_OnEnd.

- The LCID property is for setting the locale that the Active Server Pages will use. This can be helpful for presenting users with either a format that they prefer or one that your site may require.

- The Session object's Contents collection lets you store and retrieve items that are accessible from any Active Server Pages in a site for the entire duration of a session. You can also remove these items using the Contents.Remove and Contents.RemoveAll methods.

- The StaticObjects collection contains items that are read-only and do not change during the session, but that can be accessed from any Active Server Page in the same site.

- All the session variables can be cleared at once using the Session object's Abandon method. This method is actually processed by the server only at the end of the script that it is called from, but subsequent scripts will only get empty results unless session variables are reassigned first.

- Active Server Pages that use the Session object are very useful; however, in larger organizations or sites with high demand you can run into session problems if you Web farms are used for load balancing.

I n Lesson 11, "Using the Session Object," you saw how you could work with variables that were persistent throughout the entire duration of each session. In this lesson you will learn how you can work with variables that are available not just on a per user per session basis but rather are available to all users of an Active Server Pages application. By the time you complete this lesson you will have covered the following key concepts:

- The Application object has similar collections, events, and methods to the Session object, but it does not limit its accessibility to a single user and a single session at a time.

- You can use the locking mechanisms of the Application object, the Lock and Unlock methods, to prevent multiple users from trying to change a variable at the same time.

- The Contents collection stores variables, arrays, and objects created by the Server object's CreateObject method. You can change the elements of the Contents collection in your scripts.

- You can use the Contents.Remove method to remove individual items from the Contents collection or the Contents.RemoveAll method to remove all the items.

- An array is a single point of reference that can hold many values. When declaring an array you use parentheses and an integer to indicate the initial size of the array.

- The IsArray function returns True if an object is an array and False otherwise.

- The StaticObjects collection contains items that do not change during the course of the application, although you can add items to it in the global.asa file.
- You can also have subroutines for both the start and end of an application, by making use of the OnStart and OnEnd events, both of which belong in the global.asa file. You are mostly limited to the Application and Server objects in these subroutines, as the other objects are not yet available.

Introducing the Application Object

The Application object deals with storing and keeping track of items that can be used by multiple users of an application. Unlike the Session object, the variables that you use with the Application object do not disappear when a user's session ends. One of the most practical uses of the Application object is to keep statistics of your site on an Application-level basis. So you could perhaps count the number of visitors that have visited your Active Server Pages application. There are also complications with using variables that multiple users can have simultaneous access to, which you shall learn about in this lesson.

At first glance the collections, events, and methods of the Application object look exactly the same as the ones for the Session object.

Collections	Events	Methods	Properties
Contents	OnEnd	Contents.Remove	-
StaticObjects	OnStart	Contents.RemoveAll	
	Lock		
	Unlock		

Table 12.1. An overview of the Application object

The Application object is made up of a mixture of collections, events, and methods, with no properties at all.

Using Locking Mechanisms with the Application Object

Because the Application object holds variables that multiple users can access and change, you could run into difficulties if more than one user tries to change the same variable at the same time. You need to be sure the variable that you begin working with in a procedure does not change midprocedure because it will ruin the calculations being made.

For this reason, the Application object offers both the Lock method and the Unlock method. Both methods are very simple to use and do not require any arguments. So to use them simply, call the Lock method first, apply your statements, and then call the Unlock statement, as shown:

```
<%
Application.Lock
  'Statements
Application.Unlock
%>
```

When you lock the application object, you are preventing any other users from changing any of the variables in the Application object, not just the one or two variables that you might want, so it is important that you unlock the Application object as soon as you can.

If you use the Lock method without calling the Unlock method, the server will automatically unlock the Application object when the current Active Server Page ends or times out, whichever occurs first.

Working with the Contents Collection

The Contents Collection is where you can store many variables for use within the Active Server Pages application using scripting commands. Application variables are accessible from all Active Server Pages that use the application regardless of user or session, making it easy to share information. To add a variable to the Contents collection, you must use the following syntax:

```
Application("keyname") = value
```

You can use your own keynames although you should try to make them descriptive. To see the number of items, use the Count method of the Contents collection, as follows:

```
Application.Contents.Count
```

Reading the value that an item in the Contents collection holds can be either from a variable that holds the keyname or from the keyname itself, as shown:

```
Application.Contents(item)          'Using a variable
Application.Contents("keyname")    'Using a keyname string
```

To make things clearer, you should try using the Contents collection in a more complete script. The following example is very similar to what you saw when working with the Session object's Contents collection. Therefore you may want to simply open up *contents.asp* from the last lesson and modify it. The example will show you how you can assign some application variables and will also show you how to see the number of items currently stored in the Contents collection, as well as how to iterate through the collection to read back all the items. Open up Notepad and save the following script as *contents.asp*:

```
<html>
<head>
<title>Using the Contents Collection</title>
</head>

<body>
<h2>Using the Contents Collection</h2>
<hr size="4">
<%
'Declare some variables
Dim item
Dim a, b, c, d
a = "Water"
b = "Earth"
c = "Air"
d = "Fire"

'Add some variables to the Contents collection
Application.Lock
Application("first") = a
Application("second") = b
Application("third") = c
Application("fourth") = d
Application.Unlock

'Find out the number of items in Contents collection
Response.Write "The number of items in Contents " _
& "collection is: " & Application.Contents.Count
```

```
'Loop through the Contents collection and read in each item
Response.Write "<br><br>The items in the collection " _
& "are:<br>"
for each item in Application.Contents
  Response.Write "This is item " & item & ": " &
  Application.Contents(item) & "<br>"
next
%>
</body>
</html>
```

Load in the file *contents.asp* to your browser to see the results, which should appear as shown in Figure 12.1.

**Figure 12.1.
The result of loading
contents.asp in the
browser**

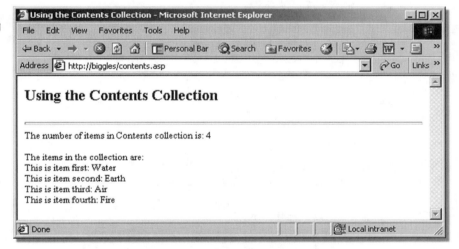

So far this has all been extremely similar to what you saw with the Session object's Contents collection. However, there are some other points that you are now ready to learn about regarding the Contents collection. You can add items to the Contents collection without declaring a variable first but rather by using a more direct approach. For example, compare the following two snippets of script which accomplish the same thing (unless you plan to use the variables by name elsewhere). The first snippet of script uses variables the way you have been writing them so far.

```
<%
'================================
'Snippet 1
'================================
Dim a, b, c, d
a = "Water"
b = "Earth"
c = "Air"
d = "Fire"

'Add some variables to the Contents collection
Application.Lock
Application("first") = a
Application("second") = b
Application("third") = c
Application("fourth") = d
Application.Unlock
%>
```

The second snippet of script uses a direct approach.

```
<%
'================================
'Snippet 2
'================================
'Add some variables to the Contents collection
Application.Lock
Application("first") = "Water"
Application("second") = "Earth"
Application("third") = "Air"
Application("fourth") = "Fire"
Application.Unlock
%>
```

You can see that the second snippet of script is obviously shorter than the first snippet. You have also not created any unnecessary variables. It is not that one approach is really any superior to the other but more importantly that you are aware of both approaches, especially when working in a team of Web developers.

There is one other important aspect about the Contents collection. Recall that in Lesson 10, "Using the Server Object," you learned about using the

CreateObject method. The Application object's Contents collection can contain instances of objects that you create with the Server object's CreateObject method, which therefore makes a single instance of the object available to all users of the Active Server Pages application. The following example shows the syntax you would use to assign an instance of an object to the Contents collection:

```
<%
set Application("myObject") = _
Server.CreateObject("myComponent")
%>
```

Removing Items from the Contents Collection

Removing items from the Contents collection in the Application object is almost exactly the same as removing items from the Session object's Contents collection. There are two ways that you can remove items from the Contents collection using the Application object. One method, the Contents.Remove method, removes a single item and the other, the Contents.RemoveAll method, removes all items. Both of these methods can be very useful when working with the Contents collection.

To remove a single item from the Contents collection, simply call the Contents.Remove method and pass it the keyname of the item you want to remove. For example, if you were to remove the item *LastVisit* from the Contents collection, you would simply write the following:

```
Application.Contents.Remove("LastVisit")
```

Similarly, to remove all the items in the Contents collection, you make use of the Contents.RemoveAll method, as shown:

```
Application.Contents.RemoveAll
```

Note that you can also use numbers to remove individual items but be careful because as you add and remove items, the place number of individual items within the Contents collection changes. So the following would remove the seventh item in the Contents collection:

```
Application.Contents.Remove(7)
```

Note: *In practice, remember to lock the Application object before removing items from the Contents collection and unlock it again afterwards.*

Working with Arrays in the Contents Collection

If you have never programmed with arrays before, they are a little trickier to work with than regular scalar variables that only store a single value. An array simply holds many values using a single point of reference. You may wonder why you would need to use an array when you already have variables. So far in this book, you have only been using a few variables at most but imagine if you had to store thousands of them! Imagine how much code would be necessary to increment each of those variables by 1. But if you were to use an array, you could increment the equivalent of thousands of variables using just three lines of code! The following code shows you an example of this, just to give you an idea of what is meant, although you have not yet seen how to declare and assign arrays:

```
for i = 0 to 9000
  myArray(i) = myArray(i) + 1
next
```

You declare an array similarly to scalar variables; however, you use parentheses and an integer value to specify the size of the array, as shown in the following example which can hold three variables:

```
Dim myArray(2)
```

To assign values to the individual elements in an array, simply specify the element number within parentheses and then use an = character followed by the value that you want to assign, as shown:

```
myArray(0) = 118
myArray(1) = 34
myArray(2) = 276
```

To add an array to the Application object's Contents collection, you simply use the same method that you use when you are adding a variable, as the following line demonstrates:

```
Application("ArrayTag") = myArray
```

To retrieve a single element from an array that is inside the Contents collection, you specify the keyname of the array and then in separate parentheses indicate the number of the array element that you want. For example, the following line outputs the array's second element:

```
<%Response.Write Application.Contents("ArrayTag")(1)%>
```

You can also retrieve all the elements of an array that is stored in the Contents collection by using either a For-Each loop or a For-Next loop. The following example shows how you might output each element of an array on a separate line using a For-Each loop:

```
for each item in Application.Contents("ArrayTag")
  Response.Write item & "<br>"
next
```

The same retrieval could be done using a For-Next loop, as shown:

```
for i=0 to 2
  Response.Write Application.Contents("ArrayTag")(i) _
  & "<br>"
next
```

To see the use of arrays with the Application object's Contents collection more clearly, the following example displays a more complete script. In the script, you declare and assign values to an array before adding it to the Contents collection. After this, you instruct the script to output all of the items in the array on a separate line. Open up Notepad and type in the following example script:

```
<%
'Declare variables
Dim item
Dim myArray(3)

'Assign values to the array
myArray(0) = "Shahla"
myArray(1) = "Martha"
myArray(2) = "Esther"
myArray(3) = "Zoe"

'Add the array to the Contents collection
Application.Lock
Application("NameArray") = myArray

'Retrieve the array from the Contents collection
for each item in Application.Contents("NameArray")
  Response.Write item & "<br>"
next
Application.Unlock
%>
```

Save the script as *array.asp* and then open it in your browser. The output from *array.asp* is shown in Figure 12.2, where you can see that the contents of the array were retrieved in order.

Figure 12.2.
The result of loading *array.asp* **in the browser**

Note that in the previous example you were aware that you were dealing with an array. However, if you were outputting all of the elements in the Contents collection, including arrays and variables, you would have to check whether you were dealing with an array first. You can check if an object is an array by using the VBScript function IsArray, which returns True if the object is an array and False otherwise. The basic syntax for using IsArray with the Contents collection is as follows:

```
IsArray(Application.Contents(item_number))
```

To give you an idea of its use, you might replace the section of script from the file *array.asp* that deals with retrieving array elements with the following:

```
'Retrieve the array from the Contents collection
For Each item in Application.Contents
   If IsArray(Application.Contents(item)) Then
     Response.Write "Here is our array...<br>"
     For Each arrayitem in Application.Contents(item)
       Response.Write arrayitem & "<br>"
     Next
   End If
Next
```

Working with Arrays and the Session Object

You now know about arrays and using them with the Application object's Contents collection. Although not covered in Lesson 11, "Using the Session Object," you can also apply what you know about working with arrays to the Session object's Contents collection, too. So you could also add an array to a session using script similar to the following example:

```
<%
'Declare variables
Dim item, arrayitem
Dim myArray(2)

'Assign values to the array
myArray(0) = 9
```

```
myArray(1) = 19
myArray(2) = 95

'Add the array to the Contents collection
Session("array") = myArray

'Retrieve the array from the Contents collection
For Each item in Session.Contents
  If IsArray(Session.Contents(item)) then
    Response.Write "Here is our array...<br>"
    For Each arrayitem in Session.Contents(item)
      Response.Write arrayitem & "<br>"
    Next
  End If
Next
%>
```

The result of this example is shown in Figure 12.3.

**Figure 12.3.
The result of
loading** *array_s.asp*

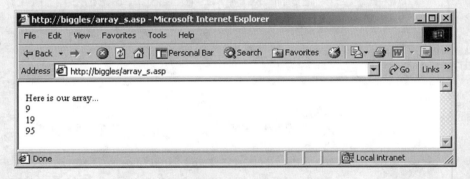

Working with the StaticObjects Collection

The StaticObjects collection holds items that are available to all Active Server Pages in an application. You do not change the items in the StaticObjects collection in your scripts; they are referenced as they are needed. The StaticObjects collection is read-only from within your usual Active Server Pages scripts. However, items in the StaticObjects collection are available to all users of an application, not just an individual user's session.

As with the StaticObjects collection in the Session object, the items within the Application object's StaticObjects collection are added only in a file known as the global.asa. You add these items using the HTML <Object> tag. You will] learn about both the *global.asa* file and the <Object> tag in Lesson 15, "Using the Global.asa, Includes and Directives."

Although, you will learn how to add the items that the StaticObjects collection contains at a later stage, you can see clearly from the following code that reading the items in is almost exactly the same as with the Session object's StaticObjects collection. The following script shows you how to read in all the items in the StaticObjects collection and display them in the browser:

```
<%
Dim item
for each item In Application.StaticObjects
  Response.Write item & "<br>"
next
%>
```

Making Use of the Application_OnStart Event

The Application object's OnStart event takes place before the first session begins and for this reason, you cannot make any reference to the Session object within the subroutine or you will get an error. Likewise, you cannot use the Response and Request objects in the OnStart event either or you will receive an error. When creating the subroutine for the OnStart event, you must place it in the global.asa file using the following syntax:

```
Sub Application_OnStart
    'Statements
End Sub
```

If you are making changes to application level variables in the OnStart event, you do not have to lock or unlock the Application object. This is basically because no other script will be running at this point that could simultaneously be using those same variables. Remember that the OnStart event for the Application object takes place even before the first OnStart event of the Session object runs.

Making Use of the Application_OnEnd Event

The Application object's OnEnd event is very similar to its cousin, the OnStart event. You cannot reference the Session object or any object for that matter, apart from the Application and Server objects. Also you cannot use the MapPath method, even though it is part of the Server object. What you can actually do in this event may seem quite restricting; however, bear in mind that at this point you are closing down the application and the Session object's OnEnd event has already run. Actually, the OnEnd event will really only be run if you purposely stop your Active Server Pages server or if something causes the server to stop.

When creating the subroutine for the OnEnd event, you must place it in the global.asa file, as you did with the OnStart event, using the following syntax:

```
Sub Application_OnEnd
    'Statements
End Sub
```

WHAT YOU MUST KNOW

The Application object allows multiple users of an Active Server Pages application to access and share variables. This is mostly useful for the purpose of gathering statistics about an application and its pages. To prevent multiple users from accessing the same application level variables at the same time, you must use the locking mechanism of the Application object to ensure that only safe changes are made. In Lesson 13, "Using the ASPError Object," you will learn about obtaining information pertaining to error conditions that occur in your Active Server Pages. Before you continue with Lesson 13, make sure that you know the following key concepts:

- The Application object has some similar collections, events, and methods to the Session object, but it contains objects that are accessible by different users of the application, which allows you to share information between multiple users.

- By using the Lock and Unlock methods, you can prevent multiple users from trying to change application-level objects at the same time.

x The Contents collection stores variables and arrays whose elements you can change in your scripts. It also stores objects that you create with the Server object's CreateObject method.

x Removing items from the Contents collection is straightforward with the Contents.Remove and the Contents.RemoveAll methods. Remove individual items with the Contents.Remove method by indicating the particular item to remove, or remove all the items with the Contents.RemoveAll method.

x An array is a single point of reference that can hold many elements. You indicate the initial size of the array by using parentheses and an integer when you declare it.

x The IsArray function lets you know if an object is an array. It returns True if an object is an array and False otherwise.

x The items that are contained in the StaticObjects collection are read-only and do not change during the course of the application. You can, however, add items to the StaticObjects collection in the global.asa file.

x Using the Application_OnStart event, you are able to have a subroutine for the start of an application, although you are mostly limited to the Application and Server objects.

x The Application object's OnEnd event, lets you create a subroutine in the global.asa file that triggers when the application is closed. As with OnStart event, you are limited mostly to the Application and Server objects and also cannot use the MapPath method.

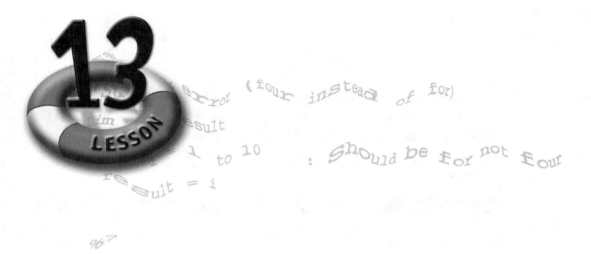

USING THE ASPError OBJECT

i n Lesson 12, "Using the Application Object," you learned about sharing variables between multiple users and preventing possible errors by using the Application object's locking mechanism. In this lesson, you will learn about how to retrieve information when an error actually does occur so that you can get to the core of the problem and fix it. By the time you finish this lesson you will know the following key concepts:

- When an error occurs in your Active Server Pages, the ASPError object can be useful in discovering what went wrong and where it went wrong to aid you in fixing the problem.
- The ASPError object is made up completely of read-only properties. There are no collections, events, or methods.
- There are three main types of errors that can occur: preprocessing errors, syntax errors, and runtime errors.
- To test your error handling capabilities, you must create error scripts than can produce each of the three main types of errors.
- Before you can use the ASPError object's properties, you must make a call to the Server object's GetLastError function.
- You use the ASPError object and its properties to customize the *500-100.asp* script, in order to present the company logo and tailor the page to your requirements or those of your client.

Introducing the ASPError Object

The ASPError object is basically for retrieving as much information as possible when an error occurs. You should use this information to try to remedy problems in your Active Server Pages. Table 13.1 shows you an overview of the ASPError object in terms of the collections, events, methods, and properties that it contains.

Collections	Events	Methods	Properties
-	-	-	ASPCode
			ASPDescription
			Category
			Column
			Description
			File
			Line
			Number
			Source

Table 13.1. An overview of the ASPError object

The ASPError object is made up entirely of properties with no collections, events, or methods.

Understanding How Active Server Pages Handle Errors

When an error occurs in your Active Server Pages scripts, the server immediately transfers control to a special page for handling the error. This page is usually the *500-100.asp* file, although you can create custom files. You will probably find that your default *500-100.asp* file is not in the same directory as your standard scripts. For instance, on Windows 2000 the default location of the *500-100.asp* file is as follows:

c:\winnt\help\iishelp\common\500-100.asp

The path on your computer may be different but you can simply perform a search or find, depending on your operating system, to discover the location of this file. The *500-100.asp* file is important because it is in this file that you will work with the ASPError object.

Causing Errors in Your Active Server Pages

There are many different possibilities that can cause an error to appear in your Active Server Pages. When an error occurs it is basically one of three main types: a preprocessing error, a syntax error, or a runtime error. To be able to test all of these types of errors you must create some scripts that are able to cause these various errors to occur. Start off by taking a look at runtime errors.

Creating a Runtime Error Script

Runtime errors are basically errors that occur when the script of the Active Server Page is being run by the server. The following example shows you a script that is correct for syntax but will cause a runtime error when you execute it. Notice that the loop includes division of a variable, *i*, minus 5. At the point in the loop where *i* is equal to 5, the script will cause a "division by zero" error. Open up Notepad and type in the following script:

```
<%
'Division by zero error
Dim i, result
for i = 1 to 10
   result = i / (i-5)
next
%>
```

Save the script as *error1.asp* and then load it into your browser to see the result. The *500-100.asp* page is loaded into the browser, and you can see from Figure 13.1 that the first part of the page really only indicates to the user that the page cannot be displayed by the server. Notice that it offers suggestions to the user to try, and also it discreetly mentions that it is an HTTP 500.100 error occurring at the server end. It also mentions ASP and Internet Information Services, although this of more limited usefulness.

Figure 13.1.
The first part of the
standard error page

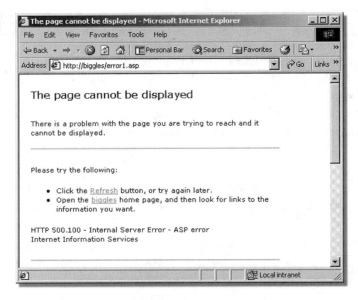

If you scroll down the default *500-100.asp* page further, you will see more details about the error, as shown in Figure 13.2. You can see all sorts of details about the error, including what type it was, what browser was used, what page caused the error and what the time was when the error occurred.

Figure 13.2.
The second part
of the standard
error page

Creating a Syntax Error Script

Syntax errors are different from runtime errors. They occur before the script is even run and are usually spelling mistakes or missing keywords. The following script contains a syntax error in its For-Next loop. The keyword For has been written as Four instead. Open up Notepad and type in the following script:

```
<%
'Syntax error (four instead of for)
Dim i, result
four i = 1 to 10      : Should be for not four
  result = i

%>
```

Save this script as *error2.asp* so that you can use it to test syntax error handling.

Creating a Preprocessing Error Script

Although you have yet to cover preprocessing directives, at this point in time you only need to be aware that a third type of error exists. Preprocessing directives will be discussed later in the book.

To create an example of a preprocessing error, you need to remove the = character that follows the word *file* within the preprocessing directive. So the following example script, *error3.asp*, would produce a preprocessing error that you can use for testing purposes:

```
<!--#include file myspecialfile.h  -->
<%
Response.Write "Hello world"
%>
```

Save the script as *error3.asp* for testing purposes.

Using the ASPError Properties

All the ASPError properties are read-only and do not take any arguments, so they are very simple to implement and require very little in the way of explanation. Recall from Lesson 10, "Using the Server Object," that you must use the Server object's GetLastError method to populate an ASPError object. You might use some script similar to the following to accomplish this:

```
<%
  Option Explicit
  Dim objASPError
  Set objASPError = Server.GetLastError
%>
```

After you have populated your ASPError object you can retrieve the value of any of its properties. To retrieve the value from any of the ASPError object's properties, simply use a dot notation between the ASPError object's variable name and the name of the property that you want to retrieve. The following line returns the Line property value:

```
<%=objASPError.Line%>
```

It is worth knowing that not all the properties will return useful information all of the time. For instance, some properties only return a value if a certain type of error occurs and are completely empty otherwise. A short overview of each of the properties will help you to understand what they potentially hold before you move on to actually customizing your own *500-100.asp* script.

The Category property lets you know if it is a runtime error, a syntax error (although it refers to this as a compilation error because it could not compile the script), or a preprocessing error (referred to simply as Active Server Pages).

The Number property is the standard COM error number, typically a long value.

The Line and Column properties indicate the line and column, respectively, of the first character that the server encountered where the error occurred. You will only get *-1* for the Column property if the error type is not a syntax error.

The File property lets you know what file the error occurred in.

The Description property gives you a short description of what the error was, while the ASPDescription property gives you a lengthier description of the problem if it was part of Active Server Pages.

The ASPCode property is simply a code number that IIS assigns to specific Active Server Pages errors.

Finally, the Source property will display the actual line of script where the error occurred. Be aware that the Source property is only useful with some syntax errors.

Customizing the *500-100.asp* Script

You would only ever use the ASPError object and its properties if you want to customize the *500-100.asp* script, so that is exactly what you will do now. You should copy the default *500-100.asp* script somewhere else so that you can restore it at a later stage or in case you want to refer to it as an example.

In this example you will create the entire *500-100.asp* page from scratch. Imagine that you are creating it for a company who wants their logo on the page. The other Active Server Pages programmers have asked if you could output the Line, Column, and Description tags in purple so that they stand out clearly. You divide the script into four neat sections: creating the ASPError object, creating a customized company heading, explaining options to the user, and outputting technical details.

Open up Notepad and enter the following script:

```
<%
   Option Explicit
   Dim objASPError
   Set objASPError = Server.GetLastError
%>
<html>
<head>
<title>Error - Unable to display the page</title>
</head>

<body>
<!--=======================================
Create a customized company heading
=========================================-->
<h4 align="center">www.ivywall.com 500-100 error page</h4>
<img src="ivy.jpg" align="left">
<h1>Sorry - Unable to display the page</h1><hr>


<!--=======================================
Explain options to user
=========================================-->
<p>Unfortunately there has been a problem trying to load
the page that you have requested. There are a number of
possibilities that can cause this to happen so you may
want to try the following:<br><br>
<ul>
```

```
<li>Click the Refresh button to attempt reloading the page
<li>Return to the home page and follow the links to the
information you require
</ul>
</p>
<hr>


<!--=======================================
    Output technical details about the error
========================================-->
<p>You can see specific technical details about the error
below:<br><br>
The error was of type:
<font color="green"><%=objASPError.Category%></font><br>

The error number is:
<font color="green"><%=objASPError.Number%></font><br>

The error occurred in the file:
<font color="green"><%=objASPError.File%></font><br>

The error occurred on: <font color="purple">line
<%=objASPError.Line%>, column <%=objASPError.Column%>
</font><br>

The short error description is:
<font color="purple"><%=objASPError.Description%></font><br>

The ASP error description is:
<font color="green"><%=objASPError.ASPDescription%></font><br>

The IIS ASP code for this error is:
<font color="green"><%=objASPError.ASPCode%></font><br>

The ASP source with syntax error is:
<font color="green"><%=objASPError.Source%></font>
</p>
</body>
</html>
```

Save the script as *500-100.asp* and save it in the default location for your setup. Now that you have your custom script, you can test it with the three error scripts that you wrote earlier. When you load in *error1.asp*, your runtime error script, you should see some output similar to Figure 13.3.

Figure 13.3.
Your custom 500-100 page after loading *error1.asp*

You can see not only a customized heading and logo but also all the details of the error neatly laid out at the end of the page. The page shows clearly that it was a VBScript runtime error in the file *error1.asp*. It also indicates that the error occurred on line 5 and describes the error as a division by zero. Less helpful were the standard error number and the column reporting –1. The other properties did not return anything, as this was not an Active Server Pages error and was not a syntactical error either.

Now try out your customized *500-100.asp* script with a syntax error. Load in *error2.asp* and you should see results in your browser similar to what is shown in Figure 13.4.

Figure 13.4.
Your custom 500-
100 page after
loading *error2.asp*

Notice that the type is reported as a VBScript compilation error, although you know that this indicates a syntax error. It also informs you that the error occurred in the file *error2.asp*, on line 4, column 11. The description given indicates that it is expecting an end to a statement. Of special value though, is the source line itself, where you can clearly see the word *four* has been used instead of the word *for*. This line of source code should alert you to the problem and is only shown with some syntax errors.

Finally, running *error3.asp* will show you how your *500-100.asp* script deals with a preprocessing error. The result that you see in your browser should look similar to that shown in Figure 13.5.

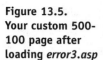

**Figure 13.5.
Your custom 500-
100 page after
loading *error3.asp***

This error seems to show a wealth of information in your *500-100.asp* page. The only properties that do not report back useful information are the Source and Column properties. You can see that the error type was dealing directly with Active Server Pages, so you can surmise that it is a preprocessing error. The file is reported as being *error3.asp*, and the line that the error occurred on is reported as being the first line. The short description informs you that an attribute value is missing, while the longer ASP description specifically states that the *file* attribute is missing a value. You are also given the standard error number, as well as the IIS code for the error. Between the Line property and the ASPDescription property alone, you have a pretty good idea of what the problem is and where it is.

WHAT YOU MUST KNOW

The ASPError object is made up entirely of read-only properties that you have access to only after using the Server object's GetLastError method. There are three main types of errors that you can encounter, namely preprocessing errors, syntax errors, and runtime errors. By customizing the *500-100.asp* script, you can create your own error handling page for reporting back what went wrong in detail. In Lesson 14, "Using the ObjectContext Object," you will learn about how to use transactions in your Active Server Pages to either commit to a block of changes or abort all the changes and roll back to a previous state.

Before you move on to Lesson 14, make sure that you have fully understood the following key concepts:

- When an error occurs in your Active Server Pages, the ASPError object can help you to find out more details about the circumstances surrounding the error and aid in fixing the problem.

- The ASPError object has nine read-only properties but no collections, events, or methods.

- The three main types of errors are preprocessing, syntax, and runtime errors. Depending on which type of error occurs, some of the ASPError object's properties do return any useful information.

- You must create error scripts than can produce each of the three main types of errors to be able to test error reporting in your Active Server Pages.

- You must make a call to the Server object's GetLastError function before you can use the ASPError object's properties.

- When you have to customize the error page either to fit in with a client's needs or to enhance the page in some way, you can use the ASPError object and its properties within the *500-100.asp* script.

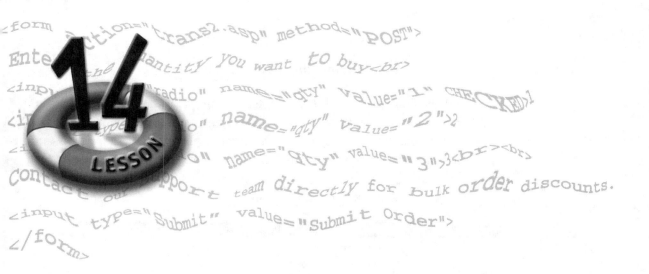

USING THE ObjectContext OBJECT

*i*n Lesson 13, "Using the ASPError Object," you learned how to extract all sorts of details about errors that can occur in your Active Server Pages so that you can fix them. In this lesson you will learn about a method of preventing errors from occurring, not with your scripts, but with the data that your scripts deal with. This method is mostly about transactions and how to implement them using the ObjectContext object. By the time you finish this lesson you will have learned the following key concepts:

- The ObjectContext object is basically for working with transactions and has only two events and two methods.
- You can commit a transaction by using the SetComplete method, which indicates that everything was successful.
- If there are any problems, you can cancel the entire transaction by using the SetAbort method, effectively rolling back any changes that were made.
- The OnTransactionCommit and OnTransactionAbort events allow you to set up subroutines that can cater to either a successful or unsuccessful transaction, respectively.
- You must make use of the transaction directive on the first line of your transaction scripts to be able to use the ObjectContext object.

Introducing the ObjectContext Object

The ObjectContext object is basically for working with transactions. By using this object you can commit an entire block of script as being a completed transaction, or if something goes awry midway through the block of script, you can abort the transaction and roll back to before any changes were made. Table 14.1 shows you an overview of the ObjectContext object in terms of the collections, events, methods, and properties that it contains.

Collections	Events	Methods	Properties
-	OnTransactionAbort	SetAbort	-
	OnTransactionCommit	SetComplete	

Table 14.1. An overview of the ObjectContext object

With only two events and two methods, the ObjectContext object is the smallest of the seven intrinsic objects and contains no properties and no methods.

Understanding the Role of Transactions

In the business world today, Active Server Pages is a very popular technology because it allows not only dynamic page content but direct access to databases. However, few businesses would trust Web pages with their live database if they did not support transactions. A transaction is really just painting several functions with the same paintbrush. These several functions either succeed as a whole or fail as a whole. The transaction binds the separate functions together ,in a sense. This is the essence of transactions and the whole concept behind it.

To understand why a business would require this sort of functionality you will first look at a simplified business model that does not use transactions. Imagine that a company called ABC sells a product online called XYZ. When a customer of ABC purchases a quantity of XYZ, there is a series of steps that are put into action.

First, the act of ordering the XYZ product triggers the step of assigning an order number. A check of the inventory database must be made to see that there is enough XYZ to fill the order. The inventory database must be reduced by the quantity in the order. Finally, the customer is sent a bill for the order and delivery of the order is arranged. Figure 14.1 shows the five separate steps that get carried out by the company, as a direct result of a customer placing an order.

**Figure 14.1.
Individual steps to
be taken without
using transactions**

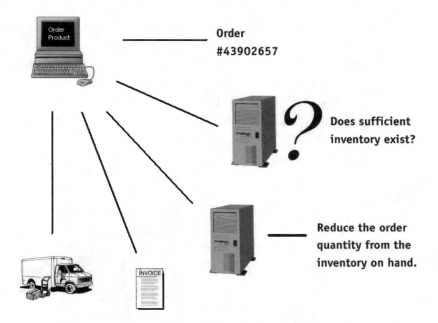

Order
#43902657

Does sufficient
inventory exist?

Reduce the order
quantity from the
inventory on hand.

If something goes wrong when trying to decrement the inventory database by the quantity in the order, it will not affect the delivery of the goods. They will still go out the door. The invoice will still be sent to the customer from the billing department. But now the database is corrupt because it no longer accurately portrays the state of inventory on hand. This is what could happen if one step did not complete. However, other steps could also cause problems if they were unable to be completed.

Imagine if the billing department was unable to send a bill. The inventory was still checked and reduced, an order number was generated originally, and the delivery was made. But no money is coming into the company for the goods! You could manually go through records trying to find problems but that can be extremely costly and time consuming, especially if sales are in the tens of thousands or more.

Enter transactions. Transactions solve this very problem by keeping all the individual steps as a related group, which either all fail or all succeed. The steps are still the same separate steps that they were before, but now there is a link between them all that causes them to succeed or fail as a whole, regardless of whether some of the steps were successful while others were not.

Going back to our example, if the inventory database had been successfully checked and reduced by the quantity of the order but there was a problem with the billing, then the transaction is aborted and everything is put back the way that it was prior to the transaction's commencement. This means that the inventory database would be restored by adding the value that was previously subtracted. Even though the step for reducing the database was successful, the transaction causes everything to fail if it is aborted.

Figure 14.2 shows the same simplistic example except that this time it is using transactions to bind the separate steps so that they either succeed or fail as a whole.

**Figure 14.2.
Combining the
individual steps
into a transaction**

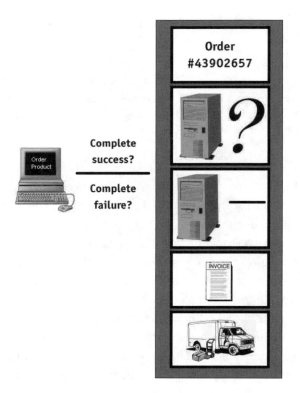

Using the SetAbort Method

In your own Active Server Pages you can abort a transaction by using the ObjectContext object's SetAbort method. The SetAbort method will abort the entire transaction. You can call the SetAbort method from anywhere in your transaction script. The SetAbort method does not take any arguments, so you simply call it on its own, as follows:

ObjectContext.SetAbort

If you call the SetAbort method, it will in turn trigger the OnTransactionAbort event, where, if you have created a subroutine for it, you can run any script that you need to run when aborting a transaction. Other lines of your script that follow the SetAbort method may still be processed by the server, even though the SetAbort method has been called.

Using the SetComplete Method

The SetComplete method simply flags the transaction as being successful. You should not call this method unless you have reached a point in your script when all steps are successful. If you are working with a database, calling the SetComplete method will indicate that all changes to the database made during the transaction can be committed. Like its counterpart, the SetAbort method, the SetComplete method takes no arguments and so is a straightforward call, as shown:

ObjectContext.SetComplete

Calling the SetComplete method will also trigger the OnTransactionCommit event.

Using the OnTransactionAbort Event

The OnTransactionAbort event is where you are able to use a subroutine that informs the user of the failure for the particular transaction that was requested. If you do not inform the user, he has no way of knowing that the transaction did not take place. The OnTransactionAbort event would have a syntax as follows:

```
<%
Sub OnTransactionAbort()
   'Statements
End Sub
%>
```

Only supported database commands will be automatically rolled back by the server when the SetAbort method is called. Therefore, within the

OnTransactionAbort event's subroutine you must undo any other changes that were made during the transaction. For instance, files that were moved should be put back, global variables that were added to should be subtracted from, and any other nondatabase changes need to be fixed.

It is important that when changing variables back in this subroutine not to assign specifically the old value again. This is because other transactions may have been made that have changed those variables since you first started using them. Of course, this does not apply to Application level variables where you have used the locking mechanisms to ensure that no other changes occur. The best way to handle variables is to subtract or add the amount that was changed. For example, if you have a variable named *myGlobal* whose value was 5000 and you subtracted 10 from it, you should not set it to 5000 again but simply add 10 to its current value, as shown:

```
myGlobal = myGlobal + 10
```

Using the OnTransactionCommit Event

The OnTransactionCommit event is where you are able to use a subroutine that informs the user of the success of the particular transaction that was requested. If you do not inform the user then, once again, he has no way of knowing that the transaction succeeded. The OnTransactionCommit event would have a syntax as follows:

```
<%
Sub OnTransactionCommit()
  'Statements
End Sub
%>
```

Within this subroutine you can make any further changes that have not already been taken care of within the transaction. However, no changes that are made in this subroutine should be critical to the success of the transaction, as by the time control reaches this subroutine the transaction is already successful. So basically, this subroutine would be useful only for tidying up and informing the user.

Knowing the Transaction Directive

Although you will be learning about preprocessing directives in the next lesson, "Using the Global.asa, Includes and Directives," you will have to make use of a directive in order to use transactions in your scripts. At the top of the page that contains the transaction script, you must put the following line:

```
<%@Transaction = "Required"%>
```

If this directive is not on the first line of your script, you will receive an error. The transaction directive informs the Active Server Pages server that this script is a transaction and must be treated with particular attention. The transaction directive will cause the script to be run to completion, although an aborted transaction can change this. For the time being, simply remember to place this directive at the top of all your transaction scripts.

Putting It All Together

Although you have learned all about the ObjectContext object and how to use its methods and events, you are still unable to really use it for what it was intended—databases. You will learn more about databases, transactions, and using the ObjectContext object later in this book in the various lessons dealing with Structured Query Language (SQL) and ActiveX Data Objects (ADO). However, the aim of this lesson is to show you how to use the ObjectContext object to prepare you for when you do learn about database work.

The following example will show you a complete example of using the ObjectContext object and its methods and events, as well as the transaction directive. You will use the example company, ABC, along with its imaginary product, XYZ, to mimic ordering a product using transactions. First, you will create an order form, which contains the company heading and basic product hype, as well as using three radio buttons to indicate the quantity of the product that the customer can order. You will also use a submit button to send the form data onto your transaction script, *trans2.asp*, which you shall create shortly. Open up Notepad and enter in the following script to create the order form:

```
<html>
<head>
<title>Transaction Form</title>
</head>

<body>
<img src="images/abc.png">
<h2>The Latest Version of XYZ!</h2>
The product that does it all.<br>
<ul>
<li>More features</li>
<li>More gizmos</li>
<li>More, more, more</li>
<li>FREE Ginsu knife thrown in</li>
</ul>
<h3>Only $99.99 while stocks last!</h3>
<form action="trans2.asp" method="POST">
Enter the quantity you want to buy<br>
<input type="radio" name="qty" value="1" CHECKED>1
<input type="radio" name="qty" value="2">2
<input type="radio" name="qty" value="3">3<br><br>
Contact our support team directly for bulk order discounts.<hr>
<input type="Submit" value="Submit Order">
</form>
</body>
</html>
```

Save this script as *trans1.asp*. Opening up *trans1.asp* in your browser should show you an order form similar to what is shown in Figure 14.3. Do not actually submit the order yet because you have not had a chance to create the transaction script for the order form.

**Figure 14.3.
The XYZ product
order form that
initiates the
transaction**

In the order form, *trans1.asp*, you have three radio buttons indicating the quantity of the product XYZ that the customer wants to buy. To allow you to see both a transaction succeeding and a transaction failing, you will simply use a conditional statement on the quantity to determine success. If the quantity is less than two then assume that the warehouse has stock and commit to the transaction, otherwise abort the transaction.

Remember that you have to use the transaction directive at the top of the page to indicate that this is a transaction script. For the two events, OnTransactionAbort and OnTransactionCommit, you will simply inform the user of the outcome of the order, letting him know if it was successful or not. Open up Notepad and type in the following transaction script:

```
<%@Transaction = "Required"%>
<html>
<head>
<title>Transaction Form</title>
</head>

<body>
<img src="images/abc.png">
<h2>The Latest Version of XYZ!</h2>
Your order is being processed. Please wait...
<%
```

```
Dim quantity
quantity = Request.Form("qty")

If quantity<2 then
  ObjectContext.SetComplete
Else
  ObjectContext.SetAbort
End If
%>
Thank you
</body>
</html>

<%
Sub OnTransactionAbort()
Response.Write "<br><br><font size=6>"
Response.Write "Unable to complete your transaction. " & _
"The transaction has been aborted due to a shortage " & _
"of inventory in our warehouse."
Response.Write "</font>"
End Sub

Sub OnTransactionCommit()
Response.Write "<br><br><font size=6>"
Response.Write "The transaction has completed " & _
successfully. " & _
"Your goods will be arriving within 2-3 working days."
Response.Write "</font>"
End Sub
%>
```

Save this script as *trans2.asp*. Open up *trans1.asp*, your order form that you created previously, and leaving it with the default quantity of one, submit an order. The transaction should succeed and give you back a response similar to that shown in Figure 14.4. Notice that the words *Thank you* appear before the output from the event, even though the SetComplete method was called first.

Figure 14.4.
The result of a successful transaction when ordering product XYZ

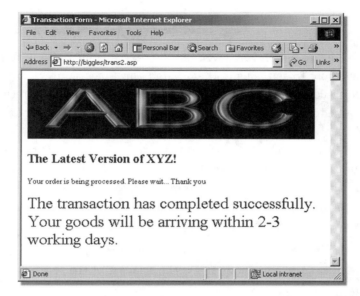

Open up *trans1.asp* in your browser again. This time select a quantity other than one and submit the order. Because you have indicated that orders with a quantity less than two do not succeed, you should see a response similar to that shown in Figure 14.5.

Figure 14.5.
The result of an unsuccessful transaction when ordering product XYZ

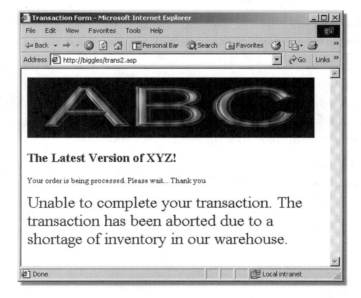

WHAT YOU MUST KNOW

In this lesson you have seen the need for transactions in Active Server Pages and how to implement them using the ObjectContext object. Although the ObjectContext object is primarily for use with databases, in this lesson you have learned how to use the object itself, and you will learn about integrating it with databases later in the book. The SetAbort and SetComplete methods, along with the OnTransactionAbort and OnTransactionCommit events, are all that comprise the ObjectContext object. In Lesson 15, "Using the Global.asa, Includes, and Directives," you will learn the final pieces concerning the basic Active Server Pages object model. Before you move on to Lesson 15, make sure that you are comfortable with the following key concepts:

- Basically for working with transactions, the ObjectContext object has only two events and two methods and is the smallest of the seven intrinsic objects.

- By using the SetAbort method, you can cancel a transaction and roll back all changes made during the aborted transaction.

- To indicate that everything in a transaction is successful, you can use the SetComplete method. The SetComplete method will commit all of the changes made within the transaction.

- The OnTransactionCommit event lets you create a subroutine that will run whenever a transaction is successful, within which you can respond to the user and perform any additional changes.

- The OnTransactionAbort event lets you create a subroutine that will run whenever a transaction is aborted. Within this subroutine you should inform the user of the failure, as well as handle the cleanup of nondatabase functions that were performed during the transaction.

- To be able to use the ObjectContext object, you must make use of the transaction directive on the first line of your transaction scripts.

*i*n Lesson 14, "Using the ObjectContext Object," you learned about working with transactions using the final object of the seven intrinsic objects. In this lesson, you will learn about some of the remaining components that make up the fundamentals of Active Server Pages. You have already seen reference to these components through learning about the intrinsic objects, but in this lesson you will go into greater detail. By the time that you complete this lesson you will know the following key concepts:

- The Global.asa is a special file that holds subroutines for four particular events—the OnStart and OnEnd events for both the Application object and the Session object.
- The Global.asa file also contains objects that you declare with the <OBJECT> tag. These objects are basically what you use with the StaticObjects collections in both the Application object and the Session object.
- There are five preprocessing directives known as @ directives.
- The @LANGUAGE directive lets you specify what scripting language to use within an Active Server Page.
- The @ENABLESESSIONSTATE directive lets you turn session tracking on or off for a page.
- The @TRANSACTION directive is what you use to indicate that the script that follows is actually a transaction.
- The @CODEPAGE directive sets the codepage for the current script.

- The @LCID directive sets the locale for the current script.
- Server-side includes allow you to add information to a Web page just prior to it being sent to the user. There are six server-side includes: #config, #echo, #exec, #flastmode, #fsize, and #include.
- The only server-side include that you can use directly in your Active Server Pages is the #include directive, which lets you import information from other files.
- The #include directive allows you to have content sharing between multiple pages, share script libraries, and use the other server-side directives in your Active Server Pages.

Working with the Global.asa

The Global.asa is an optional file but it is often employed by Active Server Pages programmers. Even though it is optional, without using the Global.asa you cannot have an OnStart or OnEnd event subroutine for the Session object. For that matter, you cannot have an OnStart or OnEnd event subroutine for the Application object either. You also use the Global.asa to store any static objects for use with session or application level scope. It is small wonder that the Global.asa file is made use of by so many programmers, and why even though it is not an intrinsic object in itself, is often listed along with the Active Server Pages object model in many discussions. The Global.asa is an important fundamental file in Active Server Pages, despite it being optional.

You can only ever have one Global.asa file per application. Also, the Global.asa file must reside in the root directory of the application.

It is important to remember that the Global.asa is not an ordinary script that you can put anything into. It is a special file that can only contain the following items:

- Session_OnStart event
- Session_OnEnd event
- Application_OnStart event
- Application_OnEnd event
- Declarations using the <OBJECT> tag

Working with the events is something that you already know about. Simply for completeness they are shown below in a skeletal form. It is not imperative that you have these events in your Global.asa. They are only an option that you can make use of when you have a need. Below are the skeletal subroutines for the four events:

```
Sub Session_OnStart
  'Statements
End Sub

Sub Session_OnEnd
  'Statements
End Sub

Sub Application_OnStart
  'Statements
End Sub

Sub Application_OnEnd
  'Statements
End Sub
```

Using the StaticObjects collections in both the Session and Application objects requires that the collections have data placed in them using the <OBJECT> tag and the Global.asa file. To use the <OBJECT> tag to add elements to the StaticObjects collections, you must assign values to several of its attributes. First, you require both a <OBJECT> tag to indicate the start of an object and a </OBJECT> to indicate the end, as follows:

```
<OBJECT></OBJECT>
```

To specify that the code is to be run at the server, you assign the value Server to the RUNAT attribute, as shown:

```
<OBJECT RUNAT=Server></OBJECT>
```

To specify that it is part of the Session object, you assign the value Session to the SCOPE attribute. You could also assign the value Application to the SCOPE attribute if you want to add something to the Application object's StaticObjects collection. So the following two lines serve as examples:

```
<OBJECT RUNAT=Server SCOPE=Session></OBJECT>
<OBJECT RUNAT=Server SCOPE=Application></OBJECT>
```

Finally, you add both the ID attribute, which basically gives the object a unique label, and the PROGID attribute, which is a string identifier of an externally creatable object. So a complete example of using the <OBJECT> tag to store a value would be as follows:

```
<OBJECT RUNAT=Server SCOPE=Session ID=MyInfo
PROGID="MSWC.MyInfo"></OBJECT>
```

If you were to place the above script in your Global.asa file (although you may find it there by default) and then you were to access the Session object's StaticObjects collection, you would be able to access properties assigned to this object. To better understand using the <OBJECT> tag in the Global.asa, change your Global.asa to the following:

```
<OBJECT RUNAT=Server SCOPE=Session ID=SessInfo1
PROGID="MSWC.MyInfo"></OBJECT>
```

Now create an Active Server Page that will initialize this static object, show the number of static objects, and finally display the static object and its property value. Open Notepad and add the following script to the document:

```
<html>
<title>Using the <OBJECT> tag in Global.asa</title>
<body>
<font size=5>
<%
'First initialize the StaticObjects
SessInfo1.UName = "Cory"
%>
The number of items in the Session object's StaticObjects
collection is <%= Session.StaticObjects.Count%>
<%
For Each item in Session.StaticObjects
  Response.Write "<br>" & item & " = "
  Response.Write Session.StaticObjects(item).UName
Next
%>
</font>
</body>
</html>
```

Save the script as *object.asp* and then load it into your browser. You should see the result shown in Figure 15.1.

Figure 15.1.
The result of loading
object.asp

Understanding Preprocessing Directives

Preprocessing directives inform your Active Server Pages server, such as IIS, about processing a script. Typically you place preprocessing directives at the top of the page containing the script that you want to affect. Active Server Pages supports the following five @ directives:

- @CODEPAGE
- @ENABLESESSIONSTATE
- @LANGUAGE
- @LCID
- @TRANSACTION

In the following sections you will learn about each of the five @ directives that Active Server Pages supports.

Using the @LANGUAGE Directive

You have already seen a little of the @LANGUAGE directive. Basically, the @LANGUAGE directive allows you to specify what scripting language to use within an Active Server Page. The default language is VBScript, although you could specify this by using the @LANGUAGE directive as follows:

```
<%@ LANGUAGE = VBScript%>
```

If you want to do some work with Microsoft's implementation of JavaScript, you might want to change some particular pages to use JScript instead of

VBScript. You can easily accomplish this by once again utilizing the @LAN-GUAGE directive, as follows:

```
<%@ LANGUAGE = JScript%>
```

You can specify other scripting languages too, although you must install support for those languages first so that Internet Information Server recognizes them. The documentation that comes with other languages should specify what is required in the @LANGUAGE directive so that you can use it in your Active Server Pages.

Using the @ENABLESESSIONSTATE Directive

The @ENABLESESSIONSTATE directive lets you turn session tracking on or off for a page. Session tracking is where information about all the requests from a single client are kept. If keeping track of this type of information does not concerned you, then turning off session tracking can be a useful practice, as it can speed up the processing of your Active Server Pages. By default, session tracking is on. To use the @ENABLESESSIONSTATE directive to turn off session tracking, you would use the following line at the top of your script:

```
<%@ ENABLESESSIONSTATE = False %>
```

The @ENABLESESSIONSTATE directive gives you flexibility on a script-by-script basis of whether you want to use session tracking. One very important point that you should be aware of before you turn off session tracking every-where is that you cannot use variables or objects that have session scope when session tracking is off. So if your script is using the Session object heavily, you may not want to use this directive.

Using the @TRANSACTION Directive

You have already seen the @TRANSACTION directive in the previous lesson when dealing with the ObjectContext object. You will recall that you must use the @TRANSACTION directive to indicate that the script that follows is actually a transaction. The server will then take the necessary steps to ensure that the individual steps that follow in the script are treated as a transaction (or a single unit), in case the transaction is aborted for any reason. The following is an example of using the @TRANSACTION directive:

```
<%@ TRANSACTION = Required %>
```

The @TRANSACTION directive can take any of the following values, shown in Table 15.1.

Value	Definition
Required	Requires a transaction
Requires_New	Requires a new transaction
Supported	No transaction is started
Not_Supported	No transaction will be started

Table 15.1. Possible values for the @TRANSACTION directive

To avoid an error being generated by your script, you must remember to place the @TRANSACTION directive on the very first line of your page. When creating multiple scripts that are all transactions, you must have the @TRANSACTION directive at the top of each script.

Normally, you would try to contain a transaction within a single page; however, if there are circumstances where you must split the transaction across multiple scripts, using the @TRANSACTION directive along with Server object's Execute or Transfer methods will enable this implementation. When trying to maintain a single transaction across multiple scripts, you must assign Required to the @TRANSACTION directive. If you were to use Requires_New, multiple transactions would be put into place rather than a single transaction.

To see a brief example of combining scripts into a single transaction, you will create a very simple transaction script from which you will call another transaction script. Open Notepad and type in the following script:

```
<%@ TRANSACTION = Required %>
<%
Response.Write "This is the first script.<br>"
Server.Execute("mtrans2.asp")
%>
<%
Sub OnTransactionAbort()
Response.Write "Unable to complete your transaction."
End Sub
```

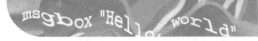

```
Sub OnTransactionCommit()
Response.Write "The transaction completed successfully."
End Sub
%>
```

Save this script as *mtrans1.asp*. Do not load it into your browser yet. First you will create the second transaction script that you are calling with the Execute method in *mtrans1.asp*. This second transaction script you will call *mtrans2.asp* and it will simply output some text and signal the transaction as complete. Control will then return to the first transaction script, where you expect the OnTransactionCommit subroutine to be run.

Open up a new document in Notepad and type in the following script:

```
<%@ TRANSACTION = Required %>
<%
Response.Write "This is the second script.<br>"
ObjectContext.SetComplete
%>
```

Save this script as *mtrans2.asp*. Now that you have made both transaction scripts, open up *mtrans1.asp* in your browser. The result is shown in Figure 15.2, where you can see clearly that the transaction completed successfully even though it was spread over multiple scripts.

**Figure 15.2.
The result of
running a
transaction across
multiple scripts**

Using the @CODEPAGE Directive

In Lesson 11, "Using the Session Object," you learned about the Session object's CodePage property and how to use it. The @CODEPAGE directive also sets the codepage for the current script. You can use the @CODEPAGE directive to set the script to use the Hebrew codepage by using the following line at the top of your page:

```
<%@ CODEPAGE = 1255 %>
```

You use the same codepage values with the @CODEPAGE directive that you saw in Lesson 11, as outlined in Table 11.3. Be aware that if you use the Session object's CodePage property that it will overwrite anything that you set with the @CODEPAGE directive.

Using the @LCID Directive

In Lesson 11, "Using the Session Object," you also learned about the Session object's LCID property and how to use it. The @LCID directive also sets the locale for the current script. You can use the @LCID directive to set the script to use the French locale by using the following line at the top of your page:

```
<%@ LCID = 1036 %>
```

You use the same locale values with the @LCID directive that you saw in Lesson 11, as outlined in Table 11.2. Besides these numeric values, there are also two built-in values that you can use with the @LCID directive: LOCALE_SYSTEM_DEFAULT and LOCALE_USER_DEFAULT. The value LOCALE_SYSTEM_DEFAULT is for the locale that is the system default and the value LOCALE_USER_DEFAULT is for the current user's locale. The following code shows how to use both of these values:

```
<%@ LCID = LOCALE_SYSTEM_DEFAULT %>
<%@ LCID = LOCALE_USER_DEFAULT %>
```

The @LCID directive directly affects how the Active Server Pages server interprets the script, whereas the Session object's LCID property affects how the HTML is processed.

Working with Server-Side Includes

You can also use server-side include directives, which are different from the @ directives that you have seen so far. Server-side includes allow you to add graphics or text to a Web page just prior to it being sent to the user. You can also use server-side includes to add application information as well. The following lists the available server-side include directives:

- #config
- #echo
- #exec
- #flastmode
- #fsize
- #include

All of these server-side include directives are used by placing them inside HTML comment tags. Depending on the particular server-side include you are using, you then assign a value to one of the valid attributes. For example, the #include directive has a file attribute and would look similar to the following when you use it:

```
<!-- #include file="myfile.inc" -->
```

Of the six server-side includes, only the #include directive can be used within an Active Server Page. For this reason, the main discussion on server-side includes deals with this directive. However, by utilizing the #include directive, you can actually use the other five server-side includes as well. Rather than using an *.asp* extension for pages that use any of the server-side includes other than the #include directive, try using *.shtm* or *.shtml* instead.

Using the #include Directive

The server-side #include directive is the only server-side include that you can use directly in your Active Server Pages. It has one function—to include an external file into your Web page. There are three main reasons for using the #include directive, as shown:

- Content sharing among multiple pages
- Sharing script libraries for reusability
- Including the other SSI directives into Active Server Pages

One very common reason to use the #include directive is for sharing content among multiple pages. For instance, if you had twenty pages in a site that all show the same information, you would have to visit twenty pages to update the same information. As an alternative, you can save yourself a lot of work by using the #include directive and simply editing the one text file that all twenty pages refer to.

To include a file using the server-side includes, you must use the #include directive within HTML comment tags. You then specify the attribute file to which you assign a text string that indicates the file you want to include, as shown:

```
<!-- #include file="license.txt" -->
```

The following example will use the #include directive within an Active Server Page. The example will load in the GNU General Public License, which is a long text file that you would not want to manually put in your Active Server Pages. This demonstrates not only how to use the #include directive but also how you can separate lengthy content from your scripts themselves. The short example script, called *include.asp*, is shown below:

```
<html>
<head>
<title>Using Include Directives</title>
</head>

<body>
<h2>Using Include Directives</h2>
This is where it all happens.<br>
The current time is <%=time%><br>
<!-- #include file="license.inc" -->
</body>
</html>
```

When you load *include.asp* into your browser and providing you have *license.txt*, the result should appear as shown in Figure 15.3. You can see that HTML, Active Server Pages script, and the server-side include all managed to co-exist together.

Figure 15.3.
Using a #include directive to insert a file into a HTML page

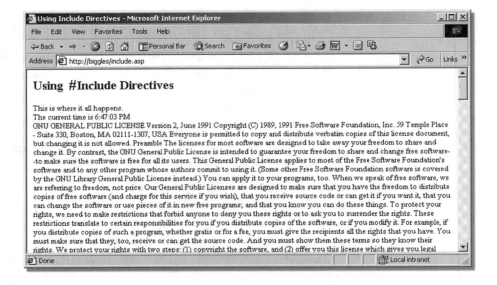

Using an Extension for Includes

Although a special file name extension is not necessary when including files, Microsoft recommends using a .inc extension to indicate clearly that it is an include file. This may not be practical if you already have thousands of documents with some other extension such as .txt and in these cases it is acceptable. However, for good coding etiquette try to use the .inc extension where you can.

Using virtual paths in your #include directive is also useful. Rather than using the *file* attribute, you use the *virtual* attribute and simply assign the path to the file. An example of using the *virtual* attribute would be as follows:

```
<!-- #include virtual="/legal/license.inc" -->
```

You can create entire script libraries full of nothing but procedures. To make use of these procedures in your Active Server Pages, you simply include the library file that you require and call the procedure wherever you need in your script. This allows an incredibly easy way to reuse code, as well as keeping the majority of your Active Server Pages small.

Using the #echo Directive

The #echo directive cannot be used in Active Server Pages directly, although you can still make use of it. To insert an #echo directive into an Active Server Page you should use the #include directive to call a separate file with the #echo directive in it. There is quite a lot of information that you can get from the #echo directive, although the Request object's ServerVariables collection also has much of this information. An example of using the #echo directive to return the remote IP address is as follows:

```
<!-- #echo var="REMOTE_ADDR" -->
```

This line of code would return the remote IP address in the form xxx.xxx.xxx.xxx. This is just one of many possible values to assign to the *var* attribute. A more complete list can be found, at the time of writing, at the following URL on the Microsoft site:

http://www.microsoft.com/WINDOWS2000/en/advanced/iis/htm/core/iisieco.htm

Using the #config Directive

The #config directive is another server-side include that you cannot use directly within your Active Server Pages. However, as with other directives of this nature you can simply use the #include directive to insert them into your Active Server Pages. The #config directive has three attributes that you can use: the *timefmt* attribute, the *sizefmt* attribute, and the *errmsg* attribute. The *errmsg* attribute is simply for configuring a text error message to display when an error occurs in any server-side include directives.

The #config directive lets you configure the date and time formats that some of the other directives use with its *timefmt* attribute. For example the following script uses the #config directive to alter the output of the #echo directive:

```
<html>
<!-- #echo var="DATE_LOCAL" --><br>
<!-- #config timefmt="%a %b %d, %Y" -->
and after using #config...<br>
<!-- #echo var="DATE_LOCAL" -->
</html>
```

If you type the above script into Notepad and save it as *config.shtm*, then loading it into your browser should give you a result similar to Figure 15.4. You can see clearly how the #config directive has abbreviated the day and month, as well as inserting a comma before the year.

Figure 15.4.
The result of loading the *config.shtm* page into the browser

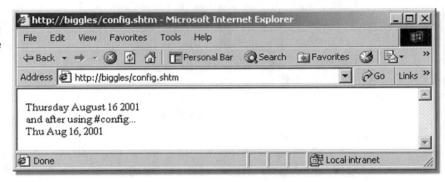

There are many parameters that you can use with the *timefmt* attribute, the full range being listed on the Microsoft site, as follows:

http://www.microsoft.com/WINDOWS2000/en/advanced/iis/htm/core/iisicfg.htm

The *sizefmt* attribute lets you specify if you want to see file sizes in bytes or kilobytes. To see it in bytes you would use the following line:

```
<!-- #config sizefmt="bytes" -->
```

To see file sizes in kilobytes you would use the following line:

```
<!-- #config sizefmt="abbrev" -->
```

Using the #exec Directive

You can use the #exec directive to execute a shell command or an application such as an Active Server Pages application, a CGI script, or an ISAPI application. Although this sounds great, most sites have now prevented the #exec directive from being run on their servers because of the huge security risk that comes with it. If you were to use the #exec directive anyway, the two attributes that it takes are *CGI* and *CMD*. *CGI* is for *CGI* scripts and other applications, while *CMD* is for shell commands. The following shows an example of using the *CGI* attribute:

```
<!- #exec cgi="/scripts/calendar.cgi" -->
```

Using the #flastmode Directive

The #flastmode directive lets you display in the browser when a specified file was last modified. You could use this directive to indicate the last time the current page was modified, as you so often see on Web pages now. You can specify the file using either the *file* attribute or the *virtual* attribute. As with the #include directive, the *virtual* attribute is for virtual paths on your Web server. You must also remember to enclose the file in quotation marks. The following is an example of using the #flastmode directive:

```
<!-- #flastmode file="index.htm" -->
```

The #flastmode directive is directly affected by the #config directive's *timefmt* attribute. So if you want to change the format of the #flastmode directive's output, use the #config directive immediately before it.

Using the #fsize Directive

The #fsize directive will display the size of a file in the browser. As with the #include directive, it takes two possible attributes: *file* or *virtual*. Both of these attributes are for indicating the path to the required file. So an example of using the #fsize directive would be as follows:

```
<!-- #fsize file="icecream.asp" -->
```

Using the #config directive's *sizefmt* attribute lets you specify whether you want the output in bytes or kilobytes.

WHAT YOU MUST KNOW

The Global.asa is a special file that contains the subroutines for the Application and Session objects' OnStart and OnEnd events, as well as any declarations with the <OBJECT> tag to define objects for use with the StaticObjects collections. There are five preprocessing directives, such as the @LANGUAGE directive for specifying the scripting language and the @TRANSACTION directive for indicating that the script is actually a transaction. Server-side includes offer a variety of means to add information to your Active Server Pages, although only the #include directive can be directly made use of. In Lesson 16, you will learn about sending e-mail from your Active Server Pages using Collaborative Data Objects (CDO), as well as learning about the CDO object model itself. Before you move on to Lesson 16 though, be sure that you have grasped fully the following key concepts:

- The OnStart and OnEnd events for both the Application object and the Session object are always in the Global.asa file, if they exist at all.

- Objects that you declare with the <OBJECT> tag are also kept in the Global.asa file. The StaticObjects collections in both the Application object and the Session object make use of these objects that you declare in the Global.asa file.

- Preprocessing directives inform your Active Server Pages server, such as IIS, about processing a script. There are five preprocessing directives known as @ directives.

- To set the codepage for the current script, you use the @CODEPAGE directive.

- To turn session tracking on or off for a page, you use the @ENABLESESSION-STATE directive.

- To specify what scripting language to use within an Active Server Page, you use the @LANGUAGE directive.

- To set the locale for the current script, you use the @LCID directive.

- To indicate that the script that follows is actually a transaction, you use the @TRANSACTION directive.

- Just prior to a Web page being sent to the user, you can make use of server-side includes to add a variety of additional information. The six server-side includes are: #config, #echo, #exec, #flastmode, #fsize, and #include.

- The #include directive, which allows you to import information from other files, is the only server-side include that you can use directly in your Active Server Pages.

- Sharing script libraries, enabling content sharing between multiple pages, and allowing you to use the other server-side directives in your Active Server Pages are the main reasons to use the #include directive.

SENDING E-MAIL WITH COLLABORATION DATA OBJECTS

In Lesson 15, "Using the Global.asa, Includes and Directives," you completed all the fundamental sections of Active Server Pages, including the Active Server Pages object model, giving you a solid foundation upon which you can add further skills. In this lesson, you will learn one of these useful skills, namely sending e-mail from your Active Server Pages using a technology known as Collaboration Data Objects (CDO). By the time you finish this lesson you will have covered the following key concepts:

- Collaboration Data Objects (CDO) is actually a variety of libraries from Microsoft dealing with messaging functionality. CDONTS (CDO for NT Server) is just one of these libraries.

- Although CDONTS supports a wide range of functionality, one separate item that it contains is the NewMail object, which is primarily for use with automated messages, such as from Active Server Pages.

- If you are running Windows NT, you must install CDONTS; however, Windows 2000 already has CDONTS installed. You must have Microsoft Internet Explorer version 4.0 or higher and the SMTP Service both installed on your server in order to utilize CDONTS.

- The NewMail object contains thirteen properties and four methods, although not all the properties and methods are required to send an e-mail message.

- You must create an instance of the NewMail object using the Set statement and the CreateObject method of the intrinsic Server object.

- The Send method takes numerous arguments, all of which are optional if you use their equivalent NewMail object properties. You can only use the Send method once per instance of the NewMail object.
- After using the Send method, you must set the instance of the NewMail object to Nothing in order to free the memory that it held.
- Use the Cc and Bcc properties to copy the e-mail message to additional recipients other than the recipient(s) that the message is actually addressed to. Use a semicolon to separate multiple recipients.
- To attach a file to the e-mail message, you must use the NewMail object's AttachFile method, indicating at least the path and filename of the file to attach.

Understanding the Collaboration Data Objects Model

The Collaboration Data Objects (CDO) model is basically a set of libraries that contain routines that enable you to easily work with messaging functions. In its very first implementation, CDO was actually called ActiveMessaging, which was really just for Exchange 5.0. There are many different versions of CDO available, which can be confusing at first, as shown in Table 16.1.

Version	File	Name
1.0	olemsg32.dll	ActiveMessaging
1.0	cdoexm.dll	CDO for Exchange Management Library
1.0	cdowf.dll	CDO Workflow objects for Exchange
1.2.1	cdo.dll	CDO (for Exchange and Outlook)
1.2.1	cdonts.dll	CDONTS (CDO for NT Server)
2.0	cdosys.dll	CDOSYS (CDO for Windows 2000)
3.0	cdoex.dll	CDOEX (CDO for Exchange)

Table 16.1. The various versions of CDO

CDOSYS is the latest version of CDO from Microsoft; however, it is not compatible with CDONTS. CDOSYS has been rewritten from the ground up to support Internet developers and has all the features of its predecessor, CDONTS, but with some additional enhancements. The main difference between CDOSYS and CDONTS as far as functionality goes is that CDOSYS supports working with newsgroups and posting messages using the protocol NNTP, as well as being able to handle protocol events.

Although, CDOSYS is newer than CDONTS, CDOSYS does not run on as many operating systems and is still establishing itself. It is more likely that you will need to be familiar with and use CDONTS in the workforce for a while yet. For this reason, you will look at CDONTS in detail throughout this lesson, and in particular only one aspect of it—the NewMail object. CDONTS is just a part of the whole Collaboration Data Objects picture, and it is a complex but powerful tool for accessing and working with messages. You will only be looking at a small part of CDONTS, which should give you an idea of the size and scope that the whole topic of Collaboration Data Objects covers. Figure 16.1 shows the CDONTS object model, where you can see how it relates to working with the folders and messages within Exchange mailboxes, as well as how disjointed and separate the NewMail object is from the rest of the objects in CDONTS.

**Figure 16.1.
The CDONTS
object model**

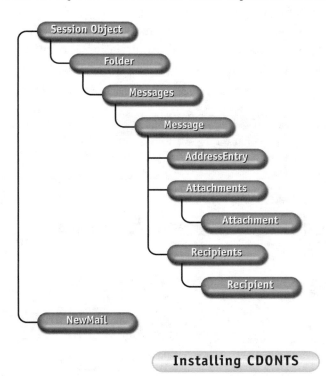

Installing CDONTS

If you are running Windows NT 4.0, you will need to install CDONTS. You can install CDONTS using one of the following methods:

- Install Microsoft® Commercial Internet System (MCIS)
- Install Microsoft® Internet Information Server (IIS) version 4.0 or 5.0
- Run the Internet Mail Service (IMS) Wizard (part of Microsoft® Exchange Server version 5.5, although you do not have to actually install Exchange itself)

CDONTS and CDOSYS both come with Windows 2000. So if you are running Windows 2000, you do not need to install CDONTS.

Regardless of operating system, you will need to have Microsoft Internet Explorer version 4.0 or higher installed on your server in order to use CDONTS.

Finally, you must have the SMTP Service running in order for CDONTS to make use of it. CDONTS relies on SMTP for sending e-mail messages. To install SMTP Service on Windows 2000, click on the Start menu, Settings item and then Control Panel item. Within the Control Panel window, double-click on the Add/Remove Programs icon. Windows will open the Add/Remove Programs dialog box, from which you must then select the Add/Remove Windows Components option from the left-hand side. Windows will then open the Windows Components Wizard dialog box in which you must select the Internet Information Services (IIS) list item and then click on the Details button. Windows will open the Internet Information Services (IIS) dialog box in which you must be sure that the SMTP Service list item has a checkmark in its check box, before clicking on the OK button, as shown in Figure 16.2.

Figure 16.2. Installing the SMTP Service from the Control Panel

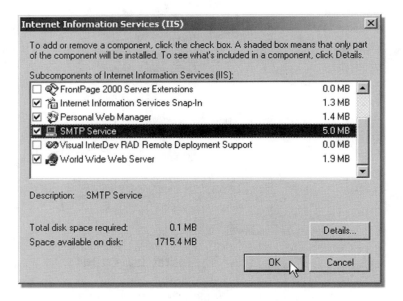

Click on the Next button in the Windows Components Wizard dialog box to install the SMTP Service. You may be required to insert the original Windows installation CD at this point. When installation is complete, click on the Close button of the Add/Remove Programs dialog box to close it. Restart your computer if the installation prompts you to.

Sending E-mail Using the NewMail Object

The NewMail object was basically designed by Microsoft to be a rapid deployment object for sending e-mails, such as notification mail. For this reason there is no interface support and no expectation of human interaction. You use the NewMail object for automated messages. The NewMail object, although simple to use, has many properties that you can make use of for creating complex e-mail messages and handling a variety of circumstances. Table 16.2 lists the various properties of the NewMail object, as well as indicating the property type and access that you have to the property.

Property	Type	Access
Bcc	String	Write-only
Body	String	Write-only
BodyFormat	Long	Write-only
Cc	String	Write-only
ContentBase	String	Write-only
ContentLocation	String	Write-only
From	String	Write-only
Importance	Long	Write-only
MailFormat	Long	Write-only
Subject	String	Write-only
To	String	Write-only
Value	String	Write-only
Version	String	Read-only

Table 16.2. Various properties of the NewMail object

The NewMail object also contains four methods that you can use in your scripts. The four methods are as follows:

- AttachFile
- AttachURL
- Send
- SetLocaleIDs

Creating an Instance of the CDONTS NewMail Object

Before you can use any of the properties or methods of the NewMail object, you must create an instance of the object in your Active Server Page. You will recall learning about the Server object's CreateObject method and how useful it was for importing other objects into your Active Server Pages. To create an instance of the NewMail object, you must use the Set statement together with the CreateObject method, as shown:

```
<%
Dim myMail
Set myMail = CreateObject("CDONTS.NewMail")
%>
```

Using the NewMail Object's Send Method

You use the Send method to send an e-mail using the NewMail object. You must have a valid instance of the NewMail object to use the Send method. The syntax of the Send method is as follows:

myMail.Send([From] [, To] [, Subject] [, Body] [, Importance])

All of the arguments for the Send method are optional. If you use the Send method together with the some of the NewMail object's properties, you do not need to supply any of the arguments to the Send method at all. You will see the Send method made use of in this fashion after you learn about each of the NewMail object's properties in detail. For now, you will learn to use the Send method with arguments.

The From argument takes a string that indicates who the message is from, for example:

```
"mickey@mousehouse.com"
```

The To argument takes a string that indicates who the message is to, for example:

```
"donald@ducksrus.net"
```

You can also have multiple recipients in the To argument by separating addresses with a semicolon, as follows:

```
"donald@ducksrus.net;goofy@dogslife.com"
```

The Subject argument also takes a string, indicating what the subject or title for the e-mail message is, for example:

```
"Greetings"
```

The Body argument takes a string that contains the actual body of the e-mail message. Although you might normally assign this string to a variable (due to its length) and then pass the variable as the Body argument, you can also directly pass a string to this argument, for example:

```
"Hello, Donald. This is the body of my test message."
```

The Importance argument will be discussed later when you look closer at the NewMail object's Importance property. For now, the default setting of Normal importance is fine.

You must set the NewMail object to Nothing after a successful call to the Send method. When you create an instance of the NewMail object, you can only use the Send method once and any subsequent calls to the Send method will result in an invalid object error.

So now that you have seen all the elements of the Send method, it is time to put everything together. Assuming that you have already installed CDONTS if you need to and that SMTP Services are running, the following script should be all that you need to send an e-mail message. Of course, you may want to try substituting real e-mail addresses that you have access to rather than the imaginary addresses before trying out the script.

```
<%
'Create an instance of the NewMail object
Dim myMail
Set myMail = CreateObject("CDONTS.NewMail")

'Send a message using only the Send method
myMail.Send _
"mickey@mousehouse.com", _
"donald@ducksrus.net", _
"Greetings", _
"Hello, Donald. This is the body of my test message."

'Clear memory for object
Set myMail = Nothing
%>
```

Using the NewMail Object's To and From Properties

The NewMail object allows you to use the To and From properties, to specify who an e-mail message is to and from, respectively. Both properties have write-only access. To actually use the properties, therefore, you can only assign values to them, as shown in the following example snippet:

```
myMail.From = "mickey@mousehouse.com"
myMail.To = "donald@ducksrus.net"
```

You can also have multiple addresses in your e-mail's To field by separating addresses with a semicolon. However, you cannot have multiple recipients in the From property—only one address is permissible. An example of having multiple recipients is as follows:

```
myMail.To = "donald@ducksrus.net;goofy@dogslife.com"
```

When you use the To and From properties, you no longer need to specify these in the optional arguments of the Send method. For example, the following script would function exactly the same:

```
<%
'Create an instance of the NewMail object
Dim myMail
Set myMail = CreateObject("CDONTS.NewMail")

'Assign message properties
myMail.From = "mickey@mousehouse.com"
myMail.To = "donald@ducksrus.net"

'Send a message using the Send method
myMail.Send ,,"Greetings", _
"Hello, Donald. This is the body of my test message."

'Clear memory for object
Set myMail = Nothing
%>
```

There is no precedence between the To property and the Send method's To argument. When you use the To property of the NewMail object to supply mail recipients, as well as supplying recipients as arguments in the Send method, the e-mail is sent to both lists of recipients. So in the following example script, the e-mail message would be sent to both donald@ducksrus.net and goofy@dogslife.com. Notice specifically that one address uses the To property and the other uses the Send method's To argument:

```
<%
'Create an instance of the NewMail object
Dim myMail
Set myMail = CreateObject("CDONTS.NewMail")

'Assign message properties
myMail.From = "mickey@mousehouse.com"
myMail.To = "donald@ducksrus.net"

'Send a message using the Send method
myMail.Send ,"goofy@dogslife.com","Greetings", _
"Hi, friends. This is the body of my test message."

'Clear memory for object
Set myMail = Nothing
%>
```

Using the NewMail Object's Subject and Body Properties

The Subject and Body properties behave similarly to the To and From properties. Basically, as an alternative to passing the subject and body of a message as arguments in the Send method, you can assign them to the Subject and Body properties, respectively. When you use the Subject and Body properties ,you do not need to specify anything for the optional arguments of the Send method. Also you can use the Subject or the Body property separately from each other. You do not need to have both of them as properties or both of them as arguments, although you can. So the following example snippet shows how to assign strings to both properties:

```
myMail.Subject = "Greetings"
myMail.Body = "This is the body of my test message."
```

Sending Carbon Copies (Cc) and Blind Carbon Copies (Bcc)

Often, you need to send an e-mail message to someone and copy it to other people that also may need to know its contents. Whether you use the Cc property to show visibly who the other recipients are or you use the Bcc property to hide the other recipients is up to you. You can assign recipients to both of these properties in the same way that you do with the To property. You can also have multiple recipients for both fields, but you must separate each address with a semicolon. Some examples of assigning recipients to the Cc and Bcc fields are as follows:

```
myMail.Cc = "minnie@mousehouse.com"
myMail.Cc = "minnie@mousehouse.com;pluto@mousehouse.com"
myMail.Bcc = "daffy@ducksrus.net"
```

Avoiding Unnecessary Use of the Blind Carbon Copy

Using the blind carbon copy may not be advisable for two main reasons. Firstly, many e-mail programs are automatically placing e-mails where your address is not clearly on the To or Cc lines into the Junk or Trash folder. This is a common method of antispamming. Secondly, when you, alone or with the help of system administrators, try to troubleshoot problems tracing mail routes from your scripts, using the blind carbon copy can significantly decrease the amount of useful information that you have to work with.

Setting Importance Levels

You can set the importance of e-mail messages using the Importance property of the NewMail object or using the Importance argument of the Send method. The various possible values of importance are shown in Table 16.3.

Constant	Value	Description
cdoLow	0	Low importance
cdoNormal	1	Normal importance (default)
cdoHigh	2	High importance

Table 16.3. The various importance levels for e-mails

You may simply want to use the value rather than the constant, as shown in the following example of setting the importance to high:

```
myMail.Importance = 2
```

If you want to use the constants, then you should place the following code in your Global.asa file first:

```
<!-- METADATA TYPE="TypeLib"
file="cdonts.dll"
version="1.2"
-->
```

This code will allow you to use the constants that are in CDONTS within your Active Server Pages. For instance, you could now assign a value of high to the Importance property using the constant cdoHigh, as shown:

```
myMail.Importance = cdoHigh
```

Note: Setting the value of the Importance argument in the Send method takes precedence over any setting of the Importance property. So if you set the Importance argument in the Send method, the Importance property is ignored by the server.

Sending HTML Messages

If you are sending HTML messages rather than plain text messages, you must be sure to set the BodyFormat property first. The default is to use plain text so you would only need to set this property when you want to send HTML. Table 16.4 shows the various settings for the BodyFormat property.

Constant	Value	Description
cdoBodyFormatHTML	0	Send body of message as HTML
cdoBodyFormatText	1	Send body of message as plain text

Table 16.4. The settings for the BodyFormat property

You must set this property before you use the Send method. It does not matter, however, if you set this property before or after the Body property, which contains the HTML or plain text message. An example of setting the BodyFormat property is as shown:

```
myMail.BodyFormat = 0
```

or

```
myMail.BodyFormat = cdoBodyFormatHTML
```

Using the ContentLocation and ContentBase Properties

Rather than specifying complete paths for every URL that you use in the message body, you can use the ContentLocation and ContentBase properties to provide the main path and shorten the amount you have to type in your message body. When multiple references are used throughout a long message body, using the ContentLocation and ContentBase properties means that there is less likelihood of errors occurring if the path is changed due to having your path reference in only one place.

For example, if you were to reference an image that had a lengthy URL in your message body, you might currently type something similar to the following:

```
<img src="http://www.financialpics.com/graphs/stocks1.png">
```

Although it may not appear to do much at first, you can add the URL path to the ContentLocation and ContentBase properties, as follows:

```
myMail.ContentBase = "http://www.financialpics.com/
myMail.ContentLocation = "graphs/"
```

By using the ContentLocation and ContentBase properties, you can drastically shorten what you have to type in your message body. So as you can see by the following example, all you have to type in your message body now is the final file name, which can be very useful when working with multiple files:

```
<img src="stocks1.png">
```

Adding Additional Headers

You can add additional headers to your e-mail messages using the NewMail object's Value property. You must place a string containing the header name within parentheses, directly after the Value property. You then assign the value of this header to the Value property. So, for example, if you were to add a header that specified the Reply-To field should go to Porky Pig, you might use something similar to the following:

```
myMail.Value("Reply-To") = "The Big Bacon<porky@pig.com>"
```

You can use any string value for a header although not all browsers will accept all headers. Some additional headers that you might use are: Reply-To, File, Reference, Date (although this is automatically added), and Keywords.

Reading the CDONTS Version

You can tell for your own purposes what version of CDONTS your scripts are running under by reading in the Version property. The Version property is the only read-only property out of the entire set of NewMail properties, of which there are thirteen. Running the following brief script should return a value of 1.2, showing that you are using version 1.2 of CDONTS:

```
<%
Dim myMail
Set myMail = CreateObject("CDONTS.NewMail")
Response.Write myMail.Version
%>
```

Setting the Message Locale

You have already seen how to set the locale for users through your Active Server Pages. You can also set the locale for users through e-mail messages by using the NewMail object's SetLocaleIDs method. This does not change the user's settings, it merely defines what locale the message should be viewed with so that if the locale is supported by the operating system it can be used. To set the NewMail object to use a specific locale, you must assign a *codepage* value as an argument to the SetLocaleIDs method, as shown in the following example that sets a Chinese locale:

```
myMail.SetLocaleID(950)
```

Setting the MailFormat Property

You can specify the encoding method that the NewMail object should use by setting the MailFormat property. The encoding method affects items such as mail attachments. There are only two settings for the MailFormat property. The default setting is for plain text and is fine for most purposes. The other setting is for MIME (Multipurpose Internet Mail Extensions). Table 16.5 shows the two settings and the constants and values that you can use with them:

Constant	Value	Description
cdoMailFormatMIME	0	Use MIME format
cdoMailFormatText	1	Use plain text (default)

Table 16.5. The settings for the MailFormat property

An example of setting the NewMail object to use MIME encoding is as shown:

```
myMail.MailFormat = 0
```

or

```
myMail.MailFormat = cdoMailFormatMIME
```

Adding File Attachments to Your E-mails

When you are adding file attachments to your e-mail messages with the NewMail object, you must use the AttachFile method. The AttachFile method takes three arguments, although two of them are optional. The syntax of the AttachFile method is as follows:

NewMailObject.AttachFile(Source [, FileName] [, EncodingMethod])

The Source argument is mandatory and is a text string that indicates the path and name of the file that you want to attach to the e-mail message. So a simple attachment example would look like:

```
myMail.AttachFile "c:\notice.doc"
```

When the recipient of the attachment receives the e-mail, the name of the attachment is given by the FileName argument. If you do not pass a string to the FileName argument then the filename in the Source argument is used. So to continue with the example, if you want the attachment to appear in the e-mail as Board Notice, you would do something similar to the following:

```
myMail.AttachFile "C:\notice.doc", "Board Notice"
```

You can choose to encode your attachment in either UUEncode format or Base64 format, as shown in Table 16.6.

Constant	Value	Description
cdoEncodingUUEncode	0	Use UUEncode format (default)
cdoEncodingBase64	1	Use Base64 format

Table 16.6. The settings for the EncodingMethod argument

Carrying the example even further, you could send your attachment in Base64 format by using code as follows:

```
myMail.AttachFile "C:\notice.doc", "Board Notice", 1
```

Note: Because the AttachFile method is closely linked to the MailFormat property, if you set encoding to Base64 format then the value of the MailFormat property is automatically set to 0 (MIME). Conversely, if the MailFormat property is set to 0 (MIME) then the default encoding format for the AttachFile method becomes Base64 format.

If your attachment contains relative links then they may well be broken when the recipient opens the attachment. For instance, if your attachment is an HTML document with a relative link to a sales chart called *chart.jpg*, then that sales chart would not be accessible unless the full path of the file is given. For example, the following would be a relative link:

```
<img src="chart.jpg">
```

You can attach the full path of files for relative links in an attachment using the NewMail object's AttachURL method. The syntax of the AttachURL method is as shown:

NewMailObject.AttachURL(Source, ContentLocation [, ContentBase] [, EncodingMethod])

The Source argument is for the full path and filename of the file that the URL refers to. However, you can simply pass the name of the file for this argument because it is then used with the ContentLocation and ContentBase arguments, which both serve a similar purpose to their property counterparts that you saw earlier. The EncodingMethod argument is exactly the same as that in the AttachFile method and, like the one in the AttachFile method, is also optional and can be used with the constants shown previously in Table 16.6.

As an example of properly handling the attachment of a file that has a relative link in it, look at a complete Active Server Page. The example shows how to construct an e-mail that attaches an HTML sales report document in it, which contains a relative link to the sales chart figure. Using the AttachURL method, the sales chart will be able to access a full URL path and therefore be shown. The example code is as follows:

```
<%
'Create an instance of the NewMail object
Dim myMail
Set myMail = CreateObject("CDONTS.NewMail")

'Assign message properties
myMail.From = "john.doe@travel.com"
myMail.To = "director@travel.com"
myMail.Subject = "Sales Report"
myMail.Body = "Here is the monthly sales report."
myMail.Importance = cdoHigh

'Attach HTML sales report
myMail.AttachFile "C:\current\sales.doc", "Sales Report"
myMail.AttachURL "chart.jpg", "august/gfx/", _
"http:/www.travel.com/sales/"

'Send a message using the Send method
```

```
myMail.Send

'Clear memory for object
Set myMail = Nothing

'Send feedback to the user
Response.Write "Your message has been sent at " & now
%>
```

WHAT YOU MUST KNOW

Actually a variety of libraries from Microsoft, Collaboration Data Objects (CDO) deal with messaging functionality. CDONTS (CDO for NT Server) is just one of these libraries. By instantiating a NewMail object and using the Send method you can send an e-mail message from your Active Server Pages in just a few lines of code. The many properties and methods of the NewMail object provide additional functionality when you need it. In Lesson 17, "Creating Page Counters," you will learn different methods of implementing page counters in your Active Server Pages, a feature that many clients will expect you to implement on their sites. Before you continue with Lesson 17, make sure that you understand the following key concepts:

- The CDONTS library's NewMail object is separate from the other items in the CDONTS object model. It is primarily for use with automated messages that do not require human interaction, and therefore is ideal for use with Active Server Pages.

- Windows 2000 already has CDONTS and CDOSYS installed; however, if you are running Windows NT, then you will need to install CDONTS. Both the SMTP Service and Microsoft Internet Explorer version 4.0 or higher must be installed on your server in order to utilize CDONTS.

- Using the Server object's CreateObject method, you must create an instance of the NewMail object before you can make use of its four methods and thirteen properties.

- To send an e-mail message with the NewMail object you must use the Send method. The Send method has five optional arguments, all of which have equivalent NewMail object properties that you can use instead.

- Only use the Send method once per instance of the NewMail object or you will receive an error.

🗴 By setting the instance of the NewMail object to Nothing after you use the Send method, you free the memory that the object was using.

🗴 You can use a semicolon to separate multiple recipients in the To, Cc, and Bcc properties, as well as the To argument of the Send method.

🗴 To copy the e-mail message to recipients in additional to those that the message is actually addressed to, use the Cc and Bcc properties.

🗴 You can use the NewMail object's AttachFile method to attach a file to an e-mail message. When using the AttachFile method you must indicate the path and filename of the file to attach in the Source argument.

I n Lesson 16, "Sending E-mail with Collaboration Data Objects," you learned a very useful technique that many clients will want you to implement in their Active Server Pages sites. In this lesson, you will learn different methods of implementing page counters in your Active Server Pages—another feature that many clients will expect you to implement in their sites. By the time that you have finished this lesson, you will have learned the following key concepts:

- Page counters can be a useful statistical tool, in addition to advertising page popularity.

- There are many different page counters available for Active Server Pages including a simple one from Microsoft, which is known as the Page Counter component and is available with IIS.

- The Page Counter component is easy to use with only three methods, each of which takes no arguments. The Hits method returns the number of page hits; the PageHit method increments the number of hits by one; and the Reset method resets the counter to zero.

- You can create your own simple counters by using the Application object and placing some script at the top of the page you want to put a counter in.

- Using an include file lets you place all of your page counter script in one place, without having to modify individual pages greatly or rewrite complex subroutines for fancier page counters.

- Working with the Session object's OnStart event, you can keep track of the number of visits to a site or an application, if there is more than one application at the site.
- Using cookies, you can distinguish between visits to a site and visitors to a site, which can be very useful statistical information for businesses.
- It is a simple matter to determine the number of active visitors that you have at any given moment by using the Application object to hold the counter, and the Session object's OnStart and OnEnd events to increment and decrement the counter as necessary.
- One of the simplest methods of maintaining page counter persistence is to save the counters to a file every given number of visitors. You can load the counter values from file when the Active Server Pages server starts.

Understanding Page Counters

Page counters were quite a novelty a few years back. Everyone wanted to have them on their site. When it came time to implement page counters, however, you would soon realize how much work was involved—requiring Java applets or special CGI applications. Most casual Web page developers were turned back at this point.

The original idea of the page counter was simply to show how popular a site is. However, it became very useful as an actual statistics tool for marketing purposes. For instance, with a properly implemented page counter, you can attain a benchmark for a company site over any given month. After holding a large advertising campaign, you can use the page counter to aid in monitoring the success of the campaign by comparing the number of visitors to the site after the campaign with the benchmark. In this manner you can help your company to ascertain which advertising media is most effective. Of course, there are many other factors that are taken into consideration, but you can see how page counters can be used. In this situation, you do not want users who simply click the Refresh button to be included in the results. Overly eager company employees checking every five minutes to see how the big campaign is doing can also offset the results somewhat.

In Active Server Pages, using a page counter is not nearly as difficult as it used to be. There are many different alternatives that you can look at to implement and use. Microsoft's Page Counter component is just one of the many alternatives to choose from.

Installing the Microsoft Page Counter Component

The Page Counter component may not be on your system by default. When you install Microsoft Internet Information Services (IIS), you can also install the Page Counter component. In Windows 2000, the Page Counter component is installed when you add the IIS components and is also part of Windows 2000 Service Pack 2 (SP2). You can check to see if the Page Counter component is already on your server by searching for the file *pagecnt.dll*.

Although in the past you could freely download this component directly from the Microsoft site, the links no longer exist. You must therefore install the component, if necessary, from one of the above mentioned Microsoft products.

Note: *If you cannot obtain the Microsoft Page Counter component for your system, simply read through the section dealing with this particular counter for general knowledge. Then in the later section on creating your own page counter, you can create a page counter without any component dependencies to worry about.*

Creating an Instance of the Page Counter Component

You must first create an instance of the Page Counter component before you can use it in your Active Server Pages. As with all nonintrinsic objects in Active Server Pages, you use the Server object's CreateObject method together with the Set statement to create an instance of the Page Counter component. The following script snippet shows an example of creating an instance of the Page Counter component:

```
<%
Dim pcnt
Set pcnt = Server.CreateObject("MSWC.PageCounter")
%>
```

Using the Microsoft Page Counter Component

The Microsoft Page Counter component only has three methods and is very simple to implement in your Active Server Pages. The three methods are shown in Table 17.1.

Method	Description
Hits	Returns the current number of page hits
PageHit	Increments the number of page hits
Reset	Resets the page counter to zero

Table 17.1. The three methods of the Page Counter component

When a user visits a page that you want to monitor a page count on, you use the PageHit method to increment the number of page hits for that particular page. The PageHit method does not take any arguments and so a simple example of using this method might look similar to the following:

```
<%
Dim pcnt
Set pcnt = Server.CreateObject("MSWC.PageCounter")
pcnt.PageHit
%>
```

The Hits method simply returns the current number of hits that the page has had. You do not supply any arguments to the Hits method, so a simple example would look something like the following:

```
<%
Dim pcnt, totalhits
Set pcnt = Server.CreateObject("MSWC.PageCounter")
totalhits = pcnt.Hits
%>
```

To see the Microsoft Page Counter component in an example, complete with an active report of current page hits, type the following short script into Notepad:

```
<html>
<head>
<title>Using Microsoft's Page Counter</title>
</head>

<body>
<h1>Using Microsoft's Page Counter</h1>
<font size=6>
<%
```

```
Dim pcnt
Set pcnt = Server.CreateObject("MSWC.PageCounter")
pcnt.PageHit
response.write "Page hits: " & pcnt.Hits
%>
</font>
</body>
</html>
```

Save the script as *pagecount1.asp* and load it into your browser. Now simply click on the Refresh button several times and watch the number of page hits increase. Figure 17.1 shows the result of loading in *pagecount1.asp* and refreshing the page six times.

**Figure 17.1.
After loading
pagecount1.asp
and refreshing the
page six times**

The final method of the Microsoft Page Counter component is the Reset method. Like its sister methods that you have just seen, the Reset method also takes no arguments. The Reset method will reset the page counter to zero. You might use the Reset method if you require a daily, weekly, or monthly style of page counter. You might also use the Reset method to flush the page count just before you make a site go live, after working with it for some time. In the following short example, the value in the variable *totalhits* will be zero at the end of the script, regardless of the number of hits that the page had previously:

```
<%
Dim pcnt, totalhits
Set pcnt = Server.CreateObject("MSWC.PageCounter")
pcnt.Reset
totalhits = pcnt.Hits
%>
```

Changing the Default Hit Count Data File

The Microsoft Page Counter component uses a Central Management object internally to handle the page count and periodically writes the output to a text file known as the Hit Count Data file. Microsoft recommends that you do not edit this file directly to prevent problems with the Page Counter component reading the file contents. You can locate the file or change the file designated as the Hit Count Data file by going to the registry and editing the following key:

HKEY_CLASSES_ROOT\MSWC.PageCounter\File_Location

The default path and filename for the Hit Count Data file is:

%windir%\system32\inetsrv\data\HitCnt.cnt

Creating Your Own Page Counter

As you read in the introduction to this lesson, constant refreshing can actually result in misleading figures for page statistics. Therefore, the Microsoft Page Counter component may not be ideal for your situation. You also may want to customize the page counter more to the specific needs of your client. In these cases you may want to consider creating your own customized page counter. This is not as difficult as you may at first think, considering the skills that you now have with Active Server Pages.

All you really need at the most basic level is a variable that you can access from different sessions and the ability to increment this variable. The Application object would seem ideal under these circumstances. All you would need to do is have the Active Server Page increment an application level variable as part of the script. The following example shows how simple this really is:

```
<%
Application.Lock
Application("PageCnt") = Application("PageCnt") + 1
Application.Unlock
%>
```

There are a few limitations to this quick and easy approach. The code above needs to be placed inside every Active Server Page that you want to have a page counter on. Therefore, you must create a unique variable name for each page.

This is a simple matter if you have a small number of pages, but if you have many pages it can get difficult.

You could always use something unique about each individual Active Server Page, such as its virtual path and filename. The Request object's ServerVariables collection can be used to retrieve the name of each page through use of its "SCRIPT_NAME" key. So by creating an include file that uses the ServerVariables collection and inserting it in all of your pages, you would accomplish the task at hand with very little effort on your part. So the following script would make up the include file:

```
<%
'Get a unique string using the virtual path and script name
Dim unique
unique = Request.ServerVariables("SCRIPT_NAME")

'Increment the counter
Application.Lock
Application(unique) = Application(unique) + 1
Application.Unlock
%>
```

You could save this script as *pagecount.inc* and then include it at the top of every Active Server Page that you want to be creating a page count for, as follows:

```
<!-- #include file="pagecount.inc" -->
```

This is all very well; however, you also require a means of extracting the actual values of each page's hit count. Once again, it is best to do this directly within the same include file, *pagecount.inc*, by simply writing a function that returns this value. You might call this function *TextPageHits*. The include file, after inserting the function, would now appear as follows:

```
<%
'Get a unique string using the virtual path and script name
Dim unique
unique = Request.ServerVariables("SCRIPT_NAME")

'Increment the counter
Application.Lock
Application(unique) = Application(unique) + 1
```

```
Application.Unlock

'Return the current page's hit count in plain text
Function TextPageHits
   TextPageHits = Application(unique)
End Function
%>
```

To get the current page counter value for a page now, you would simply call the *TextPageHits* function from within your scripts, as in the following example:

```
<head>
The total number of page visits is <%=TextPageHits%>
</head>
```

In practice, you may want to display each page's hit count with a little more than plain text. Using HTML tables is a nice way to give a better effect without resorting to a lot of little graphic files and complex script routines for displaying them. You might also show the current date and time on the server when the last hit occurred simply by adding the *Now* function to the output. You have already seen how to do all of this, so now it is just finding a practical use for implementing what you already know. You could use the following script as a starting point for creating your own table function for your custom page counter:

```
<% Sub TablePageHits %>
<html>
<head><title>My Custom Page Counter</title></head>

<body>
<table cellpadding="1" border="3">
<tr>
<td width=400 bgcolor="#dddddd">
<font color="#333333">
<b>  Total Visits:   <%=Application(unique)%>
   On Server it is   nbsp;<%=Now%></b>
</td>
</tr>
</table>
</body>
</html>
<% End Sub %>
```

To test your *TablePageHits* function, you will create a script that includes the *pagecount.inc* file and also makes a call to the *TablePageHits* subroutine. Open up Notepad and enter in the following script:

```
<!-- #include file="pagecount.inc" -->
<html>
<head><title>Test Page Counter</title></head>

<body>
<%TablePageHits%>
<p>This is where you would add lots of script and content.</p>
</body>
</html>
```

Save the script as *testpagecnt.asp* and then load it into your browser. Try refreshing the browser a few times to see that the counter changes correctly. Figure 17.2 shows the result of loading in *testpagecnt.asp*, displaying the visual result of your *TablePageHits* subroutine and page counter.

**Figure 17.2.
The result of
loading *test-
pagecnt.asp*
into a browser**

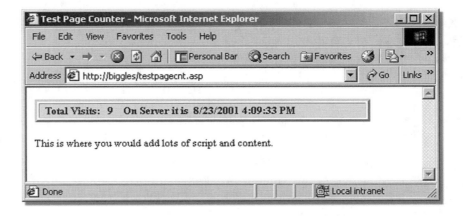

Working with Sessions

When you are working with hundreds of Active Server Pages, you may not want to see the number of visits to individual pages. It may also be preferable not to have to actually work on individual pages for your statistics but rather pull in more manageable totals of visits to your individual applications from one place. You can accomplish this by utilizing the Session object's OnStart event.

Open the Global.asa for your application. If you only have one Active Server Pages application then this is actually reflecting the entire site. Within the subroutine for the Session object's OnStart event you can place a simple call to the Application object to increment a variable (use a meaningful variable name, such as *visits*) which holds the count for the number of visits to the entire application. You do not increment the *visits* variable with every refresh of a page, as the Microsoft Page Counter component does, but instead you keep track of the number of individual visits to each application. For instance, if a user creates a session and visits twenty pages and then leaves, shortly followed by another user who also creates a session and visits seventeen pages and then leaves, the number of visits is not thirty-seven but two.

```
Sub Session_OnStart
  Application.Lock
  Application("visits") = Application("visits") + 1
  Application.Unlock
End Sub
```

Note: *It is, of course, possible to make use of the Session object's OnStart event for retrieving figures for entire Active Server Pages applications or possibly sites, as well as maintaining individual page counters. You are not limited to choosing one or the other.*

Distinguishing Visitors from Visits

You may need to collect statistics for a site where the actual number of different visitors to the site must be distinguished from the amount of visits that a site has had. For instance, if the site *http://www.abc.net* had 6,000 visits last week, you might at first think it was doing quite well. However, if you were able to see that only three visitors had been to the site in the last week, this might alter your judgment somewhat. Perhaps last week visitor 1 had been to the site 1,000 times, visitor 2 had been 2,000 times, and visitor 3 had been 3,000 times, thus making up the 6,000 visits. The key point here is that you must be able to distinguish the visitors from the visits to get more accurate statistics for companies to work with.

Checking for a cookie on the user's computer to see if that user has been to the site previously would allow you to get a better idea of the number of visitors actually visiting your site. If a cookie exists then do not increment the visitor counter; otherwise, send a cookie to the user and increment the visitor counter. The following script shows you a simple example of how to implement both visits and visitors separately in your Session object's OnStart event:

```
Sub Session_OnStart
Dim vCookie
  vCookie = Request.Cookies("visitor")
  Application.Lock
  Application("visits") = Application("visits") + 1
  if Len(vCookie) = 0 then
    Application("visitors") = Application("visitors") + 1
    Response.Cookies("visitor") = "visited"
    Response.Cookies("visitor").Domain = "192.168.7.2"
    Response.Cookies("visitor").Path = "/"
    Response.Cookies("visitor").Expires = #May 1, 2002#
    Response.Cookies("visitor").Secure = False
  end if
  Application.Unlock
End Sub
```

You may have noticed the use of the *Len* function in the script to help you determine whether the user has a cookie or not. The *Len* function is a VBScript function that returns the length of a given string. So basically, if the result of the cookie is 0 length, then the user does not have a cookie.

Facing Real World Problems

Although the code for distinguishing visitors from visits works, it is very simple and not without problems when you implement it in the real world. For instance, there is the possibility that the user does not allow cookies to be saved on his system. If no cookie gets saved, the visitor count will increase with each visit or session, marring the accuracy of the statistics.

In this case you can look at trying to keep track of visitors' IP numbers. You could combine the cookie check with the IP number check, only checking the IP number if there is no cookie. However, introducing IP numbers into the equation introduces new problems. You now run into the problem of multiple users on NAT (Network Address Translation) behind a firewall and proxy all appearing as the same visitor, when in fact they are not. This could mar the accuracy of your statistics by understating the number of visitors.

For many sites, using the cookies is sufficient.

Determining Currently Active Visitors

Using Active Server Pages, you can also easily determine the number of currently active visitors, as well. By making some changes to the Global.asa, you can increment a counter whenever sessions start and decrement the same counter when sessions end. Add the following procedures to your Global.asa if they do not already exist, otherwise just add the contents to the existing procedures:

```
Sub Application_OnStart
  Application("active") = 0
End Sub

Sub Session_OnStart
  Application.Lock
  Application("active") = Application("active") + 1
  Application.Unlock
End Sub

Sub Session_OnEnd
  Application.Lock
  Application("active") = Application("active") — 1
  Application.Unlock
End Sub
```

To show the number of currently active visitors simply use the following in one of your Active Server Pages:

```
<%Response.Write Application("active")%>
```

Creating Persistence for Your Page Counter

In all the examples so far, you have used the Application object to store the variables for the various counters. When the Active Server Pages server is shutdown, all those counters and their current values are lost. When the Active Server Pages server is started up again, the counters are all reset to default values (usually 0).

To create persistence for your page counters and various other counters, you should write out the values of the counters to a file. If you write to this file every time that the counters are modified, however, you will run into problems on big, busy sites because of all the Application object locking and disk access that is required. It is not conducive to large sites to read and write files in this manner.

There are numerous solutions to this. You could implement a timer that simply saves the script every given interval of time. You could implement a third-party persistence object to handle the task. Another solution, and probably the simplest of the lot, is to save the file when a certain number of visitors have visited your site.

For instance, every ten visitors might be sufficient for a small site to save the counters to file. A larger site might use every hundred visitors before it saves the file. To check for every ten visitors, simply use the Mod operator on the total number of visitors whenever you increment it in the Session object's OnStart event. The following snippet of code should give you an idea of when to save a file based on every ten visitors:

```
Sub Session_OnStart
    ...
    'Save counter file every 10 visitors
    vtotal = Application("visitors")
    If (vtotal Mod 10)=0 then
        'Save your file here
    End If
    ...
End Sub
```

Automatically Saving Your Page Counter File

Although saving your counter file every certain number of visitors that come to your site is a good idea, you may also want the file to be automatically saved when the Active Server Pages is being closed down by the system administrator. You can easily handle this by making use of the Application object's OnEnd event. By creating a subroutine for this event in the Global.asa file, you can insert the file saving routine code or a call to it, which will save the counters to file automatically. The other method should also be used to handle the ever common occurrence of servers crashing unexpectedly.

Saving the counters to file is all very well, but you must also remember to load them in again. Using the Application object's OnStart event, you should create a procedure in the Global.asa that reads in the counter file and assigns the last saved value to each counter appropriately.

WHAT YOU MUST KNOW

Page counters are very common and useful for both advertising a page or site's popularity, and as being a wonderful source for obtaining statistics for use with business decisions. The Microsoft Page Counter component is a very simple page counter that you can easily add to your pages. However, you already have the necessary skills and knowledge to create your own customized page counters that you can modify to meet the needs of your site or client. In Lesson 18, "Working with Files," you will learn how to use the FileSystem object to use files for reading and writing data from within your Active Server Pages. This is a natural progression for you after learning about page counter persistence through the use of files. Before you continue with Lesson 18, however, make sure that you fully understand the following key concepts:

- The Microsoft Page Counter component, available with IIS, is just one of many different page counters available for Active Server Pages.

- When you want to increment by one the number of hits, you use the Page Counter component's PageHit method. The Hits method returns the number of page hits, and the Reset method resets the counter to zero.

- By using the Application object to hold a variable and incrementing this variable at the top of an Active Server Page, you can create your own simple page counter.

- To save yourself the effort of repeating lengthy script in all of your Active Server Pages, use an include file to contain all of your page counter script in one place. You can then simply include this file and call its procedures as you required.

- The number of visits that an Active Server Pages application has can be kept track of by working with the Session object's OnStart event and incrementing an Application object variable.

- Distinguish between visitors and visits by using cookies to determine if the user has visited the site previously. This information can be very useful for businesses.

- The number of active visitors that you have at any given moment can be kept track of by incrementing and decrementing an Application object variable using the Session object's OnStart and OnEnd events.

- A simple way to maintain page counter persistence is to use a file to store all the counters in whenever a given number of new visitors begin sessions.

```
<%
Dim
cons
        nding = 8
          CreateObject("Scripting.FileSystemObject")
           enTextFile("c:\myfile.txt", ForAppending)
ts
     Wr      "Here is some important data."
ts.Close
set fs = Nothing
%>
```

18 LESSON

WORKING WITH FILES

*i*n Lesson 17, "Creating Page Counters," you learned different methods of implementing page counters in your Active Server Pages, including learning about page counter persistence through the use of files. In this lesson, you will learn how to actually implement file-based solutions by learning to use the FileSystemObject object to use files for reading and writing data from within your Active Server Pages. By the end of this lesson you will have covered the following key concepts:

- The FileSystemObject contains all the necessary objects that you require to work with files. You can use the Server object's CreateObject method to create an instance of the FileSystemObject.
- You use the CreateTextFile method to create a new text file to write to by assigning an instance of the TextStream object to a variable.
- When writing to a text file, you can make use of the TextStream object's Write, WriteLine, and WriteBlankLines methods.
- When cleaning up after writing to a file, you use the TextStream object's Close method to close the file and then set the FileSystemObject to Nothing to free memory resources.
- You can use the OpenTextFile method to write to an existing text file without creating a new file. You can also use the File object's OpenAsTextStream method. In both cases you must be sure that the *iomode* argument is set to ForAppending (8) or else you cannot add data to the existing contents of the file.

- Using either the OpenTextFile method or the OpenAsTextStream method, you can read in data from a text file. You must make sure that the *iomode* argument is set to ForReading (1).
- The three methods for reading data are the Read, ReadLine, and ReadAll methods.
- You can work with files using the FileSystemObject's CopyFile, DeleteFile, and MoveFile methods. The File object also has methods with similar functionality.
- You can work with folders using the FileSystemObject's CopyFolder, CreateFolder, DeleteFolder, and MoveFolder methods. The Folder object also has methods with similar functionality.
- Using the FileSystemObject's GetDrive method to return a Drive object, you can make use of a large variety of properties to get information about a particular drive.

Understanding the FileSystemObject Object Model

Within Active Server Pages there are many occasions where you will need to access files, either for reading in data or writing it out. You may also need to navigate folders and drives on the server, as well as supply information about their contents, size, or other relative information. You can handle these tasks by learning the FileSystemObject object model and how to utilize its many useful functions.

To access or use any of the items in the FileSystemObject object model, you must first create an instance of the FileSystemObject itself. The FileSystemObject contains three collections—Drives, Folders, and Files—all of which can hold multiple objects. For instance, the Folders collection can hold multiple Folder objects, and for each Folder object you can get information or work with given methods. The Textstream object is not a collection and does not contain other objects, but rather is used when you want to read and write from a file. Figure 18.1 shows the FileSystemObject object model and how the items within it relate to one another.

**Figure 18.1.
The FileSystemObject
object model**

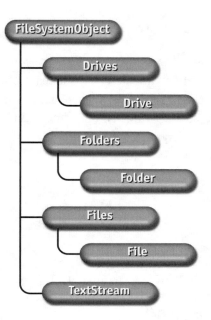

You can see an overview of what each item in the FileSystemObject object model actually is, and what it is for, in Table 18.1.

Item	Description
FileSystemObject	This is the main object for manipulating drives, folders, and files. You can create, delete, and gain information about them using the methods and properties that it contains.
Drives	This is a collection containing a list of the physical and logical drives attached to the system.
Drive	This is an object with methods and properties for getting information about a particular drive attached to the system, such as the available space.
Folders	This is a collection containing a list of all the folders within a Folder.
Folder	This is an object with methods and properties that allow you to create, delete, or move folders. Also lets you query the system for folder names, paths, and various other properties.
Files	This is a collection containing a list of all files contained within a folder.
File	This is an object with methods and properties that let you work with and get information about a file.
TextStream	This is an object that lets you read and write text files.

Table 18.1. Overview of the items in the FileSystemObject object model

Creating an Instance of the FileSystemObject

To create an instance of the FileSystemObject in your Active Server Pages, you must assign the object to a variable using the Set statement and the Server object's CreateObject method. You have already used this method of creating instances of objects several times already, and should find that you are becoming familiar with its standard syntax. An example of creating an instance of the FileSystemObject is as follows:

```
<%
Dim fs
Set fs = Server.CreateObject("Scripting.FileSystemObject")
%>
```

When you destroy an instance of an object, you free up any memory resources that it is taking. To destroy an instance of an object, you can set it to Nothing, as in the following example:

```
set fs = Nothing
```

Creating a New File

You can create a new file after you create an instance of the FileSystemObject. To create a new file, you must have an instance of a TextStream object. When creating an instance of the TextStream object, you use one of three methods, such as the CreateTextFile method, which has the following syntax:

object.CreateTextFile(filename[, overwrite] [, unicode])

The object in the syntax must be an instance of the FileSystemObject. The *filename* argument gives the path and name of the text file that you want to create. The *overwrite* argument is optional and indicates whether you want to overwrite the file if it already exists. By setting the *overwrite* argument to True, you will overwrite the file if it exists. The default is False, which is safer. The *unicode* argument is also optional. By setting the *unicode* argument to True, the text file is written as a Unicode document, which means that each character takes up two bytes. If you do not supply a *unicode* argument, the default setting of False is used, which means a regular ASCII text file that takes up one byte per character. The following example shows how to set up instances of the FileSystemObject and TextStream object, as well as creating an ASCII text file called *myfile.txt* on the C: drive and overwriting it if it already exists:

```
<%
Dim fs, ts
Set fs = Server.CreateObject("Scripting.FileSystemObject")
Set ts = fs.CreateTextFile("C:\myfile.txt", True)
%>
```

When you actually want to write data into the text file, there are three various methods that the FileSystemObject provides via the TextStream object, as shown in Table 18.2.

Method	Description
Write	Write data to an open text file without a trailing newline character.
WriteLine	Write data to an open text file with a trailing newline character.
WriteBlankLines	Write one or more blank lines to an open text file.

Table 18.2. The various writing methods of the TextStream object

To write a line of text to file using the WriteLine method, you simply supply the text as an argument. This argument is optional and if you do not supply any, a simple newline character is added to the file. To close the file once you have finished working with it, you must use the TextStream object's Close method. The Close method takes no arguments.

To try writing your own file from Active Server Pages, open up Notepad and type in the following script:

```
<%
Dim fs, ts
Set fs = Server.CreateObject("Scripting.FileSystemObject")
Set ts = fs.CreateTextFile("c:\myfile.txt", True)
ts.WriteLine "Hello world. This is a test."
ts.Close
Set fs = Nothing
Response.Write "The test file has been written."
%>
```

Save the script as *firstfile.asp* and then load it into your browser. You should see a simple message in your browser informing you that the test file has been written. There should be a file on your C: drive called *myfile.txt* that, as shown in Figure 18.2, contains the text, "Hello world. This is a test."

Figure 18.2.
The contents of
C:\myfile.txt

Writing to an Existing File

You do not always want to create a new file or overwrite an existing file when you have to write data. Sometimes you will want to append data to an existing file without losing the existing contents of the file. In these cases you can use the FileSystemObject's OpenTextFile method to open an existing file to work with. The OpenTextFile method has the following syntax:

object.OpenTextFile(filename[, iomode][, create][, format])

The object is a FileSystemObject and the *filename* argument indicates the path and name of the file that you want to open. These are the only necessary items that you must supply when using the OpenTextFile method.

When you want to open an existing file, you can specify the input/output (I/O) mode you want to open the file under. You specify the input/output (I/O) mode through the *iomode* argument. To be able to append to the file, you must specify a value of 8 for the *iomode* argument. The *iomode* is optional and if you do not specify any mode then the default mode of reading only is assumed. To make things easier to read and understand for both yourself and others, consider assigning 8 to a constant using the Const keyword and a meaningful name for the constant, such as ForAppending.

The *create* argument is also optional and, if it is set to True, will create a file if the one indicated by the *filename* argument does not exist. If you do not supply this argument, it is set to False by default and no file will be created. The last argument of the OpenTextFile method is the *format* argument, which is also optional. Basically, this argument indicates whether the file is ASCII or Unicode. You can set the *format* argument to 0 for ASCII, which is the default, -1 for Unicode, or –2 for the system default setting.

The following example shows you how to open an existing file, *myfile.txt*, write a line of text to the end of the file and then close the file again:

```
<%
Dim fs, ts
Const ForAppending = 8
Set fs = Server.CreateObject("Scripting.FileSystemObject")
Set ts = fs.OpenTextFile("C:\myfile.txt", ForAppending)
ts.WriteLine "Here is some important data."
ts.Close
Set fs = Nothing
%>
```

Another method that you can use to write text to an existing file is the File object's OpenAsTextStream method. The File object is part of the FileSystemObject object model. To assign a file to a File object, you use the GetFile method of the FileSystemObject. The GetFile method has the following simple syntax:

object.GetFile(filename)

The object is a FileSystemObject and the only argument is the path and name of the file. You must use the Set statement to assign a file to a File object. An example of using the GetFile method is as follows:

```
Set file = fs.GetFile("C:\myfile.txt")
```

The File object's OpenAsTextStream method has the following syntax:

object.OpenAsTextStream([iomode][, format])

The object is a File object and the two arguments, *iomode* and *format*, are identical to the equivalent arguments that you learned earlier for the OpenTextFile method. Instead of having an *iomode* argument with a value of 8, ForAppending, with the OpenAsTextStream method you can also have a value of 2, ForWriting. Be sure to understand the subtle difference between the two modes for writing data to text files. When you use ForAppending (8) you append or add text to the end of a file. However, when you use ForWriting (2) you completely write over the previous contents of an existing file.

In the following example, you will use the OpenAsTextStream method to open an existing file for writing. However, because you will use the ForWriting constant, your Active Server Page script will overwrite the current contents of the file. The example script is as follows:

```
<%
Dim fs, file, ts
Const ForWriting = 2
Set fs = Server.CreateObject("Scripting.FileSystemObject")
Set file = fs.GetFile("C:\myfile.txt")
Set ts = file.OpenAsTextStream(ForWriting, True)
ts.WriteLine "Whatever happened to Rocky Balboa?"
ts.Close
Set fs = Nothing
%>
```

Reading from a File

When you work with files from your Active Server Pages you do not always want to simply write data. There are many occasions when reading data is all that you must do to achieve some required goal. Before you can actually read in data from a file, you must make sure that you open the text file in the correct mode. The input/output mode that you use for reading is a value of 1. You can set up a constant to use with both the OpenTextFile method and OpenAsTextStream method, as follows:

```
'Create constant
Const ForReading = 1

'Example of use with OpenAsTextStream method
Set ts = file.OpenAsTextStream(ForReading, True)

'Example of use with OpenTextFile method
Set ts = fs.OpenTextFile("C:\myfile.txt", ForReading)
```

After opening a file in the ForReading mode, you can use the TextStream object to actually read in the data. The TextStream object has three different methods for reading data in from a file, as shown in Table 18.3.

Method	Description
Read	Read a specified number of characters from a file.
ReadLine	Read an entire line (up to, but not including, the newline character).
ReadAll	Read the entire contents of a text file.

Table 18.3. The various reading methods of the TextStream object

Using the Read method, you can read in a set number of characters. It takes a single argument, which is not optional, that indicates the number of characters that it is to read in. For example, if you want to read in eight characters from a file, you might use something similar to the following:

```
data = ts.Read(8)
```

Consequent calls to the Read method carry on from the position in the file where it left off. Imagine that you have a text file that contains the following text:

```
ABCDEFGHI
```

And then you read in data three characters at a time using the Read method, as follows:

```
data1 = ts.Read(3)
data2 = ts.Read(3)
data3 = ts.Read(3)
```

The result would be that *data1* contains "ABC," *data2* contains "DEF," and *data3* contains "GHI."

The ReadAll method reads in the entire contents of a file in one shot. Even if data is spread over multiple lines of a file, it is all read in. The ReadAll method is useful for smaller files but is a waste of memory resources when used with large files. With large files, you should use something like the ReadLine method to read a single line at a time instead. A simple example of using the ReadAll method is shown:

```
data = ts.ReadAll
```

The ReadLine method lets you read in a line of text at a time, as its name implies. This is a very useful method in many cases, as you can process individual lines of text very easily. The ReadLine method does not take any arguments, and you simply assign it to a variable to hold the result. For instance, in the following example you will use the ReadLine method to read in client information from a file and assign each line of data to an appropriate variable. The example is as follows:

```
<%
Dim fs, file, ts
Dim firstname, lastname, address, telephone
Const ForReading = 1
Set fs = Server.CreateObject("Scripting.FileSystemObject")
Set file = fs.GetFile("C:\clients\client09223.txt")
Set ts = file.OpenAsTextStream(ForReading, True)
firstname = ts.ReadLine
lastname = ts.ReadLine
address = ts.ReadLine
telephone = ts.ReadLine
ts.Close
Set fs = Nothing
%>
```

Skipping Over Lines

When you do not want to read in every line of data in a text file, you can use the TextStream object's Skip or SkipLine methods to skip areas of the file when reading. The SkipLine method will skip over a single line of the file so that all characters up until and inclusive of the next newline character are ignored. You can have multiple calls to the SkipLine method also. The SkipLine method takes no argument. If you had a TextStream object by the name of ts, an example of using the SkipLine method would simply be as follows:

```
ts.SkipLine
```

The TextStream object's Skip method will skip over a certain number of characters that you specify as its only argument. If you had a TextStream object by the name of ts, an example of using the Skip method to skip over 300 characters would be as follows:

```
ts.Skip(300)
```

Working with Files

Besides reading and writing files, you can also delete, move, copy, and rename them from within your Active Server Pages. The FileSystemObject and the File object offer similar methods that perform the same function. The reason for both objects containing similar functions is merely convenience for you as a programmer. Table 18.4 indicates the various methods for working with files.

Method	Description
File.Copy	Copies a file to another location
File.Delete	Deletes a file
File.Move	Moves a file to another location
FileSystemObject.CopyFile	Copies a file to another location
FileSystemObject.DeleteFile	Deletes a file
FileSystemObject.MoveFile	Moves a file to another location

Table 18.4. Various methods for working with files

You can delete a file using the FileSystemObject's DeleteFile method, simply by passing it the path and filename of the file you want to delete. Optionally, you can indicate if you want to force the file to be deleted if it is set to read only, by passing True for the second argument. So the following example would delete the file, *myfile.txt*, even if it is set to read-only:

```
<%
Dim fs
Set fs = Server.CreateObject("Scripting.FileSystemObject")
fs.DeleteFile "C:\myfile.txt", True
%>
```

You could also use the File object's Delete method to delete a text file. The only argument that this method takes is optional and indicates whether to force a read-only file to be deleted. So the previous example repeated but this time using the File object's Delete method to delete a file would be as follows:

```
<%
Dim fs, file
Set fs = Server.CreateObject("Scripting.FileSystemObject")
Set file = fs.GetFile("myfile.txt")
file.Delete True
%>
```

To copy a file, you can use the FileSystemObject's CopyFile method, which has the following syntax:

object.CopyFile source, destination[, overwrite]

The object is a FileSystemObject. The *source* argument is the path and filename of the file that you want to copy, while the *destination* argument is the path and filename of the file that you want to copy to. The optional *overwrite* argument is True if you want to overwrite the destination file if it already exists, which is the default setting, or False otherwise. Even if the *overwrite* argument is set to True, the CopyFile method will fail if the destination file already exists and is set to read-only. An example of using the CopyFile method to copy the file *red.txt* to *blue.txt*, only if the file does not already exist, is as follows:

```
<%
Dim fs
Set fs = Server.CreateObject("Scripting.FileSystemObject")
fs.CopyFile "C:\red.txt", "C:\blue.txt", False
%>
```

The File object's Copy method works exactly the same as the Copyfile method except that you do not have to supply the source as an argument—only the destination. The optional *overwrite* argument behaves the same in both methods. So an example of using the Copy method to repeat the previous example would be as follows:

```
<%
Dim fs, file
Set fs = Server.CreateObject("Scripting.FileSystemObject")
Set file = fs.GetFile("red.txt")
file.Copy "C:\blue.txt", False
%>
```

Moving a file is quite similar to copying a file, except that the original source file is deleted after the completion of the task. The FileSystemObject's MoveFile method takes two arguments—the source path and filename to move and the destination path and filename to move to. The File object works similarly except that it only requires the destination path and filename as an argument. An example of moving the file *red.txt* from the C: drive to the E: drive might appear as follows:

```
<%
Dim fs
Set fs = Server.CreateObject("Scripting.FileSystemObject")
fs.MoveFile "C:\red.txt", "E:\red.txt"
%>
```

To rename a file, you simply move the file to the same destination path but with a different filename. For example, to rename the file *red.txt* to *green.txt* you might do the following:

```
<%
Dim fs, file
Set fs = Server.CreateObject("Scripting.FileSystemObject")
Set file = fs.GetFile("red.txt")
file.Move "C:\green.txt"
%>
```

Using Wildcards in Your Paths

A wildcard is basically a symbol that stands for one or more unspecified characters. When working with files, you can use a * character to indicate a wildcard. You can also use a ? character to specifically indicate a single character wildcard. However, you can only use a wildcard in the filename section of a string, not in the path section. For instance, if you want to delete all document files in the accounts folder that begin with the letters "may" you might use something similar to the following:

```
<%
Dim fs
Set fs = Server.CreateObject("Scripting.FileSystemObject")
fs.DeleteFile "C:\accounts\may*.doc", True
%>
```

However, the following example would not work because the wildcard is in the path section:

```
<%
Dim fs
Set fs = Server.CreateObject("Scripting.FileSystemObject")
fs.DeleteFile "C:\*\may12.doc", True
%>
```

Working with Folders

Using the FileSystemObject, you can also work with folders from your Active Server Pages. Table 18.5 indicates some of the various methods that you can use when working with folders.

Method	Description
CopyFolder	Copy an existing folder to another destination
CreateFolder	Create a new folder
DeleteFolder	Delete an existing folder
MoveFolder	Move an existing folder to a new destination

Table 18.5. Various methods for working with folders using the FileSystemObject

Besides these methods of the FileSystemObject's, you can also use the Folder object in a similar way to how you used the File object. There are many methods in the FileSystemObject that are also available in the Folder object. Although the syntax and name may be slightly different, the functionality is the same.

Even though the folder options cannot be examined in detail, the manner of accessing them is so similar to what you have learned from working with files that a simple example should suffice to show you how to work with folders. The following example creates a folder, copies it to another location, moves the copied folder, and finally deletes the original folder:

```
<%
Dim fs
Set fs = Server.CreateObject("Scripting.FileSystemObject")
fs.CreateFolder "C:\redfolder"
fs.CopyFolder "C:\redfolder", "C:\bluefolder"
fs.MoveFolder "C:\bluefolder", "C:\greenfolder"
fs.DeleteFolder "C:\redfolder"
%>
```

Working with Drives

You can also work with drives using the FileSystemObject and the Drive object. You can obtain information about the type of drive, such as whether it is a network drive, fixed drive, CD-ROM, etc. You can find out the serial number of a drive, its drive letter, its volume name, and even its free space. Table 18.6 lists some of the FileSystemObject's methods for working with drives.

Property	Description
AvailableSpace	Returns the available space for a user on the drive
DriveLetter	Returns the drive letter
DriveType	Returns a number representing the drive type
FileSystem	Returns whether drive uses FAT, NTFS, or CDFS
FreeSpace	Returns the free space on the drive
IsReady	Returns True if drive is ready for use, False otherwise
Path	Returns the path of the drive (useful for network drives)
RootFolder	Returns a Folder object for the drive's root
SerialNumber	Returns the drive serial number
ShareName	Returns a share name if drive is a network drive
TotalSize	Returns the total size of the drive in bytes
VolumeName	Returns or sets the volume label of a drive

Table 18.6. The various properties of the Drive object

There are obviously many different properties at your disposal for working with drives and obtaining information about them. The DriveType property is a little confusing, as it only returns a number. Table 18.7 sheds some light on the values that the DriveType property returns.

Return Value	Meaning
0	Unknown drive type
1	Removeable (i.e. Floppy disk)
2	Fixed
3	Network
4	CD-ROM
5	RAM disk

Table 18.7. The various return values for the DriveType property

To obtain a Drive object, simply use the FileSystemObject's GetDrive method and pass it a drive letter as its only argument. The GetDrive method will then return a Drive object, which you assign to a variable using the Set statement. You can see an example of using some of the drive properties in an Active Server

Pages script, as well as an example of how to use the FileSystemObject, GetDrive method and Drive object in the following script. The script converts the total size and free space values from bytes to megabytes by dividing by 1048576 (1024 * 1024). Open up Notepad and type in the following short script:

```
<html>
<head>
<title>Getting Drive Information</title>
</head>

<body>
<h1>Getting Drive Information</h1>
<%
Dim fs, drive
Const path = "C:\"
Set fs = CreateObject("Scripting.FileSystemObject")
Set drive = fs.GetDrive(fs.GetDriveName(path))
Response.Write "Drive " & UCase(path) & " - " & _
drive.VolumeName & "<br>"
Response.Write "Total Space: " & _
FormatNumber(drive.TotalSize / 1048576, 0) & " MB" & "<br>"
Response.Write "Free Space: " & _ FormatNumber
(drive.FreeSpace / 1048576, 0) & " MB" & "<br>"
Response.Write "Serial # " & drive.SerialNumber
%>
</body>
</html>
```

Save the script as *driveinfo.asp* and then load it into your browser. The result of loading *driveinfo.asp* into the browser is shown in Figure 18.3.

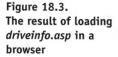

Figure 18.3.
The result of loading
driveinfo.asp **in a**
browser

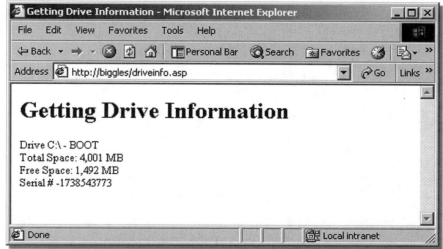

WHAT YOU MUST KNOW

The FileSystemObject contains everything that you need to work with files from within your Active Server Pages. You can create new text files or open existing files, as well as being able to copy, move, and delete files. You can also work with folders and drives using the FileSystemObject. In Lesson 19, "Working with the Content Linking Component," you will learn various methods for populating your Active Server Pages, including using approaches for minimizing the amount of work necessary to handle constantly changing content. Before you begin with Lesson 19, make sure that you know the following key concepts:

- The Server object's CreateObject method is used to create an instance of the FileSystemObject, together with the Set statement.

- You assign an instance of the TextStream object to a variable using the Set statement and the CreateTextFile method when you want to create a new text file for writing to.

- The Write, WriteLine, and WriteBlankLines methods are all part of the TextStream object. You make use of these methods when writing to a text file from within your Active Server Pages.

- To clean up after writing to a file, use the Close method with the TextStream object and then free memory resources by setting the FileSystemObject to Nothing, being sure to use the Set statement.

- To write to an existing text file, you can use the FileSystemObject's OpenTextFile method or the File object's OpenAsTextStream method. Make sure that the *iomode* argument is set to ForAppending (8) to add data to the end of the file.

- To read data in from a text file, you can use the FileSystemObject's OpenTextFile method or the File object's OpenAsTextStream method. Make sure that the *iomode* argument is set to ForReading (1) to read in data.

- The TextStream object's Read, ReadLine, and ReadAll methods allow you to easily read data in from a text file.

- The FileSystemObject and the File object provide you with methods for copying, moving, and deleting files with relative ease.

- When working with folders, you can use the FileSystemObject or the Folder object to create, copy, move, or delete folders.

- To get information about a particular drive, you can use the Drive object's many properties. Use the FileSystemObject's GetDrive method to return a Drive object, passing it the drive letter as an argument.

myValue = 7

19 LESSON

/content/index.htm Front Page This is the beginning page
/co...asp New Law Causes Stir In Parliament
...sp Religious Unity Convention Success
/co...asp Weekly Sports Events
/content/special1.asp Showing Special Advertising
/content...4.asp ...all Tower to Get Overhaul
/content/page5.asp Gardens Planned for City

WORKING WITH THE CONTENT LINKING COMPONENT

n Lesson 18, "Working with Files," you learned how to use the FileSystem object to use files for reading and writing data from within your Active Server Pages. In this lesson, you will learn about using the Content Linking component in your Active Server Pages, including minimizing the amount of work that you must do to manage previous page and next page links. By the time you complete this lesson you will have learned the following key concepts:

- To create a site that uses the book style approach in its Web pages, you can use the Microsoft Content Linking component to remove some of the burden of constantly updating each page.

- A Content Linking List must reside in a virtual directory and holds all the entries for the links in your site in the order that they are to appear.

- The entries in the Content Linking List contain a relative URL, as well as an optional description and comment, with each item being separated by a tab character.

- You must use the CreateObject method to create an instance of the NextLink object.

- Using the NextLink object's GetListCount method, you can find out the number of links that you have in the Content Linking List.

- The GetListIndex method returns the index number of the current page in your Content Linking List.

- The GetNthDescription method returns the description, if any, of the entry indicated by the index number. The GetNthURL loads the actual page for the entry.

- The GetNextDescription method returns the description of the next page and the GetPreviousDescription method returns the description of the previous page.
- The GetNextURL method loads the next page and the GetPreviousURL method loads the previous page.

Introducing the Content Linking Component

You will often see sites set up in such a manner as to imitate a common book; pages have links that connect them to the next page and the previous page. This helps to make many users feel more comfortable on a computer. Managing large amounts of page content, especially spread out over thousands of pages, and then trying to maintain all the next and previous links, can be a nightmare. This is not really a problem for small sites with only ten pages where page content and links can be updated manually, but it is not really feasible for larger sites consisting of thousands of pages.

Microsoft created the Content Linking component to let you link your Active Server Pages in a similar way to a book, yet using a simple, manageable interface. You can use the Content Linking component to generate tables of content and navigational links, which are updated automatically. When using large amounts of page content, such as online newspapers or archival resources, the Content Linking component can save an incredible amount of time or add to your site what was once an infeasible feature to implement.

Basically, the Content Linking component uses an object, the NextLink object, in your Active Server Pages to communicate with a text file known as the Content Linking List. The Content Linking List contains an entry for all the Web pages that you want to use the Content Linking component with. The pages are in the Content Linking List in the order that you want them to appear in your site.

The Content Linking component is installed with Internet Information Server and also comes with the Windows 2000 Service Pack. To check whether you already have the Content Linking component installed on your server, simply search for the file *nextlink.dll*, which is usually stored in the *C:\winnt\system32\inetsrv* folder.

Creating a Content Linking List

A Content Linking List holds all of the entries for each link in your site that you want to use with the Content Linking component. The Content Linking List must reside in a virtual directory on your Active Server Pages server. A virtual directory is not the actual root directory for your Web server but a directory that appears to be the root. Basically, a virtual directory, regardless of the actual location of that directory, is the starting point for any folders or files inside it. To create a virtual directory using Microsoft Personal Web Manager, select the Advanced icon in the left-hand panel to bring up the Advanced Options pane. Within the Advanced Options pane, click on the Add button, as shown in Figure 19.1.

**Figure 19.1.
Clicking on the
Add button**

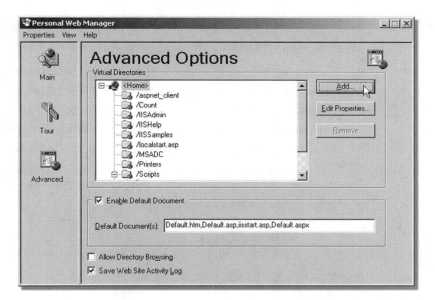

Microsoft Personal Web Manager will display the Add Directory dialog box, as shown in Figure 19.2. Inside the Add Directory dialog box, you can enter the actual path and directory that you want to use for your virtual directory in the Directory field.

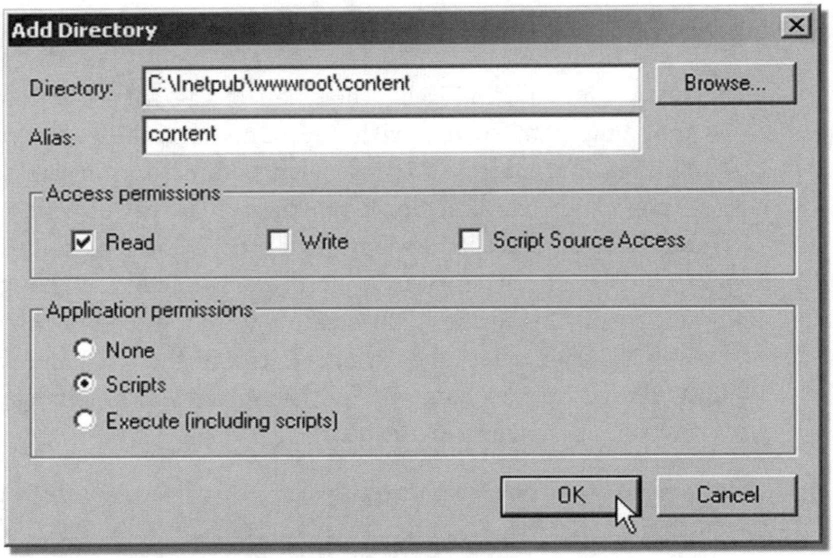

Figure 19.2.
The Add Directory
dialog box

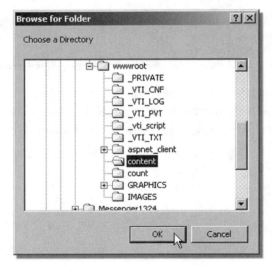

If you want to, use the Browse button situated to the right of the Directory field to browse through your current directory structure to find a folder to use as a virtual directory, as shown in Figure 19.3. The Alias field in the Add Directory dialog box is for the alias that you will use for your virtual directory. For instance, if you have a file called *test1.asp* in the virtual directory with an alias of *content*, you would access the file in your scripts by using */content/test1.asp*, rather than the entire path. Click on the OK button when you are ready to create the virtual directory.

Figure 19.3.
Browsing for a
folder to use as a
virtual directory

Within the virtual directory, create a text file and call it *links.txt*. This will be your Content Linking List. You can use another name for your Content Linking List if you want, but it must be within a virtual directory. You can also have multiple Content Linking List files for managing distinct and different areas of your site.

You add entries into your Content Linking List in a specific order, with a tab character separating items and a new line for each entry. The first item of an entry is the URL. You cannot use a full URL (one that starts with *http://*) but only a relative URL. So a URL such as the following would work because it is relative:

```
/content/page1.asp
```

However, the following URL would not work because it is a full URL:

```
http://www.mycompany.com/information/page12.asp
```

The second item in an entry is the entry description, which is optional. This is just a string of text that you can use to inform the user about the corresponding URL. For instance, if you have an entry for sport you might write *Weekly Sports Events* for the description. In the Content Linking List it would appear something similar to the following:

```
/content/page1.asp Weekly Sports Events
```

The third and final item in an entry is optional and is for comments only. This text will not be used by the Content Linking component, but is there primarily for the site manager to make comments about individual entries.

As an example of a brief Content Linking List, the following script represents ten entries that you could save in *links.txt*:

```
/content/index.htm     Front Page     This is the beginning page
/content/page1.asp     New Law Causes Stir in Parliament
/content/page2.asp     Religious Unity Convention Success
/content/page3.asp     Weekly Sports Events
/content/special1.asp  Shopping Specials 1 Advertising
/content/page4.asp     Library Tower to get Overhaul
/content/page5.asp     Gardens Planned for City
/content/page6.asp     Rare Fish Species Sighted
/content/page7.asp     Weather
/content/special2.asp  Shopping Specials 2 Advertising
```

Note: *Remember that the two or three items per entry are separated by a tab character, so you can easily have spaces within your text descriptions and comments.*

The Content Linking List cannot contain loops. It is permissible to contain the same entry twice within a Content Linking List, as long as this does not create a loop.

Creating an Instance of the NextLink Object

The Content Linking component does not actually return a Content Linking object in your Active Server Pages but what is known as a NextLink object. You must create an instance of the NextLink object before you can make use of the various methods that it contains. To create an instance of the NextLink object, use the Set statement and the Server object's CreateObject method to assign the NextLink object to a variable. In the same manner that you used the Scripting.FileSystemObject to create a FileSystemObject, use the MSWC.NextLink to create a NextLink object. An example of creating an instance of a NextLink object is as follows:

```
<%
Dim NextLink
Set NextLink = Server.CreateObject("MSWC.NextLink")
%>
```

After you create an instance of the NextLink object you will have access to the various methods that the NextLink object exposes. Table 19.1 shows the various methods of the NextLink object.

Method	Description
GetListCount	Counts the number of items linked
GetListIndex	Gets the index of the current page
GetNextDescription	Gets the description of the next page listed
GetNextURL	Gets the URL of the next page listed
GetNthDescription	Gets the description of the Nth page listed
GetNthURL	Gets the URL of the Nth page listed
GetPreviousDescription	Gets the description of the previous page listed
GetPreviousURL	Gets the URL of the previous page listed

Table 19.1. Various methods of the NextLink object

When you finish working with the NextLink object, you should use the Set statement to set the object to Nothing. This will properly free up the memory resources that the server allocates to it. The following script shows how to free the memory used by the NextLink object when you finish using it:

```
<%
Dim NextLink
Set NextLink = Server.CreateObject("MSWC.NextLink")

'Do statements here

'Free up memory
Set NextLink = Nothing
%>
```

Using the NextLink Object

You can determine the number of links that are in the Content Linking List by using the NextLink object's GetListCount method. The GetListCount method takes as its only argument the virtual path and filename of the Content Linking List. An example of using the GetListCount is as follows:

```
<%
Dim NextLink, totalcount
Set NextLink = Server.CreateObject("MSWC.NextLink")
totalcount = NextLink.GetListCount("/content/links.txt")
Response.Write "The number of links in the list is: " _
& totalcount
%>
```

The GetListIndex method returns the index number of the current page in your Content Linking List. If the current page is not in the Content Linking List then the GetListIndex method returns 0. The first page in the Content Linking List has an index value of 1. The following script creates a NextLink object, reporting back the current page out of the total number of pages, and giving you an example of using the GetListIndex method:

```
<%
Dim NextLink, totalcount, current
Set NextLink = Server.CreateObject("MSWC.NextLink")
totalcount = NextLink.GetListCount("/content/links.txt")
current = NextLink.GetListIndex("/content/links.txt")
Response.Write "The current page is " & current & _
" out of a total of " & totalcount & "."
%>
```

If you want to retrieve the description for one of the links in the Content Linking List and you already know what index number the link is, you can use the NextLink object's GetNthDescription method. The GetNthDescription method takes two arguments—the first is the location of the Content Linking List and the second is the index number of the link. For example, if you wanted to retrieve the description of the sixth link in the Content Linking List and then send this back as a response, you would use script similar to the following:

```
<%
Dim NextLink, desc
Set NextLink = Server.CreateObject("MSWC.NextLink")
desc = NextLink.GetNthDescription("/content/links.txt", 6)
Response.Write "The description for item 6 is: " & desc
%>
```

You can also easily go to the URL or link whose index number you know by using the NextLink object's GetNthURL method. The GetNthURL method also takes two arguments, which are identical to those for the GetNthDescription method. So if you put the path and filename of the Content Linking List as the first argument and the index number of the link as the second argument, the URL of that link will be loaded in the user's browser. So the following short example shows how you can display the URL of the sixth link:

```
<%
Dim NextLink
Set NextLink = Server.CreateObject("MSWC.NextLink")
NextLink.GetNthURL("/content/links.txt", 6)
%>
```

Using the skills that you now have with the Content Linking component, create an Active Server Page that will display the entire table of contents of your Content Linking List by displaying the description of each entry as an HTML link. The following script gives you a simple example of how to accomplish this:

```
<html>
<head>
<title>Table of Contents</title>
</head>

<body>
<h1>Table of Contents</h1>
<%
Dim NextLink, totalcount, i
Set NextLink = Server.CreateObject ("MSWC.NextLink")
totalcount = NextLink.GetListCount("/content/links.txt")
for i=1 to totalcount
  %>
  <a href="<%=NextLink.GetNthURL("/content/links.txt",
  i)%>">
  <%=NextLink.GetNthDescription("/content/links.txt", _
  i)%></a><br>
<%next%>
</body>
</html>
```

Save the example script as *toc.asp* and load it into your browser. Figure 19.4 shows the result of loading *toc.asp*, although the result may vary depending on the Content Linking List that you have.

Figure 19.4.
Displaying all the descriptions in the Content Linking List as links

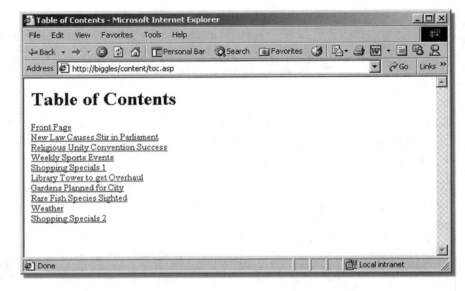

Working with the Next and Previous Links

You have already seen how to get the description of the Nth link. Using this knowledge you would have to do something similar to the following script to retrieve the description of the next link:

```
<%
Dim NextLink, totalcount, current
Dim next, s
Set NextLink = Server.CreateObject("MSWC.NextLink")
totalcount = NextLink.GetListCount("/content/links.txt")
current = NextLink.GetListIndex("/content/links.txt")
if current < totalcount then
  next = current + 1
  s = NextLink.GetNthDescription("/content/links.txt", next)
  Response.Write "The next link is: " & s
end if
%>
```

As an alternative, you can simply make use of the NextLink object's GetNextDescription method. The GetNextDescription method makes the code shorter and cleaner, as well as less prone to errors. If the current page is not in the Content Linking List, then it returns the description of the last page instead. The GetNextDescription method takes the virtual path and filename of the Content Linking List as its only argument. So the same example that you just saw, but this time using the GetNextDescription method, appears as follows:

```
<%
Dim NextLink, s
Set NextLink = Server.CreateObject("MSWC.NextLink")
s = NextLink.GetNextDescription("/content/links.txt")
Response.Write "The next link is: " & s
%>
```

You can also go to the next URL in the Content Linking List by using the NextLink object's GetNextURL method. Like the GetNextDescription method, the GetNextURL method takes the virtual path and filename of the Content Linking List as its only argument. The GetPreviousDescription method and GetPreviousURL method also take as an argument the virtual path and filename of the Content Linking List. The GetPreviousDescription method returns the description of the previous entry in the Content Linking List and the GetPreviousURL loads the URL of the previous entry.

The following snippet of code would be placed in the header section of all the Active Server Pages in the Content Linking List and shows an example of using all four of these new methods to display the previous and next pages as links:

```
<%
Dim NextLink
Set NextLink = Server.CreateObject ("MSWC.NextLink")
%>
<<< Previous page: <a href="<%=NextLink.GetPreviousURL
("/content/links.txt")%>">
<%=NextLink.GetPreviousDescription("/content/links.txt")%>
</a><br>
>>> Next page: <a href="<%=NextLink.GetNextURL
("/content/links.txt")%>">
<%=NextLink.GetNextDescription("/content/links.txt")%></a>
```

Figure 19.5 shows an example of a page using this script in the header section of the page after it is loaded into a browser.

**Figure 19.5.
An example of
having links for
the previous and
next page**

WHAT YOU MUST KNOW

In this lesson you have seen how you can use the Microsoft Content Linking component to manage a large number of page links from a central location, namely the Content Linking List. Make sure to place the Content Linking List within a virtual directory and also to use tabs and not spaces when separating items for each entry. The various methods of the NextLink object allow you to easily discover the total number of entries, retrieve descriptions, and load pages with a minimum of complexity. In Lesson 20, "Using Additional Components," you will learn how to use other Microsoft components, such as obtaining valuable information about the browser that the client is using, working with more page content tools, and dealing with ad rotations. Before you move on to Lesson 20, make sure that you understand the following key concepts:

- The Microsoft Content Linking component makes it easier to create a site that uses the book style approach in its Web pages.

- Holding all of the entries for the links in your site, a Content Linking List must reside in a virtual directory. Each line of the Content Linking List indicates a new entry, and the index order is relative to the line order.

- You cannot use full URLs in your entries in the Content Linking List but only relative URLs. You can also add an optional description and comment, separating the items with a tab character.

- Use the Server object's CreateObject method to create an instance of the NextLink object, and then set the object to Nothing to release allocated memory when you finish with the object.

-  The GetListCount method returns the number of links in the Content Linking List, while the GetListIndex method returns the index number of the current page.

-  To load a page for which you know the index number, you can use the GetNthURL method. You can also use the GetNthDescription method to return its description.

-  When working with the next page, you can use the GetNextDescription method to return the description and the GetNextURL method to actually load the next page.

-  When working with the previous page, you can use the GetPreviousDescription method to return the description and the GetPreviousURL method to actually load the next page.

USING ADDITIONAL COMPONENTS

n Lesson 19, "Working with the Content Linking Component," you learned about using the Content Linking component in your Active Server Pages, including minimizing the amount of work that you must do to manage previous page and next page links. In this lesson, you will learn more about handling page content, as you use other Microsoft components. You will also learn how to obtain valuable information about the browser that the client is using, as well as how to deal with advertisement rotations. By the time you have completed this lesson you will have learned the following key concepts:

- Using a server-side approach, the Browser Capabilities component lets you discover information about the user's browser.

- The CreateObject method, together with the Set statement, is used to create an instance of the Browser Capabilities object.

- The Browser Capabilities component relies on the *browscap.ini* file to provide its information about the various versions of browsers.

- Using a dot formation, you can access the various entries of the *browscap.ini* file as if they were read-only properties of your Browser Capabilities object.

- When you require continually changing content in your pages, you can make use of the Content Rotator component to prevent manual intervention and work.

- The CreateObject method, together with the Set statement, is used to create an instance of the ContentRotator object.

- The Content Schedule file contains all of the HTML content items and their various weightings, which determine the likelihood of being shown.

- The ContentRotator object's GetAllContent method will display all the content in the Content Schedule file, whereas the ChooseContent method will display a single content item at a time.
- The Ad Rotator component is similar to the Content Rotator component but it is geared specifically towards the needs of advertising.
- The Rotator Schedule file contains the URLs to the advertising images, as well as the advertisers' home pages. It also holds global parameters for viewing the image, optional alternative text strings, and weight values for determining likelihood of selection.
- The three properties of the AdRotator object let you set the size of the advertisement border, determine whether the advertisement is a hyperlink, and specify the name of the frame in which to display the advertisement.
- The AdRotator object's only method, the GetAdvertisement method, gets an advertisement from the Rotator Schedule file and displays it.

Working with the Browser Capabilities Component

The Browser Capabilities component lets you find out from the server all sorts of details and information about the user's browser. This can be very useful when dealing with Web content that has specific requirements, such as cookies or frames. You may need to know whether your client is using Netscape Navigator or Microsoft Internet Explorer to make changes to your response dynamically.

The Browser Capabilities component is part of Internet Information Server, which you may have to install if you do not already have the Browser Capabilities component on your server. Windows 2000 will have the Browser Capabilities component by default. To check if you already have the Browser Capabilities component, do a file search for *browscap.dll* on your server. Typically, this file is kept in the *C:\winnt\system32\inetsrv* directory.

To use the Browser Capabilities component, you must first create an instance of the Browser Capabilities object by using the Server object's CreateObject method. You must also use the Set statement to assign this new instance of the Browser Capabilities object to a variable. The following script shows an example of creating an instance of the Browser Capabilities object:

```
<%
'Create an instance of the Browser Capabilities object
Dim bc
Set bc = Server.CreateObject("MSWC.BrowserType")
%>
```

The Browser Capabilities object does not have a set of predefined properties that you can list. Instead it refers to the file *browscap.ini*, which can contain varying entries for different browsers and versions. Your Browser Capabilities object can reference the entries within the *browscap.ini* file as if they were read-only properties using the same dot formation that you normally use. For instance, the following example shows how to retrieve the tables entry from the *browscap.ini* file using the Browser Capabilities object:

```
<%
'Create an instance of the Browser Capabilities object
Dim bc
Set bc = Server.CreateObject("MSWC.BrowserType")
Response.Write "Supports tables: " & bc.tables
%>
```

There are some entries that are fairly common across all browser versions listed in the *browscap.ini* file. Table 20.1 shows a list of some of the more common entries, although this is by no means comprehensive. For a complete list, you will have to open up the *browscap.ini* file on your server and view its contents.

Entry	Description
activexcontrols	Whether browser supports ActiveX controls (True/False)
backgroundsounds	Whether browser supports background sounds (True/False)
browser	The browser type (IE, Netscape, etc.)
cookies	Whether browser supports cookies (True/False)
frames	Whether browser supports frames (True/False)
javaapplets	Whether browser supports Java applets (True/False)
javascript	Whether browser supports JavaScript (True/False)
majorver	The major version of browser
minorver	The minor version of browser
tables	Whether browser supports tables (True/False)
vbscript	Whether browser supports VBScript (True/False)
version	The version of browser

Table 20.1. The more common entries that you will find in the browscap.ini file

The following example shows how to create a page that lists some of the more common browser capabilities in a simple HTML table. First, an instance of the Browser Capabilities object is created and then the table is populated with both the browser capabilities and the status of those capabilities. Open up Notepad and type in the following short script:

```
<html>
<head>
<title>Browser Capabilities</title>
</head>

<body>
<h1>Browser Capabilities</h1>
<%
'Create an instance of the Browser Capabilities object
Dim bc
Set bc = Server.CreateObject("MSWC.BrowserType")
%>
<table border="2" cellpadding="5" align="center">
<tr><td>Browser</td><td> <%=bc.browser%>
<tr bgcolor="#e7e7e7"><td>Version</td><td> <%=bc.version%>
<tr><td>Frames</td><td> <%=bc.frames%>
<tr bgcolor="#e7e7e7"><td>Tables</td><td> <%=bc.tables%>
<tr><td>Cookies</td><td> <%=bc.cookies%>
<tr bgcolor="#e7e7e7"><td>Background sounds</td><td>
<%=bc.backgroundsounds%>
<tr><td>VBScript</td><td> <%=bc.vbscript%>
<tr bgcolor="#e7e7e7"><td>JavaScript</td><td>
<%=bc.javascript%>
<tr><td>Java Applets</td><td> <%=bc.javaapplets%>
<tr bgcolor="#e7e7e7"><td>ActiveX Controls</td><td>
<%=bc.activexcontrols%>
</table>
</body>
</html>
```

Save the script as *bc.asp* and then load it into your browser. The result should appear similar to Figure 20.1, although depending on your browser and the capabilities that it has, the table entries may appear different.

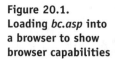

Figure 20.1.
Loading *bc.asp* into
a browser to show
browser capabilities

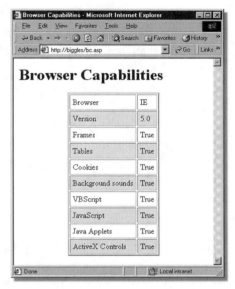

Note: *If an entry does not exist in the browscap.ini file then the result is given as Unknown. As an example, if you request <%=bc.shockwave%> you will get Unknown as the result because this is not in the typical browscap.ini.*

The most important thing to keep in mind if you are going to use the Browser Capabilities component is that you must maintain a current version of the *browscap.ini* file. You can do this yourself manually if you have plenty of time on your hands, or you can go to the following site on the Internet where a regularly updated *browscap.ini* file can be downloaded:

http://www.cyscape.com/browscap/

The company CyScape is not only home to an up-to-date *browscap.ini* file but also to its own product, BrowserHawk, which has many additional features compared to Microsoft's Browser Capabilities component. So if you are looking seriously at setting up a site that places a high importance on browser awareness, you may want to look more closely at this commercial component. You do not really have to make any changes to your scripts if you are upgrading from the Browser Capabilities component to BrowserHawk because the exact same calls are supported. The following brief list shows you some of the main additional features of BrowserHawk, which could be potentially very useful for some of your clients:

- Prevent new browsers from breaking your pages
- Perform reverse DNS lookups

- Detect when cookies, JavaScript, or Java applets are disabled
- Detect Macromedia Flash and Director plug-ins
- Detect screen size resolution and available browser window size
- Detect user connection speed in Kbps
- Detect all Netscape and IE browsers, search engines, and hundreds of others including Opera, WebTV, and older AOL browsers
- Automatically have new browser definitions downloaded and installed for you
- Detect capabilities including style sheets, file upload, DHTML, table background colors and images, SSL, SSL key size, JavaScript version, mouseover, and dozens of others
- Monitor for any unknown browsers

Working with the Content Rotator Component

When you are creating a site that you want visitors to return to time and time again, you have to continually change the content of a page. Sometimes it is not feasible to make these changes manually. For instance, if you are designing a site for a major client who has four major products that they want to promote, you do not want to manually change the advertisement on the home page every time someone connects to the site. You could change it on a daily basis manually but this is still not a good solution, especially if the advertising campaign is to last for two months.

A solution to this common type of problem is to use automated page content rotation. Microsoft provides the Content Rotator component to cater to this very need. The Content Rotator component will automatically change text, images, background colors, or anything else that you can contain in HTML code. This makes an incredibly useful tool for having a set of content that you can leave on a site for months, yet still provide a constantly changing appearance.

The Microsoft Content Rotator component comes with Internet Information Server, although you will also find it on Windows 2000. You can check to see if the Content Rotator component is already on your server by searching for the file *contort.dll* on your local hard disk. The default path that you will find *contort.dll* is *C:\winnt\system32\inetsrv\contort.dll,* although this may be different depending on the configuration of your server.

To use the Content Rotator component in your scripts, you must first create an instance of the ContentRotator object using the Server object's CreateObject method, together with the Set statement. You have used this same process many times now to create nonintrinsic objects in your Active Server Pages, and should be realizing how useful this capability is. However, in this particular call to the CreateObject method you must use the argument MSWC.ContentRotator. The following script shows an example of creating an instance of the ContentRotator object:

```
<%
'Create an instance of the ContentRotator object
Dim cr
Set cr = Server.CreateObject("MSWC.ContentRotator")
%>
```

Before you can actually use the ContentRotator object, you must decide on the content that you want to rotate and prepare it. To do this you use a Content Schedule file, which holds all the content information. You can have more than one Content Schedule file, each one holding different rotational content. The Content Schedule files are simply text files that you can create in Notepad or any other text editor. The syntax of each item in a Content Schedule file is as follows:

```
%% [#Weight] [//Comments]
ContentString
```

Each item of content begins with two % characters on a new line. This is what the Content Rotator component uses to differentiate between each of the items. On subsequent lines, you can place the HTML code that makes up the item of content. So an example of three different items might appear as follows:

```
%%
<b>This is item 1</b>

%%
<i>This is item 2<br>
This is <b>more</b> of item 2.</i>

%%
This is item 3
```

The weighting of each item determines how often it is shown, or at least the likelihood of it being shown, as random numbers are also made use of as part of the selection process. To weight an item, you simply add a # character and an integer from 0 to 10000, after the two % characters. If you use a weight of 0, then the item will not be shown. Also be aware that the sum of all your weight values cannot be more than 10000, otherwise it will cause an error when the ContentRotator object's methods access the Content Schedule file. The weight of each item is itself divided by the sum of all weight values in the Content Schedule file. Therefore, the higher the weight value, the greater the likelihood of that item being selected by the Content Rotator component. To add weighting to the previous example, you could do something like the following:

```
%% #5
<b>This is item 1</b>

%% #15
<i>This is item 2<br>
This is <b>more</b> of item 2.</i>

%% #10
This is item 3
```

From this example, item 1 has a weight of 5, item 2 has a weight of 15, and item 3 has a weight of 10. Therefore the total sum of weight values is 30. Item 1 will probably occur 5 out of 30 times, or one sixth of the time. Item 2 will probably occur 15 out of 30 times, or half the time. Item 3 will probably occur 10 out of 30 times, or one third of the time. Table 20.2 shows the weight values of the example items more clearly.

Item	Weight	Divided by Total Sum	Final Likelihood of Selection
item 1	5	5/30	one sixth of the time
item 2	15	15/30	one half of the time
item 3	10	10/30	one third of the time

Table 20.2. The weight values of three items in a Content Schedule file

You can also add comments to your Content Schedule files by using a double forward slash (//) prior to the comment. The comment comes after either the two % characters or after the weight value, if any. You can also have comments on subsequent lines before the HTML string itself. In the following example, you can see one comment for item 1, no comment for item 2, and two lines of comments for item 3:

```
%% #5 //This is a comment for item 1
<b>This is item 1</b>

%% #15
<i>This is item 2<br>
This is <b>more</b> of item 2.</i>

%% #10 //This is a comment for item 3
//This is another comment for item 3
This is item 3
```

To add content that includes hyperlinks or images, rather than simply text, all you have to do is add the appropriate HTML code. An image would use the <img> tag and a hyperlink would use an <a> tag. For example, the following script will add a hyperlink to item 1's content and an image to item 3's content:

```
%% #5 //This is a comment for item 1
<b>This is <a href="product1.htm">item 1</a></b>

%% #15
<i>This is item 2<br>
This is <b>more</b> of item 2.</i>

%% #10 //This is a comment for item 3
//This is another comment for item 3
This is item 3
<img src="hardware3.png" align="center">
```

To actually access the Content Schedule files from your Active Server Pages, you use the ContentRotator object's GetAllContent method or ChooseContent method. The ChooseContent method is probably the method that you will most often want to use, as it selects one of the items from the Content Schedule file that you pass to it as an only argument. The following shows an example of using the ChooseContent method:

```
<%
'Create an instance of the ContentRotator object
Dim cr
Set cr = Server.CreateObject("MSWC.ContentRotator")

'Output one item from the Content Schedule file
Response.Write cr.ChooseContent("content.txt")
%>
```

Note: *You cannot call the ChooseContent method from within the Global.asa file or you will get an error.*

Creating a Tip of the Day

Using the Content Rotator component to add a "Tip of the Day" feature to your site is a popular use for this component. The Content Rotator component lends itself to this by allowing you to create a simple text file, namely a Content Schedule file, which you fill with handy tips. You do not need to give any weight value to any of the tips, as they should all be equal for the most optimal rotation. As users visit the site, they will continue to see a different tip with no effort on the part of the developer or site manager.

The other method of the ContentRotator object is the GetAllContent method. The GetAllContent method takes as its only argument the path and file-name of a Content Schedule file. The output from the GetAllContent method is every item in the Content Schedule file separated by a horizontal rule or <hr> tag. You would probably only use the GetAllContent method for design times when you want to test that all of the items in a Content Schedule file appear correctly. This prevents you from having to constantly refresh your page in order to see all the items, or a particular item that you may have been working on. You might like to consider the GetAllContent method as a developer's tool rather than a feature for a site.

Working with the Ad Rotator Component

You could probably use the Content Rotator component to handle changing advertisements on your site; however, Microsoft also has a similar component that is more suited to the task. The Ad Rotator component is part of Microsoft Internet Information Services and comes with Windows 2000, as well as the Windows 2000 Service Pack. You can determine whether you have the Ad Rotator component already installed on your server by searching for the file *adrot.dll,* which is usually located in the directory *C:\winnt\system32\inetsrv*.

Like the Content Rotator component, the Ad Rotator component creates an instance of an object that references a schedule file to rotate the images. The Ad Rotator component has more options than the Content Rotator component, which are geared specifically towards the needs of advertising.

To use the Ad Rotator component in your scripts, you must first create an instance of the AdRotator object using the Server object's CreateObject method, together with the Set statement. In this particular call to the CreateObject method you must use the argument MSWC.AdRotator. The following script shows an example of creating an instance of the AdRotator object:

```
<%
'Create an instance of the AdRotator object
Dim aro
Set aro = Server.CreateObject("MSWC.AdRotator")
%>
```

Before you can make use of the AdRotator object's properties or methods, you must set up the rotation schedule that it makes use of. The Rotator Schedule file is similar in purpose to the Content Schedule file; however, it has a completely different syntax. Basically, the Rotator Schedule file is divided into two separate sections by an asterisk (*) character.

The first section applies to all the advertisements and comprises four optional global parameters—REDIRECT, WIDTH, HEIGHT, and BORDER. The REDIRECT parameter indicates an Active Server Page script that you write yourself. You can use this script as you require, such as to increment a count for statistical purposes before transferring the user to their intended destination (the advertiser's home page). The other three parameters indicate the default values for the advertising images that are used. If you do not specify any values for these parameters, then the default settings are used by the Ad Rotator component and you would start your Rotator Schedule file with the * character as the first line.

The second section applies to all of the advertisements that you want to rotate through, which really just indicates the URL of the advertisement image, the URL of the advertiser's home page, alternative text for those browsers that cannot display the image, and finally a weighting value (impressions) that is identical to the one used in the Content Schedule file. Each of these elements is on a new line.

The syntax of the Rotator Schedule file is as shown:

```
[REDIRECT URL]
[WIDTH numWidth]
[HEIGHT numHeight]
[BORDER numBorder]
*
adURL
adHomePageURL
Text
impressions
```

In the following example of a Rotator Schedule file, you can see how all these elements go together. The example portrays an imaginary sporting advertisement company, with only three different advertisements (for simplicity). The weighting for each advertisement (impressions) is 1, so they are equal in display percentages. The example of a Rotator Schedule file is as follows:

```
REDIRECT http://www.sportsads.com/redir.asp
WIDTH 400
HEIGHT 80
BORDER 5
*
http://www.surfboards.com/logo.png
http://www.surfboards.com/
Check out these surfboard specials!
1
http://www.windsurfers.com/gfx/maintitle.jpg
http://www.windsurfers.com/
Don't miss these windsurfer deals!
1
http://www.skimania.com/ultimate.png
http://www.skimania.com/
Looking for ski gear? Look no further!
1
```

Table 20.3 shows the properties of the AdRotator object and gives a brief description of what they are for.

Property	Description
Border	The size of the border around the advertisement
Clickable	Determines whether the advertisement is a hyperlink
TargetFrame	Specifies the name of the frame in which to display the advertisement

Table 20.3. Properties of the AdRotator object

To use the Border property, simply assign a size value to it, indicating the size of the border to apply. If you do not assign a size value, then the size value that you assigned (if any) with the BORDER parameter at the top of the Rotator Schedule file will be used. If you want to have no border, then set the AdRotator object's Border property to 0.

The Clickable property will determine whether the advertisement is a hyperlink or not. By setting this property to False, you can stop the advertisement from actually loading the advertiser's home page. This is a better way to handle not going to the advertiser's home page than removing the URL entirely from the Rotator Schedule file.

The TargetFrame property allows you to specify which frame you want the advertisement to be shown in. If you do not specify a value for the TargetFrame, then the default value of no frame is used. Table 20.4 shows the main values that you can use for the TargetFrame property.

Value	Description
_BLANK	Displays in a new browser window
_PARENT	Displays in the parent frame
_SELF	Displays in the current browser window
_TOP	Displays in the top level (entire browser window)

Table 20.4. The main values that you use with the TargetFrame property

Note: *On its site, Microsoft mentions two other values for use with the TargetFrame property—namely the _NEW and _CHILD values. However, these are not part of the four universal target destination values that are mentioned on W3's HTML specifications for the target destination values, so you cannot be sure of the support for these values.*

The only method that the AdRotator object has is the GetAdvertisement method. This method will get an advertisement from the Rotator Schedule file that you specify for the method's only argument.

The following example of using the AdRotator object and its properties will create an instance of the AdRotator object, set the advertisement border to 2, set the Clickable property to 0 so that it is not a link, set the target frame to itself, and finally get an advertisement from the Rotator Schedule file. The script is as follows:

```
<%
'Create an instance of the AdRotator object
Dim aro
Set aro = Server.CreateObject("MSWC.AdRotator")

'Set the advertisements properties
aro.Border=2
aro.Clickable=False
aro.TargetFrame="_SELF"

'Display an advertisement
aro.GetAdvertisement("/adrotsched.txt")
%>
```

WHAT YOU MUST KNOW

In this lesson you learned about three additional components from Microsoft. The Browser Capabilities component allows you to discover information about the user's browser. The Content Rotator component is for when you require continually changing content in your pages. The Ad Rotator component is similar to the Content Rotator component, but it is geared specifically towards the needs of advertising. In Lesson 21, "Introducing Structured Query Language (SQL)," you will start to learn about databases and how they work in preparation for using them from your Active Server Pages. Before you move on to Lesson 21, make sure that you understand the following key concepts:

- To create an instance of the Browser Capabilities object, you use the CreateObject method, together with the Set statement.

- The *browscap.ini* file contains information about the various versions of browsers which it supplies to the Browser Capabilities component as required.

- The various entries of the *browscap.ini* file are similar to read-only properties, and you can access them with a Browser Capabilities object by using the same dot formation as ordinary properties.

- To create an instance of the ContentRotator object, you use the CreateObject method, together with the Set statement.

- The Content Schedule file contains all of the content items in HTML format. It also contains a weight value that determines the likelihood of the content item being shown.

- You use the ChooseContent method to display a single content item at a time. For testing purposes, you can also use the GetAllContent method to display all the content in the Content Schedule file.

- The Rotator Schedule file contains global parameters for viewing the advertising images, as well as the URL of the images. Optionally you can have a URL to the home page of the advertiser, or alternative text when the image cannot be shown.

- The weight values for determining likelihood of selection work in the same way for the Rotator Schedule file as they do for the Content Schedule file.

- The AdRotator object has three properties: the Border property, the Clickable property, and the TargetFrame property. These properties allow you to configure the advertisement by setting the border size, stopping the hyperlink, and specifying the target frame.

- To get an advertisement from the Rotator Schedule file and display it, you would use the AdRotator object's GetAdvertisement method. The only argument of this method is the path and filename of the Rotator Schedule file.

21

LESSON

INTRODUCING STRUCTURED QUERY LANGUAGE (SQL)

n Lesson 20, "Using Additional Components," you saw how to make use of various external components to add functionality to your Active Server Pages. In this lesson, you will look at introducing database functionality to your Active Server Pages that, for many companies, is one of the most significant reasons to use Active Server Pages. Structured Query Language (SQL) deals with relational databases and the methods used to access them for both reading and writing data. By the time you finish this lesson you will have covered the following key concepts:

- Structured Query Language (SQL) is a declarative language for working with data that is stored in relational database tables. You simply declare what you want and let the database itself handle the details of how to accomplish the task.
- Rows and columns of cells containing data are known as tables and are essentially the heart of the database that SQL works with. You can manually create a table using a database management system such as Microsoft Access.
- The SELECT statement in SQL is extremely useful and enables you to query relational databases for specific information. You must specify the table that you want after the FROM keyword.
- Using the WHERE clause lets you create conditional queries in your SQL code, targeting only the data that you require.

- Logical operators, such as AND, NOT, and OR, let you combine multiple conditions in your SQL queries. You can also use the BETWEEN and IN keywords to further shorten your queries, making complex queries easier to read and write.
- The LIKE keyword, together with the % character as a wildcard, enables you to specify part of a value, which is useful to retrieve clusters of data that have a common similarity, such as beginning with the same letter.
- A primary key is a column in a table that contains some uniquely identifiable data in each cell, which you can use to identify all data in the same row. A foreign key is an ordinary column in one table that is a primary key in another.
- Using the UNION keyword between SELECT statements, you can join the results of multiple queries together.
- You can use the DISTINCT keyword to stop repetitions of data, or use the ORDER BY keywords to sort the results.
- SQL also has aggregate functions that return a summary of the data results rather than the data itself. The aggregate functions include the ADD, COUNT, MIN, MAX, and SUM functions.
- You can manipulate the actual data in tables also by using the INSERT INTO, UPDATE SET, and DELETE FROM keywords.

Learning the Basics of SQL

Structured Query Language (SQL) is a very large topic. Entire books—nay, entire careers—are wholly devoted to the subject. This lesson will introduce you to the basics of SQL only so that you can understand enough to use it from your Active Server Pages and in preparation for working with ActiveX Data Objects (ADO). The discussion of SQL in this lesson is by no means comprehensive, but it does cover enough fundamentals for you to be able to make use of SQL and understand what you are doing. To truly do the subject of SQL justice, you should consider reading books, articles, and other material that are devoted entirely to it. For specifics on all aspects of SQL, turn to the book *Rescued by SQL,* Onword Press, 2001.

Working with relational databases is an important part of many computer systems in place today. Regardless of whether they are accessed through Web pages or not, relational databases play a vital role for businesses due to their ability to hold vast amounts of data, yet maintain fast results when you search or query the database for results. Relational databases allow greater flexibility than flat databases when querying the database because you can search on individual fields or combinations of them rather than only a single field.

There are many different relational database systems to choose from, such as Microsoft SQL Server, Microsoft Access, Oracle, Sybase, and Informix to name a few. These database management systems are sometimes referred to individually as a DBMS, and you will see this acronym used in many books, articles, and newsgroups. SQL allows you to communicate with any of these database systems by providing a common language that you can access data with or manipulate that data, if need be. The generic SQL language is known as a *declarative* language rather than the more common *procedural* languages, such as Visual Basic, Modula-2, C++, and others. This is because in procedural languages you explicitly state what is to be done, whereas in a declarative language you declare what you would like and allow the database to actually formulate the step by step sequence of actions to take. The beauty of a declarative language, such as SQL, is that two databases can store information in completely different formats, yet you can obtain information from both of them.

Although, SQL is a declarative language, you may come across the acronyms DDL and DML in many references to SQL when describing the type of language. Basically, DDL stands for Data Definition Language and refers to table creation within SQL, and DML stands for Data Manipulation Language and refers to the various statements for working with databases, such as those you will learn about in this lesson.

Understanding Database Tables from an SQL Perspective

When you are working with SQL and relational databases, everything is stored in tables. Tables are really just rows and columns of cells containing data. To be able to use SQL to access the data that you require, you must have at least a general understanding of tables.

In a relational database table, related information is stored in a single row which covers multiple columns. For example, if you want to store a database of customers and their special interests, perhaps to sell to potential advertisers, you might like to store a unique customer identification number, the customer's first and last names, as well as their special interest. Table 21.1 shows the basic columns that you might use to create such a database, as well as the data for five imaginary customers.

Cust_ID	First_Name	Last_Name	Interest
379648	John	Smith	Dog breeding
735233	Margaret	Doe	Baking
899454	Jason	McDuff	Boxing
543678	Judy	Parker	Karate
543679	Andrew	Massie	Aviation

Table 21.1. Basic columns and data for five imaginary customers

Regardless of which database program you use, with the provision that it supports SQL, you can enter in data manually, which will also highlight some of the key characteristics of tables as they relate to SQL. For instance, the name of the table itself is often important, as you can have larger databases that contain multiple tables. In the following example, you will see how to implement the data from Table 21.1 using Microsoft Access 2000. If you do not have Microsoft Access 2000, a similar approach will work for any database program, due to the simplicity of the table structure. Remember that SQL is not dependent on the database program itself, so the example uses Microsoft Access 2000 purely to portray the concepts in a clear manner.

In Microsoft Access, you must create a new database by selecting the File menu New item. Access will display the New dialog box, from which you can simply select the default Database icon in the General tab. Click on the OK button to create the database. Access will then prompt you to save the new database by displaying the File New Database dialog box. Save the new database as *customer.mdb* using the File New Database dialog box. From within the Tables section, double-click on the *Create table by entering data* item, as shown in Figure 21.1. This is the same section where you will see the list of tables that are contained within the database.

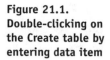

**Figure 21.1.
Double-clicking on
the Create table by
entering data item**

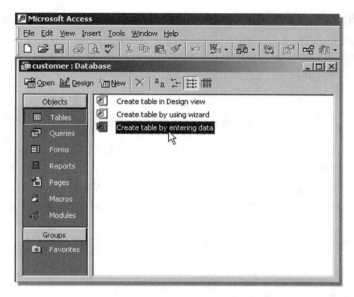

After double-clicking on the *Create table by entering data* item, Access displays
another window with a list of columns and rows, which is the table itself. Within
this new table you can simply click in a particular cell and type in the data from
Table 21.1. You can rename the column headings by right-clicking on them and
selecting Rename Column from the pop-up menu. The table will look similar to
Figure 21.2 when you finish inputting the data.

**Figure 21.2.
The Customers
table with the
data and column
headings**

Cust_ID	First_Name	Last_Name	Interest
379648	John	Smith	Dog breeding
543678	Judy	Parker	Karate
543679	Andrew	Massie	Aviation
735233	Margaret	Doe	Baking
899454	Jason	McDuff	Boxing

When you close the table, Access will prompt you to save the changes to the
table. Click on the Yes button and then save the table as *customers*. If you are
prompted about setting up a primary key, simply select the No button at this
stage. You will look at primary and foreign keys later in this lesson.

Finally, you can close Access completely, saving the *customer.mdb* database
file. For the remainder of this lesson, you will be working with this database

that you have set up. By using SQL statements and theory, you will see how to manipulate a database such as the one you have just created, and in the next two lessons you will actually put the theory into practice from your Active Server Pages.

Working with the Basics of the SELECT Statement in SQL

The SELECT statement in SQL is very different from that in VBScript. You use the SELECT statement to select data to retrieve from a database table. This type of statement is known as a query because you are querying the database for information. The most basic syntax for the SELECT statement is as follows:

```
SELECT * FROM tablename;
```

The SELECT keyword and the FROM keyword do not change. The asterisk (*) character indicates a column wildcard, which means that all columns are to be retrieved. The tablename argument indicates the name of the table in the database. Remember that a single database can contain multiple tables. The SELECT statement ends with a semicolon (;) character. In the *customer.mdb* database that you created, you gave the table the name *customers*, so to retrieve all the data from that table you would use the following SQL code:

```
SELECT * FROM customers;
```

You may not always want to retrieve all of the columns that a table contains, especially if the table is large and the number of columns quite high. In these cases you can specify the individual column header. You can have multiple column headers separated by a commas, as shown in the following syntax:

```
SELECT col1, col2, …, coln FROM tablename;
```

To show this concept more clearly, you can once again use the *customer.mdb* database. Imagine that you want to retrieve only the customer's first name, perhaps for a personalized e-mail, as well as their special interest. The table *customers* contains the column headers First_Name and Interest that meet your requirements so you could write your SQL code as follows:

```
SELECT First_Name, Interest FROM customers;
```

Working with the WHERE Clause in SQL (Conditional Queries)

You can also have conditional queries in your SELECT statements by using the WHERE clause. You use the WHERE clause in one of your SELECT statements, together with a column header and an operator, to place a conditional limit on the data that is returned. The WHERE clause makes use of one of six generic operators that are available in SQL, as shown in Table 21.2.

Operator	Description
=	Equal
<> or !=	Not equal
<	Less than
>	Greater than
<=	Less than or equal to
>=	Greater than or equal to

Table 21.2. Conditional operators in SQL

For instance, if you wanted to only see the data of customers whose customer identification numbers were less than 600,000, you would write something similar to the following:

```
SELECT * FROM customers WHERE Cust_ID < 600000;
```

This would basically return the following data from your customer database:

```
379648        John Smith          Dog breeding
543678        Judy Parker         Karate
543679        Andrew Massie       Aviation
```

You can also combine multiple conditions by using logical operators (AND, OR, NOT). The AND operator requires that both conditions are met. The OR operator requires that either one or both conditions are met. The NOT operator requires that a condition not be met.

An example of using multiple conditions might be if you require all customers whose customer identification numbers are less than 600,000, and whose interest is karate. As your SQL queries grow in size, you can split them onto multiple lines without the need for special characters, unlike VBScript, as you can see in the following example:

```
SELECT * FROM customers
WHERE Cust_ID < 600000 AND Interest = 'Karate';
```

The result of this example would simply be as follows:

```
543678          Judy Parker          Karate
```

You can also use parentheses to add clarification. If you were to query for all customers who have a customer identification number less than 600,000, but whose interest is not karate, you could use something similar to the following example:

```
SELECT * FROM customers
WHERE NOT (Interest = 'Karate') AND (Cust_ID < 600000);
```

This would basically return the following data from your customer database because customer 543678 has a karate interest and is excluded:

```
379648          John Smith           Dog breeding
543679          Andrew Massie        Aviation
```

Understanding Logical Operator Precedence

The logical operators also have their own precedence. This basically means that the NOT operator will be processed first, followed by the AND operator, and finally the OR operator. It is good to keep this in mind when working with your SQL queries, as you can sometimes overlook this fact and wonder why the results are not what you expect them to be.

You can use the IN and BETWEEN keywords instead of the combinations of the operators in Table 21.2. The IN keyword can take multiple values separated by commas and is equivalent to multiple = operator conditions. For instance, the following example shows the traditional way to retrieve all the customers who have John, Judy, or Jason as their first name:

```
SELECT * FROM customers
WHERE First_Name = 'John' OR First_Name = 'Judy' OR
First_Name = 'Jason';
```

Using the IN keyword, the exact equivalent SQL statement would be written as follows:

```
SELECT * FROM customers
WHERE First_Name IN ('John', 'Judy', 'Jason');
```

In the same way, the BETWEEN keyword can also save you work and make your SQL queries more understandable and easier to read. The BETWEEN keyword replaces a < operator and a > operator combination. For example, the following SQL query will find all customers who have a customer identification number greater than 500,000 but less than 600,000:

```
SELECT * FROM customers
WHERE (Cust_ID > 500000) AND (Cust_ID < 600000);
```

Using the BETWEEN keyword, the exact equivalent SQL statement would be written as follows:

```
SELECT * FROM customers
WHERE Cust_ID BETWEEN 500000 AND 600000;
```

When you do not want to specify the whole value of a column but merely a part of it, you can use the LIKE keyword and a % character as a wildcard. For instance, if you require all the customers whose first name begins with the letter J, you would not be able to easily do this with the = operator or the IN keyword. However, by using the LIKE keyword you can easily accomplish this feat, as shown in the following example:

```
SELECT * FROM customers
WHERE First_Name LIKE 'J%';
```

This example would return the information for the customers John, Judy, and Jason. You can also put the wildcard in other positions in relation to required characters. For example, if you want to see all the customers whose interest ends in the letter g, you could simply use the following SQL code where the wildcard precedes the character g:

```
SELECT * FROM customers
WHERE First_Name LIKE '%g';
```

This example would return the customers whose interests include dog breeding, baking, or boxing, as well as any other interests that you might add later that end in the letter g.

Understanding Primary and Foreign Keys

A primary key in a table is a column that contains unique data to identify all other data in a given row. For instance, if you had a customer table with twenty customers called John Smith, you could uniquely identify each of them using a unique customer identification number. You would make the column containing the unique customer identification number the primary key for the table.

A foreign key is nothing more than an ordinary column in a table, which happens to be a primary key in another table. In database management systems this is known as referential integrity because you can refer to the corresponding table to gather further data. As an example and to add clarity, imagine if you had a second table that was for *sales*. Each time you had a sale, you create a unique sales identification number (your primary key), as well as the amount of the sale and other information. One column may be the customer identification number (your foreign key). To find out the actual name of the customer, you could use the foreign key from the *sales* table in the *customers* table. Figure 21.3 shows you what the *sales* table might look like, where the Cust_ID column is actually a foreign key:

Figure 21.3.
A simple table
for sales, with
Cust_ID column
as a foreign key

Sale_ID	Cust_ID	Product_ID	Price	Qty
3789	379648	342	9.99	1
3790	899454	458	57.95	1
3791	899454	463	34.95	1
3792	379648	342	9.99	2
3793	543678	121	34.95	1
3794	735233	299	5.95	5

Sales : Table

Record: 1 of 6

Creating a Primary Key in Microsoft Access

Earlier, you saw how to create a table of data using Microsoft Access when you were entering in data manually. If you want to make the Cust_ID column the primary key for this table, you must change to Design View. To change to Design View, simply click on the View menu Design View item. Once you are in Design View, you can right-click on the Cust_ID item, and from the pop-up menu select Primary Key. Alternatively, you can select the Cust_ID item by clicking on it, and then selecting the Edit menu Primary Key item.

Joining Multiple SQL Queries

When you want to combine the results of two or more SQL queries you can use the UNION keyword. The UNION keyword is extremely useful when you are dealing with more than one table and require data partly from each table to provide the results. Basically, you simply type the UNION keyword between two distinct SELECT statements. The following example uses the UNION keyword to produce a single column of all the sales identification numbers and then all the customer identification numbers directly underneath them. Although not a practical use of the UNION keyword, it is simple enough for you to see its position in the SQL code and also understand the resultant union (or merging) of the two SELECT statements. The example SQL code is as follows:

```
SELECT Sale_ID
FROM sales
UNION
SELECT Cust_ID
FROM customers;
```

Eliminating Duplicates from SQL Queries

When you are dealing with multiple occurrences of the same data, such as all customers who have made purchases of a particular item, you may not want to see the same customer listed over and over again. You can avoid multiple occurrences in the results by using the DISTINCT keyword. For example, if you were to list all the customer identification numbers in the sales table, you would get one line of result for every sale, using SQL code similar to the following:

```
SELECT Cust_ID FROM sales;
```

By using the DISTINCT keyword, you can cut down the result to clearly show one distinct entry for each customer, using SQL code similar to the following:

```
SELECT DISTINCT Cust_ID FROM sales;
```

Ordering the Results of an SQL Query

You can order the results of your SQL queries alphabetically by using the ORDER BY clause. The ORDER BY clause comes at the end of your SELECT statement and is followed immediately by the column header that you want to order the results by. For example, if you want to order all of the data in the customers table by each customer's last name, you would use code similar to the following:

```
SELECT * FROM customers ORDER BY Last_Name;
```

The result of this code would be as follows:

```
735233 Margaret     Doe         Baking
543679 Andrew       Massie      Aviation
899454 Jason        McDuff      Boxing
543678 Judy         Parker      Karate
379648 John         Smith       Dog breeding
```

Working with Aggregate Functions in SQL

SQL also provides aggregate functions to use with your queries. Aggregate functions do not return the actual data but rather they summarize the results of the data. For instance, if you did a query that returns four rows of data, using the COUNT function would return simply 4, but not the actual data contained in the four rows. The five aggregate functions available to you in SQL are shown in Table 21.3.

Function	Description
AVG	Returns the average of the given column
COUNT	Returns the number of rows satisfying the conditions
MAX	Returns the largest figure in the given column
MIN	Returns the smallest figure in the given column
SUM	Returns the total of all the rows, satisfying any conditions, of the given column, where the given column is numeric

Table 21.3. The five aggregate functions in SQL

As an example, to return the total number of customers in the *customers* table, you could use the COUNT function on the Cust_ID column, as follows:

```
SELECT COUNT(Cust_ID) FROM customers;
```

The result of this SQL code would be 5, using the example table you created earlier.

Working with Table Data from SQL

You can use SQL to manipulate the tables in a database to a great extent. Some of the more useful functionality is the ability to add data to a table, delete data from a table, and change or update data in a table.

To add data to a table, you can use the INSERT INTO keywords, followed immediately by the name of the table. You then follow this by the VALUES keyword and then supply each value for the columns, separated by commas. For example, if you want to add a new customer by the name of Frank Diaz who has an interest in music, you could use the following code:

```
INSERT INTO customers VALUES (902767, 'Frank', 'Diaz', 'Music');
```

Deleting data from a table is just as easy. To delete data from a table, you use the DELETE FROM keywords followed immediately by the name of the table. You then use the WHERE clause in the same way as you do with a SELECT statement. You must be careful not to delete too much data. For instance, if you want to delete the entry for Frank Diaz, you cannot simply delete customers whose first name is Frank because there may be many of them in the database. Remember to be precise when you are deleting data, or you may cause yourself a lot of unnecessary grief.

The correct code for deleting the entry for the customer Frank Diaz would be to use the customer identification number, as this is unique. In this manner even if there were multiple customers by the name of Frank Diaz, only the one that you want to remove will be deleted. The following example shows how you might write SQL code to delete the customer Frank Diaz from the database:

```
DELETE FROM customers WHERE Cust_ID = 902767;
```

To change data that is in a table, you use the UPDATE keyword. The UPDATE keyword is followed by the name of the table that you are updating. You then use the SET keyword and follow it with the column header that you are changing, followed in turn by the equals sign (=) character and the value of the data that you want the column to now hold. You should now use the WHERE clause again to be more specific about which items to change, otherwise all the items in that column will be changed. An example will make this clearer.

Imagine that you want to change all the entries in the *customers* database from the first name of John to Joe. You could write SQL code such as the following to accomplish this:

```
UPDATE customers SET First_Name = 'Joe'
WHERE First_Name = 'John';
```

WHAT YOU MUST KNOW

Structured Query Language (SQL) is a declarative language for working with data that is stored in relational database tables. You simply declare what you want and let the database itself handle the details of how to accomplish the task. You can use the SELECT statement along with optional keywords to query databases for specific information. You can also use SQL to manipulate the data in tables as well, although this is not the limit of what you can do with SQL. You should really look more deeply into this subject if you plan to use it to its capacity. In Lesson 22, "Using ActiveX Data Objects (ADO) Part I," you will learn to break out of theory and actually apply your knowledge of SQL by using it with ActiveX Data Objects (ADO) to manipulate databases from within your Active Server Pages. Before you begin with Lesson 22, make sure that you are very familiar with all of the following concepts, as you will be directly using these concepts in the lessons to come:

- SQL works with tables that are stored in relational database management systems, such as Oracle, SQL Server, and Access. Tables are made up of rows and columns of cells containing data.

- You use the SELECT statement in SQL to query relational databases for specific information.

- After the SELECT keyword, you indicate either specific table column headers or an asterisk character for all columns. You then use the FROM keyword and specify the table that you want the information from.

- To use conditional queries in your SQL code, you use the WHERE clause. This allows you to target specific data only, rather than returning every single row.

- To combine multiple conditions in your SQL queries, you use logical operators, such as AND, NOT, and OR. However, as your queries become increasingly complex, you can use the BETWEEN and IN keywords to help shorten them.

- When you want to retrieve all data that has something specific in common but may not be exactly the same, you can use the LIKE keyword, together with the % character as a wildcard.

- A primary key indicates unique data that you can use to identify all data in the same row. A foreign key is a primary key in a table other than the current table.

- To join the results of multiple queries together, you use the UNION keyword between SELECT statements.

- The DISTINCT keyword allows you to stop repetitions of data occurring in your SQL query results.

- To sort the results of your SQL queries, you must use the ORDER BY keywords.

- The ADD, COUNT, MIN, MAX, and SUM functions are SQL's aggregate functions. These aggregate functions return a summary of the data results rather than the data itself.

- To manipulate the actual data in tables, you can use the INSERT INTO keywords to add new data, the DELETE FROM keywords to delete data, and the UPDATE SET keywords to change existing data.

myValue = 7

Set ... ver.CreateObject("ADODBConnection")
adoCon.open "provider=Microsoft.Jet.OLEDB.4.0;Data source="
& dbPath

USING ACTIVEX DATA OBJECTS (ADO), PART 1

n Lesson 21, "Introducing Structured Query Language (SQL)," you learned how to work with databases using generic structured query language, as well as all the theory behind the various commands and options that SQL presents. In this lesson, you will learn about putting your knowledge of SQL into practice by using it with ActiveX Data Objects (ADO) to manipulate databases from within your Active Server Pages. By the time you complete this lesson you will have covered the following key concepts:

- ActiveX Data Objects (ADO) is just one part of Microsoft's Universal Data Access strategy.
- The ADO programming model comprises nine different objects including the Command, Connection, Error, Field, Parameter, Property, Record, Recordset, and Stream objects.
- The ADO Connection object provides a connection to the data source through which the other objects can access data or pass commands. You create an instance of the ADO Connection object using the CreateObject method and the Set statement.
- The Recordset object is a complex object that contains many methods and properties to work with. The Recordset object also contains a Fields collection and a Properties collection.
- The Connection object's Execute method returns an instance of the Recordset object. The Execute method takes a single text string for an argument, in which you can place your SQL query.

- You can also create an instance of the Recordset object using the CreateObject method and the Set statement. In this case, you would associate the Recordset with a Connection object using the ActiveConnection property.
- You can set a string of text holding an SQL query in the Source property of the Recordset object and then action this query when you use the Open method.
- You can limit the amount of data that is sent back to the user's browser in the response either by placing conditions on the loop when creating the table or by using SQL commands to actually reduce the results themselves.

Introducing ActiveX Data Objects (ADO)

You have already seen the theory for using some basic but functional SQL commands to work with databases. Using SQL from your Active Server Pages was not always easy and so Microsoft released Universal Data Access. The main goal of Universal Data Access is to make things easier and simpler for developers to access data without being language dependent or having to lock themselves into a particular vendor's software. Its focus is targeted for enterprise situations and large-scale corporations using open industry specifications with broad industry support for all major database platforms.

To enable Universal Data Access, you use the Microsoft Data Access Components (MDAC), which comprise of the following technologies shown in Table 22.1.

Technology	Description
ActiveX Data Objects (ADO)	A single, high-level interface that provides consistent, high-performance access to data
Remote Data Services (RDS)	For creating data-centric applications within ActiveX-enabled browsers (and technically a feature of ADO)
OLE DB	An open specification providing a low-level interface for accessing data
Open Database Connectivity (ODBC)	A widely accepted application programming interface that uses SQL as its database access language

Table 22.1. The main technologies comprising the Microsoft Data Access Components

In this lesson and the next, you will be working with ActiveX Data Objects (ADO), which is just a part of the Microsoft Data Access Components (MDAC) and Microsoft's overall Universal Data Access strategy. The most common version of MDAC in use at the time of writing is version 2.6, which is available from the following Microsoft site:

http://www.microsoft.com/data/download.htm

Note: *Microsoft also has some information concerning MDAC version 2.7 on their site, as well as ADO 2.7, and you will learn more about this later in the book in sections dealing with ADO.NET and ASP.NET.*

ADO is not really a new technology but more of an amalgamation of earlier data models from Microsoft, coupled with some new features. Microsoft basically used the best parts of Data Access Objects (DAO) and Remote Data Objects (RDO) to create ADO. From this standpoint, if you are familiar with either of these two older data models then you will immediately feel comfortable with ADO. ADO has also been restructured and modified by Microsoft throughout its lifetime, evolving into a useful and up-to-date technology. You can program using ADO from languages such as Visual Basic, VBScript, C++, Java, and JScript, due to ADO not being language specific. This makes ADO very attractive to developers, including Active Server Pages application developers.

The ADO object model comprises nine different objects, as shown in Table 22.2.

Object	Identifier	Description
Command	ADODB.Command	Takes a query or statement, such as SQL, that is sent to the data source for processing
Connection	ADODB.Connection	Provides a connection to the data source
Error	ADODB.Error	Handles multiple errors returned by the data source
Field	ADODB.Field	An interface to a column in the Recordset for obtaining and modifying values
Parameter	ADODB.Parameter	An optional object for holding parameters of the query in the Command object
Property	ADODB.Property	Custom property for handling properties from data source that are not already in ADO
Record	ADODB.Record	An interface to a row in the Recordset for obtaining and modifying values
Recordset	ADODB.Recordset	A complex object that provides a large interface of methods and properties for data manipulation
Stream	ADODB.Stream	An interface providing stream access, such as saving data to disk

Table 22.2. The objects in the ADO programming model

One of the most important of these objects is the Recordset object, which you will work with quite a lot throughout this lesson and the next. You can see graphically how the objects in the ADO programming model relate to each other in Figure 22.1. Although there are only nine objects in the ADO object model, some of them are associated with multiple objects and therefore appear numerous times. There are also items that represent collections of singular objects. For instance, the item Fields in the ADO object model is really just a collection of singular Field objects.

**Figure 22.1.
The ADO object
model**

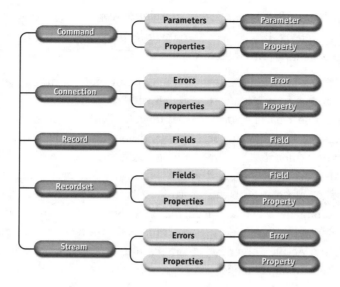

Creating an Instance of the ADO Connection Object

The ADO Connection object is necessary for any work with ADO from your Active Server Pages, and is probably the first object that you will require an instance of. The ADO Connection object provides a connection to the data source through which the other objects can access data or pass commands. To create an instance of the ADO Connection object, you use the CreateObject method from the intrinsic Server object, together with the Set statement. You must also use the string *ADODB.Connection*, as the CreateObject method's argument, as shown in the following example:

```
<%
'Create an instance of ADO Connection object
Dim adocon
Set adocon = Server.CreateObject("ADODB.Connection")
%>
```

Introducing the Recordset Object in ADO

The Recordset object can basically be thought of as a big container for all the records that come back as a result of a query. The Recordset object is a complex object that contains many methods and properties. These methods and properties are basically provided to make common tasks that developers perform easier. Due to the large number of these properties and methods, they cannot all be discussed here in detail, although you can purchase quite a number of books that are dedicated entirely to ADO, which do go into detail. Using the Recordset in ADO can be a challenging task; however, this book will keep the use of the Recordset fairly straightforward, explaining the methods and properties as you come to them.

From Active Server Pages, you create an instance of the Recordset object with which you can pass SQL statements and queries to the data source via the Connection object. Simply put, the ADO Connection object's Execute method will return a Recordset object, on the provision that rows of data result from the Execute method.

For an overview of the methods that are contained in the Recordset object, you can browse over the list in Table 22.3.

Name	Description
AddNew	Creates a new record for an updateable Recordset object
Cancel	Cancels a pending Open operation
CancelBatch	Cancels a pending batch update
CancelUpdate	Cancels any changes made to a current record or to a new record prior to calling the UpdateBatch method
Clone	Creates a duplicate Recordset object from an existing Recordset object
Close	Closes an open object and any dependent objects
CompareBookmarks	Checks equality of two bookmarks
Delete	Deletes the current record in an open Recordset object or an object from a collection
Find	Searches for records that match search criteria
GetRows	Retrieves multiple records of a Recordset into an array
GetString	Retrieves a Recordset as a string
Move	Moves the position of the current record in a Recordset object
MoveFirst	Moves to the first record in a specified Recordset
MoveLast	Moves to the last record in a specified Recordset
MoveNext	Moves to the next record in a specified Recordset
MovePrevious	Moves to the previous record in a specified Recordset
NextRecordset	Moves to next Recordset for when multiple commands return Recordsets
Open	Opens a cursor on a Recordset
Requery	Updates the data in a Recordset object by reexecuting the query on which the object is based (equivalent to calling the Close and Open methods in succession)
Resync	Refresh data in current Recordset
Save	Saves a Recordset in a file or Stream object
Seek	Search current index for specified value
Supports	Determines whether a specified Recordset object supports a particular type of function
Update	Saves any changes you make to the current record of a Recordset object
UpdateBatch	Writes all pending batch updates to disk

Table 22.3. Various methods of the Recordset object

For an overview of the properties that are contained in the Recordset object, you can browse over the list in Table 22.4.

Name	Description
AbsolutePage	Determines the exact page in which a record resides
AbsolutePosition	Ordinal number of current record in Recordset
ActiveCommand	Returns the Command object that created the specified Recordset
ActiveConnection	Sets or returns the Connection object that the specified Recordset object currently belongs
BOF	Indicates whether the current record position is before the first record in a Recordset object
Bookmark	Returns a bookmark that uniquely identifies the current record in a Recordset object or sets the current record in a Recordset object identified by a valid bookmark
CacheSize	Sets or returns the number of records from a Recordset object that are cached locally in memory
CursorLocation	Sets or returns the location of the cursor (whether the cursor is on the client or the server side)
CursorType	Sets or returns the type of cursor used in a Recordset object. Only the adOpenForwardOnly CursorType is supported by the current version of the OLE DB Provider for DB2
DataMember	Specifies the name of the data member to obtain from the data source
DataSource	A source of data that can return data in the form of a Recordset
EditMode	Indicates the editing status of the current record type
EOF	Indicates whether the current record position is after the last record in a Recordset object
Filter	Filter records within Recordset so they are not seen
Index	Name of index that Recordset is using with data
LockType	Sets or returns the types of locks placed on records during editing. The OLE DB Provider for DB2 supports locks of type adLockReadOnly and adLockPessimistic
MarshalOptions	Settings for packaging and sending records between server and client
MaxRecords	Sets or returns the maximum number of records to return to a Recordset from a query
PageCount	Determines the number of pages of records returned
PageSize	Total number of records that make up one page
RecordCount	Exact number of records returned or −1 if unsupported
Sort	Specifies a field name to sort records by
Source	Sets or returns the source (table name or command object) for the data in a Recordset
State	Describes the current state of an object
Status	Indicates the status of the current record with respect to batch updates or other bulk operations
StayInSync	Boolean value indicating if parent row changes when child row changes

Table 22.4. Various properties of the Recordset object

The Recordset also contains two collections—the Fields collection and the Properties collection. The Fields collection contains a Field object for each data column in the Recordset object. The Properties collection contains any properties that are custom (nonstandard) that you can use with the specific data provider (i.e. Oracle, SQL Server, etc.).

Working with Simple Queries

You now have enough theory about using SQL, as well as a fairly basic understanding of ADO and where it fits into the grand scheme of things within Microsoft Data Access Components (MDAC) and Microsoft's overall Universal Data Access strategy. To add some clarity to all this theory, you will now return to Active Server Pages scripts and actually start working with a database.

The first example you should try is a simple query to a database, the result of which you will display in a table within the user's browser. You will use the Microsoft Access Customer database that you created in the previous lesson, although if you are confident, feel free to try similar Active Server Pages scripts with other databases, using this example as a guide.

To understand the script you should break it down into sections, stating what each section is doing. Firstly you state that you are using VBScript and that variables must be declared. You then create a standard HTML header section. In the body section you create some variables. So far this is standard script that you already know.

Now you want to assign the full path and filename of the Microsoft Access database file to a variable. For trouble-shooting purposes you might like to output the result of this variable assignment to the browser. The code snippet would appear as follows:

```
'Assign the database physical path to a variable
dbPath = Server.MapPath("customer.mdb")
Response.Write dbPath & "<br><br>"
```

Next you create an instance of the ADO Connection object. This is something that you have already seen earlier in this lesson. However, you will also use the Connection object's Open method, stating that you want to use a Microsoft Jet database. You will also pass as an argument the path to the database file itself, which you prepared earlier. The Open method takes a single text string as its argument, so all this information must go into a string. The snippet of code for accomplishing all this is as follows:

```
'Create an instance of ADO Connection object
Set adocon = Server.CreateObject("ADODB.Connection")
adocon.Open "Provider=Microsoft.Jet.OLEDB.4.0;Data Source="
& dbPath
```

You now use the Connection object's Execute method to pass an SQL query to the data source for processing. The result of the Execute method is an instance of the Recordset object, so you use the Set statement. The SQL code that you supply as an argument to the Execute method must be a string. The closing semicolon is optional in this string, although you may want to use it with SQL queries of greater complexity. This code snippet shows you all that is necessary to query the database for everything in the customers table:

```
'Execute a SQL query and store the results within a Recordset
Set recset = adocon.Execute("SELECT * FROM customers")
```

Now that you have an instance of the Recordset object containing the results of your query, you need some method of relating that information back to the user. Creating an HTML table is a simple yet effective option for accomplishing this task, and you will probably agree that it is an obvious choice, as you are displaying table data after all. Combining your HTML table with a loop that iterates through each row of the Recordset object completes the picture, one might say.

Although you know about loops in Active Server Pages, you have to learn about iterating through rows of the Recordset object. Moving the focus to the next row (or record) is made simple for you by the Recordset object's MoveNext method. The focus is on the first record automatically, although you can use the Recordset object's MoveFirst to explicitly state this if you want to. You do not have to supply any argument to either of these methods. You can continue looping until the end of the file is reached, which is the last record as far as the Recordset object is concerned. The Recordset object's EOF property is True if the last record has already been reached and False otherwise.

The Recordset also accesses the Fields collection, as you saw from the ADO programming model earlier in this lesson. It might help you to think of the fields as representing columns from the original table. Remember that you do not always retrieve all the columns from the table whenever you make an SQL query. For this reason, you should use the Fields collection's Count property rather than a hard-coded value in your loop. To actually display the value of an individual cell from the table, you simply indicate after the Recordset object the number of the field from the current row in parentheses.

The code that displays the HTML table, looping through each field of each row, is as follows:

```
<!-- Create a table to display the results in -->
<table border = 1>
<%Do while (Not recset.eof) %>
<tr><%
For i=0 to (recset.fields.count-1)
   Response.Write "<td>" & recset(i) & "</td>"
Next
%></tr>
<%recset.MoveNext
Loop%>
</table>
```

Finally, you use the Close method of the Recordset object, as well as the Close method of the ADO Connection object, to cleanly free up any resources held by the objects or their dependents, although the objects themselves remain in memory. You should close the objects in the reverse order that they were created. For example, you would always close the Recordset object before the Connection object. The following script snippet shows you an example of closing the objects:

```
<%
'Clean up the objects
recset.close
adocon.close
%>
```

The entire example script appears as follows:

```
<%@ LANGUAGE = VBScript %>
<% Option Explicit %>

<html>
<head>
<title>Simple Database Query</title>
</head>
```

```
<body>
<h1>Simple Database Query</h1>
<b>Database path: </b>

<%
Dim adocon
Dim recset
Dim dbPath
Dim i

'Assign the database physical path to a variable
dbPath = Server.MapPath("customer.mdb")
Response.Write dbPath & "<br><br>"

'Create an instance of ADO Connection object
Set adocon = Server.CreateObject("ADODB.Connection")
adocon.Open "Provider=Microsoft.Jet.OLEDB.4.0;
Data Source=" & dbPath

'Execute a SQL query and store the results within a Recordset
Set recset = adocon.Execute("SELECT * FROM customers")
%>

<!-- Create a table to display the results in -->
<table border = 1>
<%Do while (Not recset.eof) %>
<tr><%
For i=0 to (recset.fields.count-1)
  Response.Write "<td>" & recset(i) & "</td>"
Next
%></tr>
<%recset.MoveNext
Loop%>
</table>

<%
'Clean up the objects
recset.close
adocon.close
%>
</body>
</html>
```

Save the script as *simplequery.asp* and make sure that the *customer.mdb* database file is in the same directory. Loading the file *simplequery.asp* into your browser will result in a display similar to that shown in Figure 22.2. You can see the path of your database and all of the contents of the *customers* table.

Figure 22.2.
The result of loading the *simplequery.asp* file in a browser

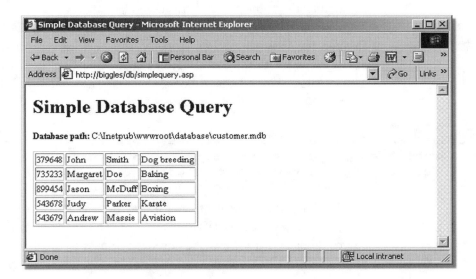

Note: *So as not to detract from the overall focus of the lesson, the HTML table that is used in the examples is extremely simple and straightforward. This makes the code that you are learning easier to spot by not burying it amongst screens of code. In practice you might put a lot more effort into your visual display, adding consistent shading, column headings, alternate shades for each row, or many other finishing touches.*

Limiting Query Results for Manageability

So far you have been working with a very small database and set of data. However, you can imagine that in the real world you will have to deal with huge volumes of data. You probably do not want to send all the data in a response to the user, as this could be extremely slow and time consuming, especially if the Active Server Page is accessed from the Internet over a slow dial-up modem. This can often be the case when company sales representatives are on the road and in need of information, or corporate consultants need to access data to work from home.

The next example will show you an example of reading in a database but limiting the response to the user to only three rows of data. The code in this example is very similar to the previous example, however, there is a change to the manner of creating an instance of the Recordset object. Rather than relying on the Connection object's Execute method, you will this time create your own instance of the Recordset object and then associate it to the Connection object through the ActiveConnection property. You must use the string *ADODB.Recordset* as the argument to the Server object's CreateObject method. The following script snippet shows how all this is done, assuming that you already have an instance of the ADO Connection object assigned to the variable *adocon*:

```
'Create an instance of ADO Recordset object
Set recset = Server.CreateObject("ADODB.Recordset")
Set recset.ActiveConnection = adocon
```

You limit the number of rows by setting the PageSize property to specific value. In the example PageSize is set to 3. You then set the Recordset object's Source property with the string you want to pass to the data source provider for processing. In this example, you will use an SQL query and pass it to Microsoft Access for processing. Finally, you simply increment a variable during the loop and conditionally continue while this variable is less than the PageSize property.

The entire example is as follows:

```
<%@ LANGUAGE = VBScript %>
<%  Option Explicit       %>

<html>
<head>
<title>Limit Database Query</title>
</head>

<body>
<h1>Limit Database Query</h1>
<b>Database path: </b>

<%
Dim adocon
Dim recset
Dim dbPath
Dim i
```

```
'Assign the database physical path to a variable
dbPath = Server.MapPath("customer.mdb")
Response.Write dbPath & "<br><br>"

'Create an instance of ADO Connection object
Set adocon = Server.CreateObject("ADODB.Connection")
adocon.Open "Provider=Microsoft.Jet.OLEDB.4.0;
Data Source=" & dbPath

'Create an instance of ADO Recordset object
Set recset = Server.CreateObject("ADODB.Recordset")
Set recset.ActiveConnection = adocon

'Limit the number of rows
recset.PageSize = 3

'Set the SQL query and open the Recordset
recset.Source = "SELECT * FROM customers"
recset.Open
%>

<!-- Create a table to display the results in -->
<table border = 1>
<%
Dim limit
Do while (Not recset.eof) and (limit < recset.PageSize) %>
  <tr>
  <%
  For i=0 to (recset.fields.count - 1)
    Response.Write "<td>" & recset(i) & "</td>"
  Next
  %>
  </tr>
  <%
  recset.MoveNext
  limit = limit + 1
Loop
%>
</table>

<%
'Clean up the objects
```

```
recset.close
adocon.close
%>
</body>
</html>
```

Save the script as *limitquery.asp* and then load it into your browser. Providing the *customer.mdb* database file is in the same directory, the result should look something similar to Figure 22.3.

Figure 22.3. Loading the *limitquery.asp* file into a browser

Using Other Methods to Limit Query Results

There is an old saying that all paths lead to Rome. There are many methods that you can use, all of which result in limiting the results of your database queries. You have seen one so far in detail using loops to limit the response. Shortly, you will see another method that uses SQL commands to limit the results of a database query. In the next lesson you will learn about limiting the results to a scrollable view in the user's browser.

If you are interested in yet further methods to limit the results of a database query, you can look into the Recordset object's MaxRecords property. Basically the MaxRecords property is 0 by default, meaning that all results are returned from the query. However, when the Recordset is closed, you can set the MaxRecords property to a specific value, thereby limiting the number of records that are returned from the database query.

If you look closely at the previous example, you can see that although you successfully send only a limited amount of rows in the response to the user, you were actually still querying for the entire data (all columns), as can be seen by the asterisk (*) character. As an alternative, you can also use your SQL skills to limit the result of the database query and then respond to the user with all of the results. In other words, instead of retrieving the whole table and then limiting what is sent back to the user, make use of your SQL skills when possible to retrieve just the portion that you want when you send your query.

The following example also limits the query; however, this time using SQL code to achieve the limitation. Although simple, you can see how to limit the results to just those rows where the customer's identification number is between 500,000 and 800,000 largely brought about by the following snippet of code:

```
SELECT * FROM customers
WHERE Cust_ID BETWEEN 500000 AND 800000
```

The example also shows how to use a variable for holding your SQL commands as they get larger and more complex, which adds to the clarity of your code. Basically, by assigning the entire SQL command to a string, you can then simply pass that string as the argument to the Execute method of the ADO Connection object. The example shown in its entirety is as follows:

```
<%@ LANGUAGE = VBScript %>
<%   Option Explicit        %>

<html>
<head>
<title>Limit Database Query Using SQL</title>
</head>

<body>
<h1>Limit Database Query Using SQL</h1>
<b>Database path: </b>

<%
Dim adocon
Dim recset
Dim dbPath
Dim sqlCmd
Dim i
```

```
'Assign the database physical path to a variable
dbPath = Server.MapPath("customer.mdb")
Response.Write dbPath & "<br><br>"

'Create an instance of ADO Connection object
Set adocon = Server.CreateObject("ADODB.Connection")
adocon.Open "Provider=Microsoft.Jet.OLEDB.4.0;
Data Source=" & dbPath

'Prepare the SQL command as a string
sqlCmd ="SELECT * FROM customers "
sqlCmd = sqlCmd & "WHERE Cust_ID BETWEEN 500000 AND 800000"

'Execute a SQL query and store the results within a recordset
Set recset = adocon.Execute(sqlCmd)
%>

<!-- Create a table to display the results in -->
<table border = 1>
<%Do while (Not recset.eof) %>
<tr>
<%
For i=0 to (recset.fields.count-1)
  Response.Write "<td>" & recset(i) & "</td>"
Next
%>
</tr>
<%
recset.MoveNext
Loop
%>
</table>

<%
'Clean up the objects
recset.close
adocon.close
%>
</body>
</html>
```

Save the script as *limitquery_sql.asp* and then load it into your browser. The result should appear similar to Figure 22.4 showing a limited set of the entire table. You can see clearly that only the rows where the customer identification number was between 500,000 and 800,000 were returned.

Figure 22.4.
The result of loading
limitquery_sql.asp
into your browser

WHAT YOU MUST KNOW

ActiveX Data Objects (ADO) is just one part of Microsoft's Universal Data Access strategy and is actually a component of Microsoft Data Access Components (MDAC). You can use the simple interface of ADO to connect to different types of data sources without having to change your approach and can also use different languages with ADO, as well. Using an ADO Connection object and a Recordset object, you can easily connect to a database, query it, and then display the results to the user. In the Lesson 23, "Using ActiveX Data Objects (ADO), Part 2," you will continue working with ADO, looking at more of the properties and methods of the Recordset object, adding rows to tables, deleting rows from tables, and changing rows in tables. Before you continue with Lesson 23, make sure you fully understand the following key concepts:

- There are nine different objects in the ADO object model. Although there are some repetitions, the main objects are the Command, Connection, Error, Field, Parameter, Property, Record, Recordset, and Stream objects.

- Objects can access data from the data source by going through the ADO Connection object. Using the CreateObject method and the Set statement, you can create an instance of the ADO Connection object.

- The Recordset object has lots of methods and properties that you can use to manipulate the data source. The Recordset object is basically a big container for holding the results of your data source queries and through it you can also access those results for displaying or other purposes.

- You can use the Connection object's Execute method, together with the Set statement, to return an instance of the Recordset object. You can pass an SQL query in the form of a string as an argument to the Execute method.

- Using the intrinsic Server object's CreateObject method together with the Set statement, you can create an instance of the Recordset object.

- You associate the Recordset with a Connection object using the ActiveConnection property.

- The Recordset object's Source property can take an SQL query as an argument, but this is only processed when you call the Recordset's Open method.

- By placing conditions on the loop that is used when creating the table of results, you can limit the amount of data that is sent back in the response. Alternatively, you can use SQL commands to actually reduce the results themselves.

USING ACTIVEX DATA OBJECTS (ADO), PART 2

*i*n Lesson 22, "Using ActiveX Data Objects (ADO), Part 1," you learned about combining ActiveX Data Objects (ADO) with your Structured Query Language (SQL) skills to access a database from within your Active Server Pages. In this lesson, you will continue directly with this line of functionality, adding to your repertoire of ADO skills by learning about adding, deleting, and changing data, as well as working with a few other properties and methods in the Recordset object. By the time you finish this lesson, you will have covered the following key concepts:

- You can use generic SQL code, making use of the INSERT INTO keywords, that you assign to a variable in the form of a string, together with the ADO Connection object's Execute method, to simply add data to a table.
- You can use the Recordset object's AddNew method to add data to a database table. You can supply no arguments and specify the values for the individual column headers or you can use arrays as arguments.
- Depending on whether you are using immediate mode or batch mode, the actual point at which the data is written to the database is different. During immediate mode, the data is written immediately after the call to the Update method, whereas in batch mode it is not written until the call to the BatchUpdate method.
- The Supports method lets you determine whether a particular method or property is supported by the underlying data source provider. You should include the *adovbs.inc* file into your Active Server Pages to make use of all the constants available to you.

- The LockType property locks records in various ways allowing you to make changes without corrupting the data. The LockType property is different depending on the mode that you are using.
- The Recordset object's Open method has five optional arguments, which you can use to shorten your code. The fifth argument is important as it informs the Open method of how to handle the Source property or argument.
- You can use generic SQL code, making use of the DELETE FROM keywords, that you assign to a variable in the form of a string, together with the ADO Connection object's Execute method, to simply delete data from a table.
- You can use generic SQL code, making use of the UPDATE SET keywords, that you assign to a variable in the form of a string, together with the ADO Connection object's Execute method, to simply change existing data in a table.

Adding Data to Database Tables

In Lesson 21, "Introducing Structured Query Language (SQL)," you learned about using SQL commands to add data to a database in theory. From Active Server Pages, you can use the objects of ADO to actually put this theory into practice. You already know how to use the Recordset to execute a string that contains SQL code. However, when you add data to a database, you will not actually receive any records. It is not a query so you do not expect any results to be returned in the form of rows of data. It is a straight SQL command. Therefore you will not use the Recordset but merely the ADO Connection object and its Execute method to add data to a database.

In the following example, you will use the Microsoft Access database that you created earlier in Lesson 21. You will be adding an imaginary customer to the *customers* table. You have already seen the SQL code for doing this, and now you only have to assign this SQL code to a variable as a string. You can then use this variable as the argument to the ADO Connection object's Execute method. Assuming you have an instance of the ADO Connection object assigned to a variable called *adocon* and another empty variable called sqlcmd to hold the SQL command string, the main code to accomplish adding a row of data to the *customers* table would appear as follows:

```
'Prepare SQL command and add new customer to database table
sqlcmd = "INSERT INTO customers VALUES "
sqlcmd = sqlcmd & "(902767, 'Frank', 'Diaz', 'Music');"
adocon.Execute(sqlcmd)
```

The Active Server Pages script to add a row of data to the *customers* table and then display the results, is shown in its entirety, as follows:

```
<%@ LANGUAGE = VBScript %>
<%   Option Explicit      %>

<html>
<head>
<title>Add Data to Database</title>
</head>

<body>
<h1>Add Data to Database</h1>
<b>Database path: </b>

<%
Dim adocon
Dim recset
Dim dbPath
Dim sqlcmd
Dim i

'Assign the database physical path to a variable
dbPath = Server.MapPath("customer.mdb")
Response.Write dbPath & "<br><br>"

'Create an instance of ADO Connection object
Set adocon = Server.CreateObject("ADODB.Connection")
adocon.Open "Provider=Microsoft.Jet.OLEDB.4.0;
Data Source=" & dbPath

'Prepare SQL command and add new customer to database table
sqlcmd = "INSERT INTO customers VALUES "
sqlcmd = sqlcmd & "(902767, 'Frank', 'Diaz', 'Music');"
adocon.Execute(sqlcmd)

'Execute an SQL query and store the results within a recordset
Set recset = adocon.Execute("SELECT * FROM customers")
%>

<!-- Create a table to display the results in -->
<table border = 1>
<%Do while (Not recset.eof) %>
<tr><%
For i=0 to (recset.fields.count-1)
  Response.Write "<td>" & recset(i) & "</td>"
Next
%></tr>
```

```
<%
recset.MoveNext
Loop
%>
</table>

<%
'Clean up the objects
recset.close
adocon.close
%>
</body>
</html>
```

Save the script as *add.asp* and then load it into your browser. Providing that the *customer.mdb* database is in the same directory as *add.asp,* the result should look similar to Figure 23.1.

**Figure 23.1.
The result of
loading *add.asp*
into a browser
to add another
customer**

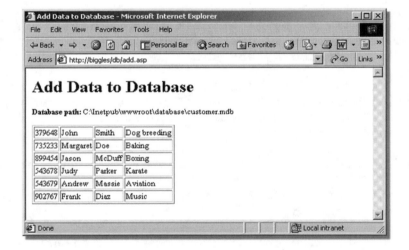

Introducing the Recordset Object's AddNew Method

Although Microsoft recommends using the SQL method of adding data to a table in a database, they have also supplied you with another method to accomplish this. The Recordset object's AddNew method also allows you to add data to a table in a database, providing that the database supports this. The syntax for the AddNew method is as follows:

recordset.AddNew fields, values

The two arguments for the AddNew method basically represent the name of the column in the table that you want to add data to (fields), as well as the value that you want to assign (values). However, working with the AddNew method introduces many other components of ADO that you must be aware of to be able to implement it confidently, so you will have to cover at least some of these sections before actually putting the AddNew method into practice.

Checking Support for Methods and Properties

Not all database management systems support all new methods of the Recordset object, including the AddNew method. You can use the Recordset object's Supports method to discover if a particular property or method is currently supported.

Constant	Value	Description
adAddNew	0x1000400	Determines whether the AddNew method can be used to add new records
adApproxPosition	0x4000	Determines whether the AbsolutePosition and AbsolutePage properties can read and set
adBookMark	0x2000	Determines whether the Bookmark property can be used to access specific records
adDelete	0x1000800	Determines whether the Delete method can be used to delete records
adFind	0x80000	Determines whether the Find method can be used to locate a row in a Recordset
adHoldRecords	0x100	Determines whether you can retrieve more records or change the next retrieve position without committing all pending changes
adIndex	0x100000	Determines whether the Index property is supported
adMovePrevious	0x200	Determines whether the current record position can be moved backwards without requiring bookmarks, which in turn indicates whether you can use the MoveFirst, MovePrevious, Move or GetRows methods
adNotify	0x40000	Determines that the underlying data provider supports notifications, which in turn determines whether Recordset events are supported
adResync	0x20000	Determines whether the Recordset cursor can be updated with the data visible in the underlying database using the Resync method
adSeek	0x200000	Determines whether the Seek method can be used to locate a row in a Recordset
adUpdate	0x1008000	Determines whether the Update method can be used to modify existing data
adUpdateBatch	0x10000	Determines whether the UpdateBatch and CancelBatch methods can be used to transmit changes to the data provider in groups (batches)

Table 23.1. The list of constants that you can use with the Supports method

To be able to make use of the constants in Table 23.1, you must include the file *adovbs.inc,* or explicitly create your own constants. You should find the *adovbs.inc* file in the following directory on your hard drive, although you can simply do a search to discover its location:

"C:\Program Files\Common Files\System\ado"

You can copy the *adovbs.inc* file to your root directory (or some other location if you want). To include the *adovbs.inc* file you would then simply add the following line to your Active Server Pages scripts:

```
<!-- #include virtual="\adovbs.inc" -->
```

To test whether there is support for the AddNew method, you can use the following pattern of code:

```
'Test first before adding data
If recset.Supports(adAddNew) then
    'You can add to the database
Else
    'AddNew method not supported by database
End If
```

Placing Your Supports Method Calls

It may not be obvious at first, but if you try to call a Recordset object's Supports method with the adAddNew constant before you have actually made a call to the Recordset object's Open method, the result will always be False and will give you the impression that the AddNew method is not supported by the data source provider when this may in fact not be the case. Even setting the ActiveConnection property to a Connection object will not change this. You must use the Supports method with the adAddNew constant only on open Recordset objects for a correct result.

Using the Recordset Object's LockType Property

When you are going to alter information held in a database using the AddNew method, you should first lock the records to prevent multiple changes from clashing or corrupting data. You lock the records in various ways using the Recordset object's LockType property. The Recordset object must be closed in order to set the LockType property. Table 23.2 lists the constants that you can use with the LockType property.

Constant	Value	Description
adLockReadOnly	1	Read-only records where the data cannot be changed
adLockPessimistic	2	Locks each record immediately after editing
adLockOptimistic	3	Locks each record only after the Update method is called
adLockBatchOptimistic	4	Locks each record only after the BatchUpdate method is called

Table 23.2. The list of constants that you can use with the LockType property

As with the Supports method, the constants that you require for the LockType property are in the *adovbs.inc* file, or you can create explicit constants if you do not want to include this file. The following one-line examples show how to set the LockType to adLockOptimistic, as well as adLockBatchOptimistic:

```
recset.LockType = adLockOptimistic
recset.LockType = adLockBatchOptimistic
```

Understanding Modes Using the EditMode Property

When you actually want to use the AddNew method, there are a number of different options and factors to take into account. The updating mode of the Recordset will determine the behavior of the AddNew method. The updating mode can be either in immediate mode or batch mode. Depending on how you use the AddNew method, you will automatically change the EditMode property. The Recordset object's EditMode property is an indication of whether the editing buffer held by the ADO has been changed. You can see in Table 23.3 the constants that the EditMode property uses to indicate what the current state of the ADO editing buffer for the current record is.

Constant	Value	Description
adEditNone	0	Specifies that no editing operation is in progress
adEditInProgress	1	Specifies that data in the current record has been modified but not saved
adEditAdd	2	Specifies that the current record in the copy buffer is a new record from a call to the AddNew method, and that it has not been saved in the database
adEditDelete	4	Specifies that the current record has been deleted

Table 23.3. The list of constants that you can use with the EditMode property

As with the Supports method and LockType property, the constants that you require for the EditMode property are in the *adovbs.inc* file, or you can create explicit constants if you do not want to include this file. Basically, you would use the EditMode property by reading its current value to give you an indication of the current state of the ADO editing buffer for the current record. You will see more of how this property is affected when you learn about the different forms of using the AddNew method.

Utilizing the Recordset Object's Open Method

You have been using the Recordset object's Open method so far with no arguments. However, the Open method can actually take arguments and has the following syntax:

recordset.Open [Source][, ActiveConnection][, CursorType][, LockType][, Options]

You can use the arguments of the Open method rather than specifying each property on an individual line. For instance, take a look at the following code:

```
recset.Source = "customers"
recset.ActiveConnection = adocon
recset.LockType = adLockOptimistic
recset.Open ,,,,adCmdTable
```

The same code can be written as:

```
recset.Open "customers",adocon,,adLockOptimistic,adCmdTable
```

The last argument of the Open method is particularly important and also catches many unwary developers. It represents the manner in which the Open method is to interpret the Source argument and is especially important if the Source argument is not a command. In the case which you will be using the Open method for here, you will set Options argument to the constant adCmdTable because you want the Open method to interpret the Source as being a table name not a command. Table 23.4 shows you the various constants that you can use with the Options argument. These particular constants are known as CommandType constants.

Constant	Value	Definition
adCmdText	1	Evaluate Source as a textual definition of a command
adCmdTable	2	Evaluate Source as a table name
adCmdStoredProc	4	Evaluate Source as a stored procedure
adCmdUnknown	8	The type of command in the Source argument is not known

Table 23.4. The CommandType constants that you can use with the Options argument

Using the AddNew Method in Immediate Mode

In immediate mode, the database is changed as soon as you call the AddNew method with arguments. If you do not supply arguments to the AddNew method, then the change occurs when you call the Recordset object's Update method. Calling the AddNew method without arguments sets the EditMode property to the constant adEditAdd. If you want to, you can use the Supports method to see if the current data provider supports the Update method, just as you did with the AddNew method.

After the call to the AddNew method, you can call the Recordset object followed by the name of the column in quotation marks and parentheses, to which you can assign the cell data. The following example should help make things a little clearer, as it is easier to see than to explain:

```
recset.AddNew
recset("Cust_ID") = 902767
recset("First_Name") = "Frank"
recset("Last_Name") = "Diaz"
recset("Interest") = "Music"
recset.Update
```

Note: You may see a variation of the standard format, as shown in this book, where an exclamation character (!) is used instead. This as a shorthand style and you may come across it in other people's scripts. So basically, the following two lines would be equivalent:

```
recset("Interest") = "Music"
recset!Interest = "Music"
```

Once again, you have covered a lot of ground with theory intermingled with snippets of code. To get a better understanding, you must see how all of these pieces go together to make the whole process work. The following example shows you a larger picture of how you can add a customer to the database, although HTML header code and tables have been omitted:

```
<%
Dim adocon
Dim recset
Dim dbPath
Dim i

'Assign the database physical path to a variable
dbPath = Server.MapPath("customer.mdb")

'Create an instance of ADO Connection object
Set adocon = Server.CreateObject("ADODB.Connection")
adocon.Open "Provider=Microsoft.Jet.OLEDB.4.0;
Data Source=" & dbPath

'Create an instance of ADO recordset object
Set recset = Server.CreateObject("ADODB.Recordset")

'Open the Recordset
recset.Open "customers",adocon,,adLockOptimistic,adCmdTable

'Test first before adding data
If recset.Supports(adAddNew) Then

    recset.AddNew
    recset("Cust_ID") = 902767
    recset("First_Name") = "Frank"
    recset("Last_Name") = "Diaz"
    recset("Interest") = "Music"
    recset.Update

    'Send user some feedback
    Response.Write "Success adding customer to database."
Else
    Response.Write "AddNew method not supported."
End If
```

```
'Close the Recordset and Connection objects
recset.Close
adocon.Close
%>
```

You can also add multiple rows by using arrays. The two basic steps are to first prepare the array and then assign them as arguments to the AddNew method. When you use arrays with the AddNew method, you do not use the Update method, as the database will automatically be updated. The following code snippet shows the slight differences that you require to add multiple customers to the database:

```
Dim cnames(3)
Dim values(3)

'Set column names array
cnames(0) = "Cust_ID"
cnames(0) = "First_Name"
cnames(0) = "Last_Name"
cnames(0) = "Interest"

'Set cell values for each column
values(0) = 902768
values(1) = "Vahid"
values(2) = "Qualls"
values(3) = "Music"

recset.AddNew cnames, values
```

The code above could be placed into the larger example seen just prior to it. It would basically replace all the code from the recset.AddNew line down to the recset.Update line.

Using the AddNew Method in Batch Mode

Batch mode means that you can add multiple rows of data, but they are all stored by ADO in a buffer and then written to the database together, as a batch. Batch mode is similar to immediate mode, in the sense that you still use the AddNew method—you still assign values by column headers, and you still use the Update method. Batch mode differs from immediate mode in the sense that when the Update method is called in batch mode, the changes are kept in a buffer, whereas in immediate mode the changes are immediately written to the database. You only write to the database whatever is in the buffer when you explicitly call the UpdateBatch method. In batch mode, you must also remember to set the LockType property to adLockBatchOptimistic. So the basic pattern that you would use would be as follows:

```
<%
...

'Open the Recordset
recset.Open "customers",adocon,, _
adLockBatchOptimistic,adCmdTable

'Test first before adding data
If recset.Supports(adAddNew) Then

    recset.AddNew
        'insert details of entry
    recset.Update

    recset.AddNew
        'insert details of entry
    recset.Update

    recset.AddNew
        'insert details of entry
    recset.Update

    recset.BatchUpdate
End If
...
%>
```

Deleting Data from Database Tables

You have seen earlier in this lesson, two different methods for adding data to a table in a database. One method made use of direct SQL commands while the other used methods from the Recordset object. You can delete data from tables using almost identical methods as what you learned when adding data.

With just a slight modification to the script in *add.asp,* you could easily delete the table entry for the customer Frank Diaz, whose unique customer identification number is 902767. Remember that a primary key in a table is always unique, so it is often good practice to make use of this key when deleting data to avoid accidentally removing more than intended. The following script snippet would replace the section where you were adding data earlier:

```
'Prepare SQL command and delete customer from table
sqlcmd = "DELETE FROM customers WHERE Cust_ID = 902767;"
adocon.Execute(sqlcmd)
```

Actually modify the Active Server Pages script in *add.asp* with this code snippet and then save the file as *delete.asp*. You might also like to change the header of *delete.asp* to reflect that you are actually deleting from a database and not adding to one. Loading this file in your browser will delete the entry for Frank Diaz (customer 902767), before querying the database for a complete listing in the table and displaying the results, which should look similar to what is shown in Figure 23.2.

Figure 23.2.
After deleting data from the database, the customer Frank Diaz is gone

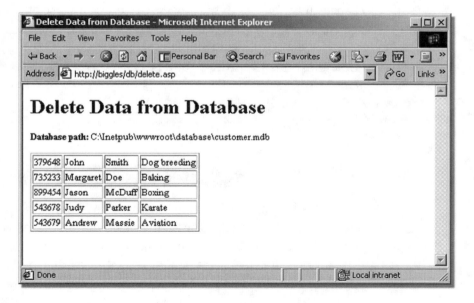

Using the Recordset to Delete Data

As you saw earlier in this lesson, you can add data to a database table using straight SQL code or you can use the Recordset object. You have already seen how to delete data from a database table using SQL, but you can also use the Recordset object's Delete method to delete the current record. In immediate mode, the current record is deleted immediately, but in batch mode, the current record is only deleted when the UpdateBatch method is called.

Changing Existing Data in Database Tables

You can also change existing data in database tables quite easily. You actually saw how to do this in theory using SQL code, which is probably the best and most straightforward method to use. You would have to modify the *add.asp* script that you created at the start of this lesson, replacing the section where you prepare an SQL statement. The following example shows how you could change all the entries in the table where the first name of the customer is Frank and change this to Francisco.

```
'Prepare SQL command and change table
sqlcmd = "UPDATE customers SET First_Name = 'Francisco'"
sqlcmd = sqlcmd & "WHERE First_Name = 'Frank';"
adocon.Execute(sqlcmd)
```

WHAT YOU MUST KNOW

In this lesson, you continued with your ADO skills, covering aspects of the Recordset object, including the AddNew method and the many methods, properties, and constants that you can work with to complete the task. You also saw methods to add, delete, and change data without using the Recordset at all, but relying on SQL code and the ADO Connection object's Execute method. Standing back and taking a broader view, this lesson completes a three-lesson course on working with databases from Active Server Pages. You now have a reasonable foundation upon which you can build working database applications, although there is much more to learn about databases, as it is a huge topic in its own right. In Lesson 24, "Being Security Conscious with Active Server Pages," you will look at security related issues within your Active Server Pages, including authentication for protecting data and obtaining information for security analysis purposes. Before you move on to Lesson 24, make sure that you are comfortable with the following key concepts and understand them clearly:

- By using the INSERT INTO keywords, that you can construct an SQL statement that you can pass as an argument to the ADO Connection object's Execute method, to add data to a table.

- To add data to a database table, you can use the Recordset object's AddNew method, supplying arrays that contain the columns and data as arguments. You can also supply no arguments and specify the values for the individual column headers.

- When you use immediate mode, the data is written immediately after the call to the Update method, unless arrays are used in which case it is written immediately after the AddNew method call itself.

- In batch mode you can add multiple rows and make multiple calls to the Update method, but the data is not actually written until the call to the BatchUpdate method.

- The Supports method allows you to determine whether the underlying data source provider supports a particular method or property.

- Due to the large number of constants that you often require access to, you should include the *adovbs.inc* file in your Active Server Pages.

- To prevent problems occurring when you are changing data, you should lock the records using the LockType property. Depending on whether you are using immediate mode or batch mode, you will require a different value for the LockType property.

- You can shorten your code considerably by using the Recordset object's Open method five optional arguments. The last argument indicates, by means of a CommandType constant, how the Open method is to handle the Source property or argument.

- By using the DELETE FROM keywords, you can construct an SQL statement that you can pass as an argument to the ADO Connection object's Execute method, to delete data from a table.

- By using the UPDATE SET keywords, you can construct an SQL statement that you can pass as an argument to the ADO Connection object's Execute method, to change existing data in a table.

BEING SECURITY CONSCIOUS WITH ACTIVE SERVER PAGES

i n Lesson 23, "Using ActiveX Data Objects (ADO) Part II," you completed a three-lesson course on working with databases from Active Server Pages. Just as being able to access important data stores is a great feature, there is also a security hole that opens up, which it is your responsibility to act upon. In this lesson you will look at security related issues within your Active Server Pages, including authentication for protecting data, as well as obtaining information for security analysis purposes. By the time you complete this lesson, you will have covered the following key concepts:

- Authentication allows you to limit access to Active Server Pages to only those who have permission to view them. Authentication can often involve supplying a username and password.
- You can use the Microsoft Management Console to set the properties and settings for security in IIS.
- The Request object's ServerVariables collection provides a huge variety of environment variables, including many that you can utilize for security information.
- When working with digital certificates, you can use the ClientCertificate collection of the Request object to gather specific information about the certificate.
- You can use NTFS permissions to provide effective authentication that makes use of an existing database of usernames and passwords.
- You can use cookies via the Session object, together with your own database of usernames and passwords, to provide simple authentication with persistence for protected Active Server Pages.

- As an alternative to usernames and passwords, you can use IP numbers as the authentication for protected pages.
- You can further protect the security of database files by storing them in folders that are not Web accessible, locking them with a password, and using encryption algorithms on the passwords contained within them.

Introducing Authentication

There are many reasons why authentication may be necessary in your Active Server Pages. You may have certain pages that contain confidential data or sensitive information. These may include company data that could jeopardize potential contracts or give an opposing company the upper hand if they could get hold of the data. You may have administration scripts that provide site information or allow remote changes to the site. For security reasons, you do not want everyone to have access to any of these Active Server Pages.

Authentication forces users to supply some form of valid credentials before they can access secure pages. There are many different ways to go about implementing authentication. Microsoft's Internet Information Server (IIS) has numerous settings that you can look into, you can use NTFS permissions on the server, you can use cookies, you can use IP numbers, or you can use third-party tools and components.

Setting Security Properties in IIS

You can set many properties and default settings for IIS that determine how to handle various security situations. Practical uses include setting anonymous access, basic authentication, and integrated Windows authentication, as well as settings for server and client certificates. You will learn more about each of these elements of security throughout this lesson. At this stage, it is best for you simply to learn how to change properties and settings for IIS security so that as you come across a security type that requires a particular setting, you will know already where to make the appropriate changes. If you are using Windows 2000 or Windows XP, Personal Web Manager makes use of IIS behind the scenes, so this is relevant to many Personal Web Manager users also.

From the Start menu, select the Settings submenu and then the Control Panel item. Windows will then present the Control Panel window from which you must double-click on the Administrative Tools folder to open it. Within the Administrative Tools window, double-click on the Computer Management icon.

This will open the Microsoft Management Console with the Computer Management items. Under the Services and Applications item is the Internet Information Services item, and underneath this is the Default Web Site item. Click once on the Default Web Site item so that it is highlighted, as shown in Figure 24.1.

**Figure 24.1.
Highlighting
the Default Web
Site item**

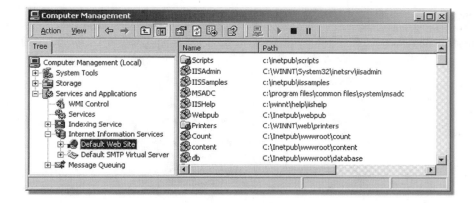

With the Default Web Site item highlighted, click on the Action menu and select the Properties item. From the Default Web Site Properties dialog box, select the Directory Security tab, as shown in Figure 24.2.

**Figure 24.2.
The Directory
Security tab**

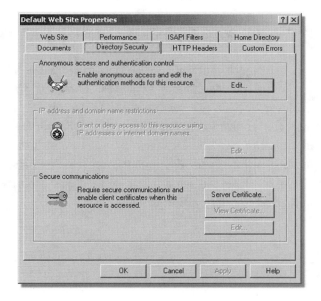

Under the section for Anonymous access and authentication control, click on the Edit button to open the Authentication Methods dialog box. Within the Authentication Methods dialog box, as shown in Figure 24.3, you can allow or deny anonymous access, basic authentication, or integrated Windows authentication, simply by using the check boxes associated with each security type.

**Figure 24.3.
The Authentication
Methods dialog box**

Using the Request Object Collections for Security

The Request object contains a collection called ServerVariables which contains many useful environment variables. Some of the variables of the ServerVariables collection are very relevant to security in Active Server Pages. You can use the environment variables of the ServerVariables collection to retrieve information on different security aspects, including usernames, passwords, authentication types, IP numbers, secure channels, secure ports, and more. In Table 24.1 you can see a list of some of the security-relevant variable keys, as well as a brief description of each.

Variable Key	Description
AUTH_PASSWORD	The password entered by the user for authentication
AUTH_TYPE	The authentication method that the server uses
AUTH_USER	The username entered by the user for authentication
CERT_COOKIE	Unique ID for client certificate
CERT_FLAGS	Indicates if the client certificate is present and valid
CERT_ISSUER	Indicates the client certificate issuer
CERT_KEYSIZE	Number of bits in Secure Sockets Layer connection key size
CERT_SECRETKEYSIZE	Number of bits in server certificate private key
CERT_SERIALNUMBER	Serial number field of the client certificate
CERT_SERVER_ISSUER	Issuer field of the server certificate
CERT_SERVER_SUBJECT	Subject field of the server certificate
CERT_SUBJECT	Subject field of the client certificate
HTTPS	Returns on if the request came in through a secure channel (SSL), otherwise off
HTTPS_KEYSIZE	Number of bits in Secure Sockets Layer connection key size
HTTPS_SECRETKEYSIZE	Number of bits in server certificate private key
HTTPS_SERVER_ISSUER	Issuer field of the server certificate
HTTPS_SERVER_SUBJECT	Subject field of the server certificate
LOCAL_ADDR	The IP address of the server accepting the request
LOGON_USER	The Windows account that the user is logged into
REMOTE_ADDR	The IP address of the remote host making the request
REMOTE_HOST	The name of the remote host making the request
REMOTE_USER	The username string as sent by the user
SERVER_PORT	The port number to which the request was sent
SERVER_PORT_SECURE	A string that contains 1 if the request is being handled on the secure port, otherwise 0

Table 24.1. Some of the security-related variable keys of the ServerVariables collection

There are 47 different variable keys that you can access, one of which breaks down further into subkeys. You can see a complete list on Microsoft's Web site at the following URL:

http://www.microsoft.com/windows2000/en/advanced/iis/htm/asp/vbob5vsj.htm

Alternatively, you can simply search for *servervariables* on search engines to discover a wealth of information on the many environment variables of this collection.

To use a ServerVariables collection environment variable, you simply place the variable name in quotation marks and within parentheses after a call to ServerVariables itself. For example, the following simple line of code shows how you might determine whether a secure channel was used when the request came in (meaning https:// was used instead of http://):

```
<%= Request.ServerVariables("HTTPS") %>
```

The ClientCertificate collection of the Request object is available you for working with digital certificates in your Active Server Pages. Certificates use the Secure Sockets Layer (SSL) connection and are deemed secure encryption for data on the Web. For this reason, you will find that most sites on the Web that accept credit card information use SSL, and you will notice that the URL begins with https:// rather than http://.

You can set the Server certificate in the Secure communications section of the Directory Security tab, which you saw earlier in Figure 24.2. There are a number of environment variables in the ServerVariables collection that deal with certificate usage, all of which begin with CERT_.

The ClientCertificate collection really just makes use of the seven keys shown in Table 24.2.

Key	Description
Certificate	A binary string containing the digital certificate
Flags	Additional certificate information
Issuer	Additional information about the certificate issuer
SerialNumber	The serial number of the certificate
Subject	Additional information about the certificate subject
ValidFrom	The date that the certificate becomes valid
ValidUntil	The date that the certificate expires

Table 24.2. The available keys in the ClientCertificate collection

The following example shows how to use the ClientCertificate collection to return the serial number of the certificate:

```
<%=Request.ClientCertificate("SerialNumber")%>
```

The ClientCertificate collection cannot be explained thoroughly and in minute detail in this lesson; however, the following points may be of additional use to you if you are interested in using digital certificates. To use the ClientCertificate collection in your Active Server Pages, you should also include the Microsoft file *cervbs.inc*. This file contains the constants that you use for the Flags key. Both the Issuer key and the Subject key can make use of subkeys for additional information. If there was no digital certificate received from the client, the ClientCertificate will be empty for any key.

If you are interested in doing more with digital certificates, the following URL on the Microsoft site will provide you with further details and examples:

http://www.microsoft.com/windows2000/en/advanced/iis/htm/asp/vbob8q5h.htm

Applying Basic NTFS Authentication

One of the simplest methods of implementing authentication is to use NTFS permissions. Although simple, this is very secure and also saves you from having to create and maintain a database of users and passwords, utilizing the list in Active Directory or the SAM instead. As its name implies, you use NTFS on Windows NT operating systems, which includes Windows 2000 and Windows XP, as well.

An example should show you how simple NTFS authentication is. In the following example you will create a script that uses some of the ServerVariables collection variables and displays their values. You will first load the script in your browser and observe the result, and then apply NTFS permissions before repeating the process. Enter the following simple example script into Notepad:

```html
<html>
<head>
<title>Using ServerVariables and NTFS Security</title>
</head>

<body>
<h1>Using ServerVariables and NTFS Security</h1>
AUTH_PASSWORD server variable =
<%= Request.ServerVariables("AUTH_PASSWORD") %><br>
AUTH_TYPE server variable =
<%= Request.ServerVariables("AUTH_TYPE") %><br>
AUTH_USER server variable =
<%= Request.ServerVariables("AUTH_USER") %><br>
HTTPS server variable =
<%= Request.ServerVariables("HTTPS") %><br>
REMOTE_ADDR server variable =
<%= Request.ServerVariables("REMOTE_ADDR") %><br>
REMOTE_HOST server variable =
<%= Request.ServerVariables("REMOTE_HOST") %><br>
REMOTE_USER server variable =
<%= Request.ServerVariables("REMOTE_USER") %>
</body>
</html>
```

Save the script as *sv.asp* on an NTFS partition and then load it into your browser. You will notice that all of the authentication fields (those starting with AUTH_) are empty. Using Windows Explorer, navigate to the file *sv.asp*, wherever you saved it. Right-click on the file and select Properties from the pop-up menu. Windows will then present the properties dialog box for the file *sv.asp* from which you must select the Security tab. Add some users or groups to the list that have permission and then click on the OK button, as shown in Figure 24.4.

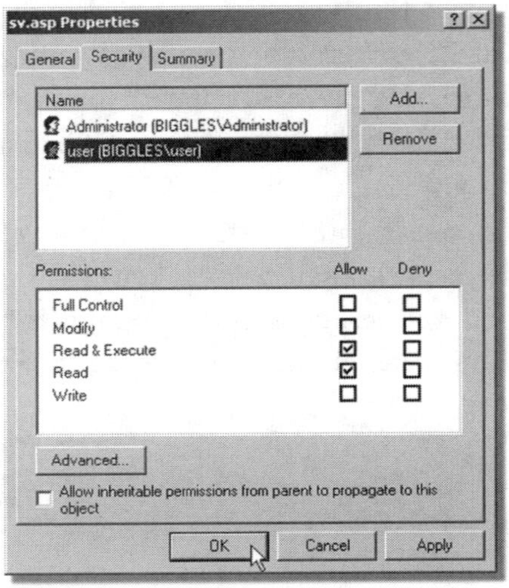

Figure 24.4. Adding some users to the permission list

Now when you load *sv.asp* in your browser, you will be prompted for a Windows NT username and password where you can enter data and authenticate, as shown in Figure 24.5.

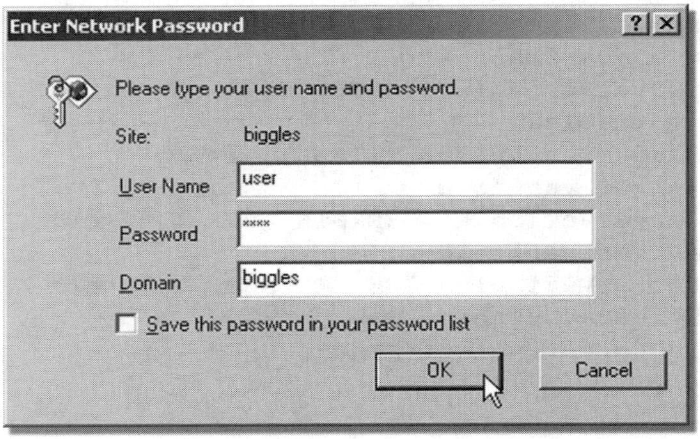

Figure 24.5. Being prompted for a Windows NT username and password

When you authenticate using NTFS, the Active Server Page *sv.asp* loads and displays results similar to those shown in 24.6.

**Figure 24.6.
The result of
loading *sv.asp*
after authenticating
with NTFS**

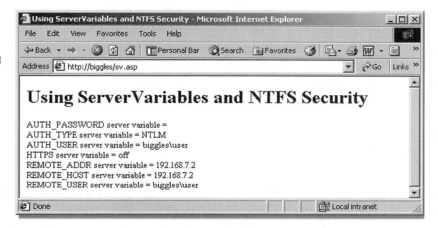

You can see clearly that using NTFS is not only simple but allows you to make use of some of the ServerVariables collection's environment variables. The AUTH_PASSWORD is only shown if you are using basic authentication, which is not recommended because the passwords are in clear text and not highly secure. You saw earlier in this lesson where you could set basic authentication if you need to.

Your system administrators may not want to use NTFS permissions for the pages on the site, especially as there may be a huge number of pages, each requiring special permissions that change constantly. There are many reasons that the authentication for Web pages is better kept separate from an internal company's Active Directory or SAM. In these cases you must provide another means of authentication using your own database.

Applying Simple Authentication Using Cookies

The reason for using cookies for authentication is not obvious at first. You think about authentication and supplying valid credentials and wonder what need there is for cookies. You will soon realize that without some form of persistence, authentication would be practically infeasible. You would have to supply a username and password on every secure page, which you can imagine would soon get rather annoying for legitimate users working with confidential data. You can use cookies to provide the persistence for a user after he authenticates, allowing him to then go to other secure Active Server Pages without having to reauthenticate.

Before you get into the code, you should first get an overview of what you are trying to do. The following steps show clearly what your overall plan should be:

1. User logs into the site
2. Session-level variables are created and initialized
3. User attempts to gain access to a protected page
4. Protected page assigns itself to one of the Session-level variables, as the intended destination page of the user
5. Protected page redirects user to a login page for authentication
6. User enters credentials in a form, which is then submitted to a validation page
7. Validation page checks credentials against a database and either 1) redirects user to the intended page, or 2) redirects user to a failure page

When the user connects to the site for the first time, the Session object's OnStart event fires. This is the perfect time to create and initialize any security variables that you might require for the duration of the session. You will require one variable to determine whether the user is authenticated or not, which you can call *UserAuthenticated*. You will also require a second variable, which you can call *Intended*.

You do not technically require the second variable; however, it is a nice touch on the developer's part to redirect a successfully authenticated user back to the page that he was intending to open. Otherwise, he would be left facing a page that simply told him that he had authenticated successfully, which is not much good to him if he had gotten to the protected page via a link and did not know the actual URL. So the second variable, *Intended*, is for holding the URL of the page that the user was intending to open.

Inside the Global.asa file, add the following script:

```
Sub Session_OnStart
  Session("UserAuthenticated") = False
  Session("Intended") = ""
End Sub
```

You have covered steps 1 and 2 of your overall plan. You now have to deal with step 3. When a user actually attempts to open a protected Active Server Page, the page itself must determine whether the user is authenticated before taking any appropriate action. You can use a conditional If-Then statement that checks the Session-level variable UserAuthenticated to see if it is True, in which case you simply continue on with the script. However, if it is False then you have to handle steps 4 and 5.

For step 4, you can use the Request object's ServerVariables method and the SCRIPT_NAME option to assign the URL of the current Active Server Page to the Session-level variable, *Intended*. Step 5 is simply a redirect to the login.asp file. You should keep your login page in your root directory rather than in the same directory as the protected files; you may want to use more secure permissions on the directories containing sensitive data, yet everyone should be able to access the login page.

At the top of every protected Active Server Page, you can place the following script, which covers all of the steps from 3 to 5:

```
<%
'Check for authentication first otherwise redirect
If Session("UserAuthenticated") = False Then
   'Take note of the intended destination page
   Session("Intended") = _
   Request.ServerVariables("SCRIPT_NAME")

   'Redirect user to the authentication page
   Response.Redirect "/login.asp"
End If
%>
```

Remembering Old Tricks

When you think about the fact that this script is quite simple and still takes up quite a few lines, not to mention the possibility of having hundreds of protected pages, you will probably start to wonder if this is the best way to go about this. If you have to modify this piece of authentication script, you will have to visit all of those hundreds of pages. Remember back to earlier in the book when you were learning about include files. This scenario with the authentication having to go at the top of each of your protected pages is ideal for making use of an include file. Basically, you can add the authentication code into a separate include file, such as authenticate.inc, *and then at the top of each protected page, simply add the following single line of code:*

```
<!-- #include virtual="/authenticate.inc" -->
```

In this way, you can update or modify your authentication script in one place without having to visit hundreds or thousands of protected pages.

Step 6 is very straightforward. All you need is a standard HTML page that contains a form with fields that lets a user authenticate. One field for a username and another for a password will normally suffice. Notice that the form's *action* attribute is sending the data on to an Active Server Page for validation. Make sure that the method is POST and not GET because you do not want the username and password appended to the URL. The following script is a simple example that you could easily elaborate on, but which gets across the main points:

```html
<html>
<head>
<title>The Login Page</title>
</head>

<body>
<h1>Welcome to the Login Page!</h1><hr>
<p>The page you were attempting to access is <b>protected</b>.
To gain access to the page, please supply a valid username and
password using the fields below:</p>
<form action="validate.asp" method="POST">
Enter your username here:<br>
<input type="text" name="username" value=""><br><br>
Enter your password here:<br>
<input type="password" name="password" value=""><br><br>
<input type="reset">
<input type="submit" value="Login">
<hr></form>
</body>
</html>
```

Save the script as *login.asp* and place it in the root directory of your Web server. Figure 24.7 shows the result of loading *login.asp* into a browser, as well as how the password field is shown with asterisk characters (*) rather than the actual letters of the password.

Figure 24.7.
The login page
that is presented
to the user

You will require a database of unique usernames and passwords. In practice, you will undoubtedly store a lot of other information as well, but to maintain focus on what you are doing, just keep things simple for now. Later in this lesson you will look at various ways to increase the security of this type of database.

You do not have to use Microsoft Access as a database provider. You can use SQL Server, Oracle, or whatever else you want, as long as it is a secure storage for usernames and passwords. The database for the example is a Microsoft Access database called *private.mdb*. The table containing all the usernames and passwords is called *passwords*. Figure 24.8 shows a table in an Access database that contains some usernames and passwords.

Figure 24.8.
An Access database
table that contains
some usernames
and passwords

Username	Password	Full_Name
cory	hair4me	Cory Ferlazzo
justin	stillgrowing	Justin Miller
rob	km2(6,aD	Rob Francis
shahla	teddybear	Shahla Francis
stuart	iextend	Stuart Hall

Step 7 of your overall plan is covered by the validation page itself. First, you have to retrieve the username and password that has been submitted by the user. Next, you have to open up a connection to your database. Using your SQL skills, you can create an SQL query that only returns the results where the username in the table matches the username from the form, and the password in the table matches the password from the form. Basically, if the resulting Recordset object is empty, then there was no match.

You can use the Recordset object's EOF property to determine whether there is a record in the results or not. If the EOF property is not True, then you have to set the Session-level variable *UserAuthenticated* to be True. This will allow the user to access other protected pages without having to reauthenticate. You also have to redirect the user to his intended destination page by using the Session-level variable, *Intended*.

If the EOF property is True then the Recordset contains no records and this means that the authentication failed. You should redirect the user to a failure page, explaining the situation and offering him a chance to login again.

The following script makes up the file *validate.asp*, which covers all the requirements of step 7:

```
<%
Dim adocon
Dim recset
Dim sqlCmd
Dim usrName
Dim pwd

'Get the username and password
usrName = "'" & Request.form("username") & "'"
pwd = "'" & Request.form("password") & "'"

'Assign the database physical path to a variable
dbPath = "C:\private.mdb"

'Create an instance of ADO Connection object
Set adocon = Server.CreateObject("ADODB.Connection")
adocon.Open "Provider=Microsoft.Jet.OLEDB.4.0;
Data Source=" & dbPath

'Prepare the SQL query
sqlCmd = "SELECT * FROM passwords " & _
```

```
"WHERE Username = " & usrName & " AND Password = " & pwd

'Execute a SQL query and store the results within a recordset
Set recset = adocon.Execute(sqlCmd)

'Check for successful validation
If Not recset.EOF Then
  'Success!
  Session("UserAuthenticated") = True

  'Redirect the user to the intended page
  Response.Redirect Session("Intended")
Else
  'Failure!
  Response.Redirect "failure.asp"
End If
%>
```

That is all there is to it. If you try protecting a few pages and then accessing them from various computers and operating systems, you will see that the security does not break and that the pages remain protected unless an authentication is given.

If a user does not have cookies enabled or does not accept the cookie from your site when prompted by his browser, that user will not be able to authenticate regardless of the fact that he may supply valid credentials. You will recall from earlier in the book that the Session object relies heavily on cookies being enabled at the client side to be able to function properly. Most users have cookies enabled, so this is usually not a problem in practice.

Although this simple example is sufficient for intranets or keeping out overly curious colleagues, it is not without security holes which clever hackers could compromise. Cookies are not a very secure means of storage at all; however, as this example shows, they can be made use of to provide a measure of security that was not available previously. For many small companies, this is enough, as the data is not that sensitive that a hacker would bother going to all the trouble to access.

Applying Simple Authentication Using IP Numbers

Rather than relying only on usernames and passwords for authentication, you can use IP numbers as an alternative. Using IP numbers means that you could implement security that did not rely on cookies and did not frustrate the users with constant authentication. For additional security, you could implement IP numbers in addition to usernames and passwords. In this way, a client would need to have not only the correct username and password, but also be connecting with a valid IP number.

You can make use of the Request object's ServerVariables collection to obtain the IP address of the remote host or user's computer. Basically, you can use the following snippet of script in your Active Server Pages to return the IP address:

```
Request.ServerVariables("REMOTE_ADDR")
```

You simply match the IP address against a list of permitted IP numbers, which you can store in a secure database. In the same manner as you saw earlier, the code for this would be at the top of every protected page (preferably using an include file) and either lets the user see the contents of the page if the IP address is in the database or redirects the user to a failure page.

Creating Administrator-Only Pages

You can make use of the fact that there may be only one machine that you administer your pages from. This computer's IP number will be unique, and without need for any database, username, or password, you can simply check the IP number against a hard-coded number that you place in an administration include file. In any administrator-only Active Server Pages you would, of course, include this file. The downside or security hole with this method is if someone gets access to your computer. Also, if your computer's IP number goes to another computer (perhaps you are on DHCP rather than static), then this could compromise security.

Protecting Databases Which Contain Passwords

You must take preventative measures to protect your password databases from being compromised by unscrupulous people. There are three simple steps that you can take to increase the security of your database files.

First, you should move the database file to a location that is not part of your Web accessible directory. For example, if your Web accessible directory is *C:\inetpub\wwwroot* then you do not want to place your database anywhere under this directory. Rather try placing it on another partition or drive if possible. If you cannot do this then put it in a different directory than your Active Server Pages on the same partition, such as *C:\special*. You might also use NTFS permissions to limit access to the database file so that only the system can access it.

Next, you should look at password protecting the database file. This is usually very simple to do and is a feature of most database programs today. For example, in a Microsoft Access database, you would select the Tools menu Security submenu and then the Set Database Password option. This means that even if someone found your database, he would have to crack a password before he could access the data contained within.

Thirdly, you should encrypt the passwords that are contained within the database. This would add yet another layer of difficulty to anyone trying to compromise your password database. It also addresses the issue of user privacy by hiding the clear text passwords from all of the administrators that have the database password. Using Data Encryption Standard or other key-based ciphers will be secure, except that if the key is compromised then the database is also compromised.

Using a one-way hash function with salt (No, not like salt and pepper) is probably the best method to take for the passwords in a database file. One-way hash functions are sometimes known as fingerprints or compression functions. Basically, they take a string and produce a fixed-length binary value, also known as a hash. There are many such hash functions; however, not all of them are equally secure. Using the SHA algorithm provides 160-bit hash values, which is considered highly secure. To crack a single hash value of this magnitude would take about 2^{160} different strings. This is about 14,615,016,373,309, 029,182,036,848,327,163 (or 1.46e48) which is a massive number of strings to generate and compare even for cracking programs and computers. Salt is a randomly generated string which is joined to the password before applying the hash algorithm. This basically defeats dictionary attacks where programs use precompiled lists of common words.

There are numerous third-party Active Server Pages components available that supply encrypting and ciphers, such as those discussed here, which you can use in your own Active Server Pages if security is a serious matter.

WHAT YOU MUST KNOW

You can use authentication with Active Server Pages, limiting access to only those who have permission. Authentication, although often involving the user supplying a username and password, can also be based on IP numbers. Working with databases rather than plain text files increases the security of the passwords and usernames, and is very secure when coupled with password protection and encryption. In Lesson 25, "Introducing and Running ASP.NET," you will begin your journey into the latest incarnation of Active Server Pages from Microsoft—ASP.NET. So far, all of your skills have been building on what you have covered in previous lessons. However, ASP.NET is different from classic Active Server Pages, so you will be learning new methods of accomplishing some tasks, which do not build on your existing skills. Before you begin with ASP.NET, make sure you understand fully the following key concepts:

- You set the properties and settings for security in IIS through the Directory Security tab in the Microsoft Management Console.

- You can make use of the many environment variables of the Request object's ServerVariables collection for obtaining security information.

- The Request object's ClientCertificate collection enables you to obtain specific information about digital certificates.

- NTFS permissions give you solid authentication that is simple to implement. NTFS has the added benefit of no maintenance, as it makes use of the NT domain's database of usernames and passwords.

- You can provide simple authentication for protected Active Server Pages, as well as preventing users from having to continuously authenticate, by using cookies via the Session object.

- IP numbers make a good alternative, or addition, to usernames and passwords for authentication purposes. When using cookies is not an option, IP number authentication provides a good solution.

- To help prevent any unscrupulous people from compromising your database files, you can store them in folders that are not Web accessible, lock them with a password, and use encryption algorithms on the passwords contained within them.

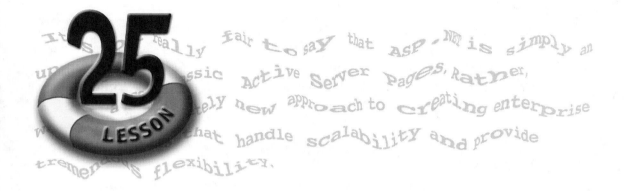

It's not really fair to say that ASP.NET is simply an update to classic Active Server Pages. Rather, it's a completely new approach to creating enterprise web sites that handle scalability and provide tremendous flexibility.

INTRODUCING AND RUNNING ASP.NET

So far in the book you have been learning about the classic aspects of Active Server Pages. This is useful, of course, because classic Active Server Pages are what most sites that are using Active Server Pages contain. However, Microsoft has released beta versions of a new style of Active Server Pages, which was originally known as ASP+ and then changed to ASP.NET for marketing reasons. By the time you finish this lesson you will have covered the following key concepts:

- Microsoft Active Server Pages has gone through many different versions, with ASP.NET being the most recent version. ASP.NET is a completely new approach to creating Active Server Pages.
- ASP.NET pages are compiled rather than interpreted, making them faster than previous Active Server Pages.
- ASP.NET requires special software to run, including a Visual Studio.NET compiler.
- To start a new project, you can select an ASP.NET Web Application from the Visual Basic.NET project type list within the New Project dialog box.
- You can select controls from the Toolbox panel, such as a Label control, and add that to your ASP.NET page using a visual approach by simply dragging the intended size and position with your mouse.
- You can edit properties of an object, such as a label, through the Properties panel. Many properties, such as the Font Size, contain further drop-down menus to make value selection even easier for you.

- You can set the name of the ASP.NET page by using the File menu's Save CurrentPageName.aspx As option and using the Save File As dialog box.
- To actually create your ASP.NET Web application, including your ASP.NET page, you use the Build menu Build option, which will compile your new page.

Introducing ASP.NET

ASP.NET is the latest version of Active Server Pages from Microsoft. It requires some special software to be able to run. This is similar to the way classic Active Server Pages require an Active Server Pages server in order to be translated into HTML pages—you cannot simply open an Active Server Page in a browser.

It is not really correct to say that ASP.NET is simply an upgrade to classic Active Server Pages. Rather, it is a completely new approach to creating enterprise Web solutions that handle scalability and provide tremendous flexibility. ASP.NET does support classic Active Server Pages so you can still make use of existing scripts and even integrate them in your applications with pages created in ASP.NET.

In ASP.NET, your Active Server Pages are no longer written as an interpreted scripting language but rather they are written as code that you compile using one of the Visual Studio.NET languages. Because the code is compiled and not interpreted at runtime, ASP.NET pages are faster than traditional Active Server Pages.

Before you can work with ASP.NET, you must install it on your system. The ASP.NET Premium Edition will run on all versions of Windows 2000 and Windows XP, so be aware that this may deter many developers working from other versions of Windows, or indeed other platforms, from using ASP.NET at this stage. For information on acquiring ASP.NET and Visual Studio.NET, visit the Microsoft Web site.

Creating an ASP.NET Application Using a Visual Interface

If you are familiar with Visual Basic.NET then creating ASP.NET applications will be very easy for you. Visual Basic.NET has almost the same syntax as VBScript, which you have been using throughout this book until now. However, there are differences in the syntax, and you will realize that VBScript only offers a fraction of the features that you can have with Visual Basic.NET. The most striking difference for those who have only used VBScript before is that you can now create Active Server Pages using a visual interface and simple drag and drop techniques. This not only allows you to shorten the time expended creating Active Server Pages, but also allows a greater flexibility in what you, as the Web developer, can accomplish.

The example in this section assumes that you have Visual Basic.NET already installed on your computer. You can also compile ASP.NET pages using a command line compiler; however, the aim here is to show you the direction that Active Server Pages are taking. Microsoft provides incredible support and features for ASP.NET from within their flagship development environment, Visual Studio.NET, and almost all Active Server Pages developers are embracing this. If you decide that the visual approach is not to your liking, you can look at using the command line approach.

Using the Command Line Compiler

If you do not have access to Visual Basic.NET or you prefer to use the command line tools, you can also create ASP.NET pages using any of the command line compilers that came with the ASP.NET SDK. You can open a command line shell by clicking on the Start menu Run command and typing **cmd** *followed by the Enter key. Windows will open a new command line shell for you to work in. Within the command line shell you will need to navigate to the following directory to find the compilers:*

c:\winnt\microsoft.net\framework\v1.0.2914

**Figure 25.1.
The Visual Basic compiler in a command line window**

```
Select C:\WINNT\System32\cmd.exe

C:\WINNT\Microsoft.NET\Framework\v1.0.2914>vbc.exe /?
Microsoft (R) Visual Basic.NET Compiler version 7.00.9254
for Microsoft (R) .NET CLR version 1.00.2914.16
Copyright (C) Microsoft Corp 2001. All rights reserved.

              Visual Basic Compiler Options

                              - OUTPUT FILE -
/out:<file>                   Specifies the output file name.
/target:exe                   Create a console application (default). (Short form: /t)
/target:winexe                Create a Windows application.
/target:library               Create a library assembly.
/target:module                Create a module that can be added to an assembly.

                              - INPUT FILES -
/addmodule:<file>             Reference metadata from the specified module.
/recurse:<wildcard>           Include all files in the current directory and
                              subdirectories according to the wildcard specifications.
/reference:<file_list>        Reference metadata from the specified assembly. (Short
                              form: /r)
```

The aim of the following example is for you to create your first ASP.NET page using a visual approach. You will not be directly coding in this particular example, but will learn to use visual techniques instead. Basically, you will create a blank page with a single text label on it, for which you have set the font, the size, and the text. You will enhance this ASP.NET Web application further in subsequent lessons.

Open up Visual Basic.NET and from the New Project dialog box, select Visual Basic Projects from the Project Types list. You can open the New Project dialog box at any time by selecting File menu New submenu and then the Project item. Under the Templates list select ASP.NET Web Application so that it is highlighted, as shown in Figure 25.2.

Figure 25.2. Select ASP.NET Web Application from the Templates list

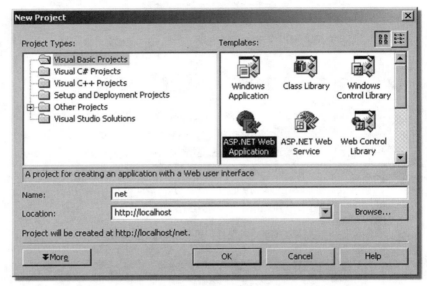

Change the text in the Name field to *net*. You can leave the Location field as *http://localhost*. The Location field and the Name field basically make up the path of the URL that appears prior to any Active Server Pages within the application. Click on the OK button to continue.

Visual Basic.NET now presents you with the main IDE (Integrated Development Environment), which can appear a little daunting at first until you become more comfortable with everything. To begin, click on the Label icon, which is situated under the Web Forms section of the Toolbox panel, as shown in Figure 25.3.

**Figure 25.3.
The Label icon is
situated under the
Web Forms section**

When you have selected the Label icon from the Toolbox panel, you can simply click somewhere in the center form and then hold down your left mouse button and drag the size of the label from any corner of the label to its diagonal opposite, as shown in Figure 25.4.

**Figure 25.4.
Dragging a
rectangle for
your label**

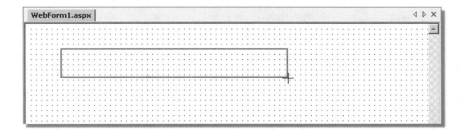

With your new label still selected, you will see its properties in the Properties panel. Within the Properties panel, click the Text property and change its value to PaperWorks. By clicking on the small plus or minus sign next to the Font property, you can toggle showing and hiding its subproperties. Click on the plus sign next to the Font property to show all the subproperties. Change the Size property to X-Large using the drop-down box and selecting from the menu. Change the Name property to Lucida Calligraphy or a font of your own choosing, using the drop-down box and selecting from the menu. After you finish making the changes, your Properties panel should appear similar to the one shown in Figure 25.5.

**Figure 25.5.
The Properties
panel after
changes**

You may not want your first ASP.NET page to be called *WebForm1.aspx*. Click on the File menu and select the Save WebForm1.aspx As . . . item. Visual Basic.NET will present the Save File As dialog box, allowing you to change the name of the ASP.NET page. Change the name of the page to *paper1.aspx* and then click on the Save button.

From the Build menu select the Build item. Visual Basic.NET will compile the ASP.NET application, including the file *paper1.aspx*. From your browser, enter in the name of your Web server in the Address field (URL) and follow it with */net/paper1.aspx*. So for example, if your Web server is named *biggles,* you would enter the following URL:

http://biggles/net/paper1.aspx

Loading *paper1.aspx* into your browser should result in a page similar to that shown in Figure 25.6.

**Figure 25.6.
After loading
paper1.aspx into
your browser**

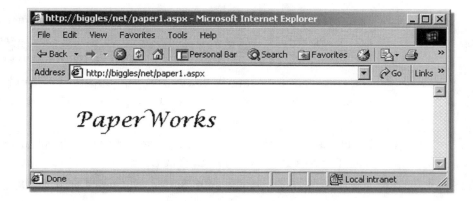

You have now created your first ASP.NET page! Although it may not be very exciting, you have seen that it is a very different approach to creating Active Server Pages and this is what is important.

WHAT YOU MUST KNOW

ASP.NET is the latest version of Active Server Pages from Microsoft, supporting all existing (classic) Active Server Pages, as well as offering a huge range of new features and options. You must compile ASP.NET pages using one of the Visual Studio.NET compilers. In Lesson 26, "Comparing Differences in Syntax," you will look in more detail at some of the fundamental changes from classic Active Server Pages to the new ASP.NET from the perspective of code creation. Before you begin Lesson 26, make sure that you fully understand the following key concepts:

- Classic Active Server Pages are basically just scripts that are interpreted by the Active Server Pages server at runtime, which can be slow. ASP.NET pages, on the other hand, are precompiled using one of the new Visual Studio.NET compilers, which can dramatically increase speed for large sites.

- ASP.NET pages end with the *.aspx* extension, rather than the traditional *.asp* extension of classic Active Server Pages.

- Within the New Project dialog box, you can select Visual Basic Projects from the Project Types list and then select the ASP.NET Web Application icon from the Templates list, to start a new ASP.NET Web application.

- The Toolbox panel offers a range of various controls that you can add to your Active Server Pages using a visual approach by simply selecting one of them and then dragging the intended size and position on the Web form with your mouse.

- The Properties panel allows you to edit the properties of an object on your Web form. Some properties contain additional options such as subproperties or drop-down menus.

- Using the File menu's Save CurrentPageName.aspx As option together with the Save File As dialog box, you can set the name of an ASP.NET page.

- You use the Build menu's Build option to compile your ASP.NET Web application and create your ASP.NET pages.

COMPARING DIFFERENCES IN SYNTAX

n Lesson 25, "Introducing and Running ASP.NET," you were introduced to ASP.NET, including creating an ASP.NET page using a visual approach. In this lesson, you will look in more detail at some of the fundamental changes from classic Active Server Pages to the new ASP.NET, mostly from the perspective of code creation. In particular, these syntax changes will be taking into consideration VBScript and Visual Basic.NET. By the time you have completed this lesson you will have covered the following key concepts:

- Due to conformity with other .NET languages to make use of the Common Language Runtime, Visual Basic.NET has some differences in syntax to Visual Basic 6 and its subset VBScript.

- Many inconsistencies and obsolete commands have been removed from Visual Basic and are no longer supported in the language.

- There is a basic interface improvement with ASP.NET, although optional, allowing you to visually create the graphic interface for Web forms using drag and drop functionality, as well as providing color-coding of the actual code itself.

- In Visual Basic.NET you must explicitly declare your variable data types unlike VBScript where all variables are of type Variant. There are some new data types in Visual Basic.NET that were not in earlier versions of Visual Basic.

- You can only use one language per page in ASP.NET unlike classic Active Server Pages where multiple scripts could be mixed on the same page.

- You must use parentheses around the argument to the Response object's Write method in ASP.NET.

- Arrays in Visual Basic.NET are zero-based rather than one-based. To access individual array elements in ASP.NET, you must use an explicit call, such as to the GetValues method.
- Procedures use the Return statement in place of Exit Sub and Exit Function statements. The Return statement also replaces the return value being assigned to the function name.
- Procedures can no longer be declared or created within <% and %> tags, but must be within a <script> block instead.
- Boolean expressions now support short-circuiting by way of the new AndAlso and OrElse operators.
- The Wend keyword is no longer supported in Visual Basic.NET. Instead you end your While loops with the End While statement.
- In Visual Basic.NET the Set command is no longer supported. Instead you simply assign the value or object to the variable.

Understanding Why Active Server Pages Syntax Has Changed

As you are already aware, VBScript is basically a subset of Visual Basic. There are differences between the two, but generally speaking VBScript and Visual Basic share the same syntax. To work with ASP.NET you must use one of Microsoft's new Visual Studio.NET languages, such as Visual Basic.NET. Visual Basic.NET syntax is not the same as Visual Basic 6 syntax. The VBScript that you use for classic Active Server Pages is based on Visual Basic 6 syntax. Here lies the key reason for differences in syntax between classic Active Server Pages and the new ASP.NET pages. Figure 26.1 portrays the language usage for both styles of Active Server Pages.

**Figure 26.1.
The language usage
for both styles of
Active Server Pages**

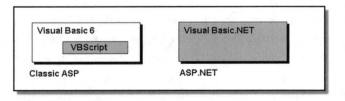

Although Visual Basic 6 and Visual Basic.NET have differences in syntax, they still share quite a lot in common. This also applies to VBScript, which is a subset of Visual Basic, having many things in common with Visual Basic.NET. The Venn diagram in Figure 26.2 shows how these three variations of Visual Basic interact and relate to each other.

**Figure 26.2.
Venn diagram
showing relation of
the old languages
with the new**

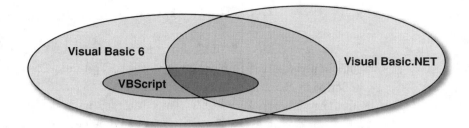

Prior to Visual Basic.NET, Visual Basic was mostly for Microsoft Windows applications and therefore VBScript was used for working with Web pages, such as classic Active Server Pages. However, Microsoft has now changed the focus of their development languages, including Visual Basic.NET, to be used for Web development and Web applications. These new languages in Visual Studio.NET share quite a lot in common, including the Common Language Runtime, unified programming classes, and an integrated development environment. To be able to conform to these new, shared components and libraries, inevitably Visual Basic had to change.

There are many old commands and keywords that Visual Basic has, until now, retained like so much old baggage. Some of these commands date back to early versions of *spaghetti* Basic, such as the GoSub and Goto commands. The continual addition of commands and features has further complicated the language, leading to inconsistencies in syntax. The current requirement to modify Visual Basic's syntax to conform to the shared common libraries in Visual Studio.NET provides the perfect opportunity to remove these outdated and redundant commands.

According to the Microsoft site and documentation, the changes in Visual Basic.NET are intended to:

• Simplify the language and make it more consistent
• Add new features requested by users
• Make code easy to read and maintain
• Help programmers to avoid introducing coding errors
• Make applications more robust and easy to debug

Overall, you can see that there is a commonality between versions of the language. Although you will need to learn some of the differences in syntax in this lesson, generally the skills that you have already with VBScript and classic Active Server Pages will continue to serve as a foundation for working with ASP.NET.

Looking at Basic Interface Differences

When working with VBScript from a standard text editor, such as Notepad, you simply see text in black on a white background. There are ways of viewing VBScript in color, but many people dabble with classic Active Server Pages without fancy editors. Microsoft's Visual Interdev is probably the most well-known editor for classic Active Server Pages that provides color-coding of the script.

Visual Studio.NET provides color distinction between HTML keywords, comments, code, and other elements of your Active Server Pages, as well as automatic indenting for added clarity of reading. You can change these colors to suit your own preferences or simply stick with the defaults.

The VBScript interface is purely coding and straight text input, whereas Visual Basic.NET, within the Visual Studio.NET integrated developer environment, offers visual coding elements, such as dragging and dropping controls into place and editing object properties from a list. Having a list of all the controls and properties visibly in front of you not only makes it easier to create a page, but also prevents you from having to recall what a particular property is called or from making typos and spelling mistakes. You also become aware of controls and properties that you may not have been aware of before, opening up new areas of learning.

Note: You can work with ASP.NET and Visual Basic.NET completely from a command line interface if you want to, although the vast majority of developers will be embracing the visual approach.

Declaring Variables in Active Server Pages

In classic Active Server Pages, using VBScript, you declare variables using the Dim keyword. Also, all variables are of type Variant. A declaration of two variables, one for holding an integer and one for holding a text string, would appear as follows:

```
Dim myInt
Dim myString
```

You could also declare these same variables on the same line, as follows:

```
Dim myInt, myString
```

However, using Visual Basic.NET to write your ASP.NET pages requires you to modify your method of declaring variables. You no longer declare all variables as Variant but rather you must specify the type explicitly using the As keyword followed by the type name. For instance, the following example shows how you would declare the same two variables as before, but this time using Visual Basic.NET:

```
Dim myInt As Integer
Dim myString As String
```

Table 26.1 shows you the main type names that are available in Visual Basic.NET. Integral numbers are whole numbers, such as integers, whereas nonintegral numbers are combinations of whole numbers and fractions or real numbers.

Type Name	Brief Description
Boolean	True or False
Byte	8 bit whole number
Char	Holds a single Unicode character
Date	Holds date and time values
Decimal	128 bit nonintegral number
Double	64 bit nonintegral number
Integer	32 bit whole number
Long	64 bit whole number
Object	32 bit address (pointer) to an object
Short	16 bit whole number
Single	32 bit nonintegral number
String	Holds multiple Unicode characters

Table 26.1. The main data types available in Visual Basic.NET

Using the Correct NonIntegral Data Type

You can see that there are three nonintegral data types of 32, 64, and 128 bit variations. These are the Single, Double, and Decimal data types, respectively. However, it is worth noting a few points about these data types, as they can be somewhat confusing at first. The Decimal data type has the smallest range of possible values despite being 128 bits. The Double actually has the largest range, with the Single data type coming next. This is because the Decimal data type is never rounded off like the Double and Single data types, which means that the Decimal is exact and precise, whereas the Double and Single are subject to error due to approximations.

Does this mean that the Decimal data type is not a good choice for large numbers? No indeed, as the range for the Decimal data type is +/- 79,228,162,514, 264,337,593,543,950,335, which is a massive number for many requirements. It does point out, though, that the Decimal data type is most suitable for financial requirements, or similar tasks, where precision of numerical figures is paramount.

Using Multiple Languages in Your Active Server Pages

In classic Active Server Pages you can have multiple scripting languages within the same page. For instance, you might use JavaScript for some of your client-side scripting code and VBScript for your server-side scripting code, mixed together with standard HTML. In ASP.NET using Visual Basic.NET, multiple languages used in this way are no longer supported. The reason for this is that each page is compiled, not interpreted, on a line-by-line basis. You can create a single Web application in ASP.NET that uses multiple languages, but each page must consist of a single language.

You can make use of controls that have been written in a language other than the language being used in a particular Active Server Page. This allows a lot of flexibility because controls will be a major factor in ASP.NET. Therefore, the ability to use them in any Active Server Pages regardless of the language that they are written in is both useful and very welcome.

Using the Response Object's Write Method

Although very simple, you must remember to always use parentheses with the Response object's Write method in ASP.NET. In classic Active Server Pages you could get away with the following code:

```
<%Response.Write "Hello World"%>
```

However, this will not work in ASP.NET. In ASP.NET you must use parentheses, as follows:

```
<%Response.Write("Hello World")%>
```

Working with Arrays

In VBScript and classic Active Server Pages you could have arrays that begin with 1 as the first element. In Visual Basic.NET and ASP.NET, all arrays have a base number of 0. So loops that use arrays should go from 0 to count - 1, instead of 1 to count.

Request object's Form and QueryString collections return a string array in classic Active Server Pages; however, in ASP.NET they return an individual string. To refresh your memory, you use the Form collection with the POST method and you use the QueryString with the GET method. For instance, if you had a form with multiple check boxes with the same name that the user could select from, you could get the results using classic Active Server Pages as follows:

```
<%Response.Write Request.QueryString("cboxes")%>
```

If the check boxes that the user checked on the form had values of red, white, and blue, then the result of the line of code above would be as follows:

```
red, white, blue
```

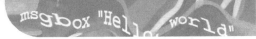

To give you an idea of the values when sent with the Get method, the Address field or URL might hold text similar to the following:

```
http://biggles/forms/colors.asp?cboxes=red&cboxes=
white&cboxes=blue
```

You could get the same result using ASP.NET by simply using parentheses with the Write method, as follows:

```
<%Response.Write(Request.QueryString("cboxes"))%>
```

However, to access an individual element of the array shows a bigger difference. In classic Active Server Pages you would access the first element of the array, which would return red in the example, as follows:

```
<%Response.Write Request.QueryString("cboxes")(1)%>
```

In ASP.NET you must use an explicit call to the GetValues method to access individual elements of an array. You must also remember that arrays in ASP.NET have a base of 0. Therefore the code to access the first element in an array using ASP.NET, as in the previous example, would appear as follows:

```
<%Response.Write(Request.QueryString.GetValues("cboxes")(0))%>
```

Working with Procedures

In VBScript when you want to return a value from a function that you have written, you assign the value to the name of the function itself. For example, if you were to write a function that takes two numbers as arguments and then adds them together and returns the result, your code might look similar to the following:

```
Function AddMe(a, b)
  AddMe = a + b
End Function
```

In Visual Basic.NET you return the value from a function by using the Return statement followed by the value that you want to return. Also, you must explicitly state the data type of each argument, as well as the data type being returned by the function itself. You explicitly state the data type of an argument by following it with the As keyword and the name of the data type. To explicitly state the data type being returned by the function itself, you use the As keyword and the name of the data type after the closing parentheses. The same example you saw before would be written in Visual Basic.NET as follows:

```
Function AddMe(a As Integer, b As Integer) As Integer
   Return a + b
End Function
```

The Return statement is new in Visual Basic.NET and it combines and replaces older statements from earlier versions of Visual Basic. Specifically, the Return statement replaces the Exit Sub command when it is used in a subroutine. Within a function, the Return statement replaces not only the Exit Function command but also the assignment of the return value to the function name, as you have just seen. Regardless of whether it is in a subroutine or a function, any lines of code within a procedure that come after a Return statement will not be executed.

The other very important aspect of working with procedures in Visual Basic.NET is where you can and cannot write them. In classic Active Server Pages using VBScript, you could declare subroutines and functions between a <% and a %> tag, as follows:

```
<%
Sub SayHello()
   Response.Write "Hello World"
End Sub

Dim counter
For counter=1 To 5
   SayHello
Next
%>
```

Using Visual Basic.NET with ASP.NET, you can no longer write code in this manner. All procedures that you write must be between a <script> tag and a </script> tag. For Visual Basic.NET, you should set the *language* attribute to VB and the *runat* attribute to server. This lets the server know that the code contained within the <script> block is Visual Basic and must be run on the server not the client. So the same example you saw previously would appear as follows when using Visual Basic.NET with ASP.NET:

```
<script language="VB" runat=server>
Sub SayHello()
  Response.Write("Hello World")
End Sub
</script>

<%
Dim counter
For counter=1 To 5
  SayHello
Next
%>
```

Due to the fact that procedures are now enclosed within a <script> block, you cannot have combinations of code and HTML mixed together in your functions in the same manner as with classic Active Server Pages. For example, the following example shows a subroutine that shows the current date and time on the server, using classic Active Server Page style:

```
<%Sub ShowDT()%>
The <b>current</b> date and time on the server is: <%=Now%>
<br>Have a nice day.
<%End Sub%>
```

Using ASP.NET you would have to place any of the HTML code within a call to the Response object's Write method, as follows:

```
<script language="VB" runat=server>
Sub ShowDT()
Dim s1 As String
  s1 = "The <b>current</b> date and time on the server is: "
  s1 = s1 & Now & "<br>Have a nice day."
```

```
  Response.Write s1
End Sub
</script>
```

Short-Circuiting Boolean Expressions

Although Boolean expressions have not changed dramatically in syntax, there is a difference in the way Visual Basic.NET handles Boolean expressions in comparison to VBScript. VBScript handles Boolean expressions by evaluating each part of an expression. Visual Basic.NET performs what is known as short-circuiting on Boolean expressions. This basically means that if one of the parts already determines the outcome of the expression, then the remaining parts of the expression are not evaluated.

To accomplish short-circuiting, Visual Basic.NET includes two new keywords —AndAlso and OrElse. AndAlso is the same as And except with short-circuiting behavior. OrElse is the same as Or except with short-circuiting behavior.

The difference is quite subtle; however, a simple example should add clarity. The following examples show the difference between using the traditional And operator and the AndAlso operator. The example uses a loop and a function, with the function being called as the second part of a Boolean expression. Each time the function is called, it increments the value of the variable counter by 1. The example is as follows:

```
<script language=vb runat=server>
Dim counter As Integer
Function NewValue() As Integer
  counter = counter + 1
  Return counter
End Function
</script>

<%
counter = 0
While counter < 5 And NewValue < 7
  Response.Write(counter & "<br>")
End While
Response.Write("Final value of counter: " & counter)
%>
```

The result of running this code is as follows:

```
1
2
3
4
5
Final value of counter: 6
```

The same example code but replacing the And operator with the AndAlso operator will enter the loop exactly the same number of times. The While loop and its Boolean expression would now appear as follows:

```
While counter < 5 AndAlso NewValue < 7
   Response.Write(counter & "<br>")
End While
```

The difference occurs when the counter reaches 5 and the first part of the Boolean expression fails. At this point the AndElse operator will short-circuit and not even call the NewValue function in the second part of the expression. You can see from the result that the second part of the script was never evaluated at all, as the result is only 5 rather than 6, as it was previously. The results are as follows:

```
1
2
3
4
5
Final value of counter: 5
```

Using While Loops

While loops are still supported in Visual Basic.NET, there has been a change to the ending of the loop. In Visual Basic 6 and VBScript, you end the While loop with the Wend keyword. For example, the following While loop continues while the variable i is less than 10:

```
Dim i
i = 1
While i < 10
  i = i + 1
Wend
```

In Visual Basic.NET, the Wend keyword is no longer supported. It has been replaced by the End While statement. So the same example you just saw but this time using Visual Basic.NET, would appear as follows:

```
Dim i
i = 1
While i < 10
  i = i + 1
End While
```

Losing the Set Command

In VBScript with classic Active Server Pages, you use the Set command quite often, such as when you are creating an instance of an object, as follows:

```
Set adocon = Server.CreateObject("ADODB.Connection")
```

In Visual Basic.NET with ASP.NET, the Set command is no longer supported. For greater language clarity and consistency, you simply assign the value in the same way as any normal variable assignment. So the above example using Visual Basic.NET would appear as follows:

```
adocon = Server.CreateObject("ADODB.Connection")
```

Note: *There is also a Let command in Visual Basic 6, which is also no longer supported in Visual Basic.NET.*

WHAT YOU MUST KNOW

Visual Basic.NET has some differences in syntax from Visual Basic 6 and its subset VBScript due to conformity with other .NET languages in making use of the Common Language Runtime. There are still many similarities between classic Active Server Pages and ASP.NET. In Lesson 27, "Working with Web Forms," you will begin to work with one of the bigger improvements to Active Server Pages, concentrating on becoming more familiar with creating graphical Web pages using a visual approach. Before you move on to Lesson 27, be sure that you are very confident with the following key concepts, as you will require these skills in your continuing endeavors with ASP.NET:

- To help with the clarity of the language, a lot of inconsistencies and obsolete commands have been removed from Visual Basic.NET and are no longer supported.

- ASP.NET provides a basic interface improvement, combining visual creation of Web forms through drag and drop functionality, as well as intelligent colored text for the code itself.

- In VBScript all variables are of type Variant. However, in Visual Basic.NET you must explicitly declare your variable data types. New data types, such as the Decimal, Short, and Char, are also in Visual Basic.NET.

- In classic Active Server Pages, you can have multiple scripts mixed on the same page. In ASP.NET, you can only use one language per page.

- In ASP.NET, you must always use parentheses with the Response object's Write method.

- In Visual Basic.NET, arrays use a 0 base. Individual array elements can be accessed in ASP.NET using an explicit method call.

- In place of Exit Sub and Exit Function statements, procedures use the new Return statement. Functions no longer assign the return value to the function name, but use the Return statement instead.

- You must declare procedures within a <script> tag and a </script> tag.

- Using the AndAlso and OrElse operators, you can use short-circuiting in your Boolean expressions.

- You must end your While loops with the End While statement because the Wend keyword is no longer supported in Visual Basic.NET.

- In Visual Basic.NET you simply assign the value or object to the variable instead of using the Set command.

WORKING WITH WEB FORMS

n Lesson 26, "Comparing Differences in Syntax," you looked at some of the differences in syntax between using VBScript with classic Active Server Pages and using Visual Basic.NET with ASP.NET. In this lesson, you will look at one of the more fascinating features of ASP.NET—Web Forms. Finally, Active Server Pages developers can use visual tools to create the visual interface for Web pages, although you will also look at the code that these tools create. By the time you complete this lesson you will know the following key concepts:

- Web Forms process the script in ASP.NET pages on the server, including when events are triggered by the user. Web Forms replace HTML forms and client-side scripting in Active Server Pages providing greater reliability and functionality.
- Clicking on the HTML button at the base of the main window in Visual Studio.NET, allows you to view the source code for any ASP.NET page.
- You can use the TextBox control and the Button control to add basic text field and button form functionality to your Web Forms. The properties of these controls allow greater ease and flexibility for customization than traditional HTML forms.
- The subroutine of the Button object's Click event is created as wrapper code by Visual Studio.NET when you double-click on the Button object in Design view.
- Using the Image control to add graphics to your Web Forms requires that you set the ImageUrl property either by hand or using the Select Image dialog box.
- When working with the ListBox or DropDownList controls, you can use either handwritten code or the ListItem Collection Editor dialog box to add individual items to the Items collection.

Understanding Web Forms in ASP.NET

To understand the need for Web Forms, you must look at what is currently available in Active Server Pages. Basically, whenever the user causes an event to occur, such as clicking on a button, client-side scripting had to come into play. This often caused problems if the scripting was done in VBScript because not everyone had the right browser or operating system. Often this meant that Active Server Pages developers ended up using JavaScript as well, to handle client-side scripting because it was more portable and compatible than VBScript. Even then, correct handling of the client-side script could not be ensured.

HTML forms, while useful, still lacked some of the niceties and flair of their Windows-based cousins (non-Web applications) when it came to displaying data and information. They were also a lot of work to create and maintain, especially when dealing with large numbers of forms and layouts of high complexity.

The demand for greater reliability, as well as an updated means of displaying data, was heard from professional sites that each represent a company presence online. To meet this demand, Microsoft came out with Web Forms in ASP.NET.

Web Forms replace traditional HTML forms, as well as the client-side scripting that is often necessary and goes hand-in-hand with those HTML forms. They accomplish this by handling code on the server. When the user triggers an event on an ASP.NET Web Form, it is no longer handled by the client but is sent back to the server for processing, as shown in Figure 27.1.

**Figure 27.1.
The Web Forms
cycle**

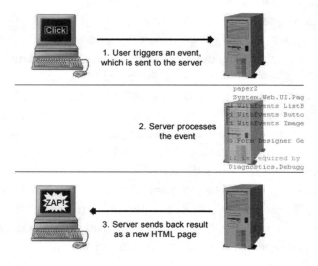

1. User triggers an event,
which is sent to the server

2. Server processes
the event

3. Server sends back result
as a new HTML page

While this means more round trips between the user's browser and the server, it also means that a powerful server (or server farm) can reliably make use of the latest software and technology and share it with users, regardless of their browser limitations. It is easier to update a single server than to wait until all of your customers are up to a certain standard of browser in order to provide new functionality.

Because all of the processing is done on the server, ASP.NET can make use of new controls for Web Forms, some of which you shall look at later in this lesson. Third parties can provide more controls for your use or you can even write your own. Basically, this means you are no longer limited to only what an HTML form can provide.

Viewing the Source Code of ASP.NET Pages

Open up the *net* project that you started in Lesson 25. You can see that you have already added a Label control to the Web Form, *paper1.aspx*. Visually you can see where this Label object is positioned on the Web Form, but you may want to look at the source code.

To discover what the source code looks like for any ASP.NET Web Form from within Visual Studio.NET, you can select the HTML button at the base of the main window, as shown in Figure 27.2. The source code in Figure 27.2 is the default source code prior to adding any controls to your Web Form. Notice that the <title> tags are inserted but empty of any actual title text.

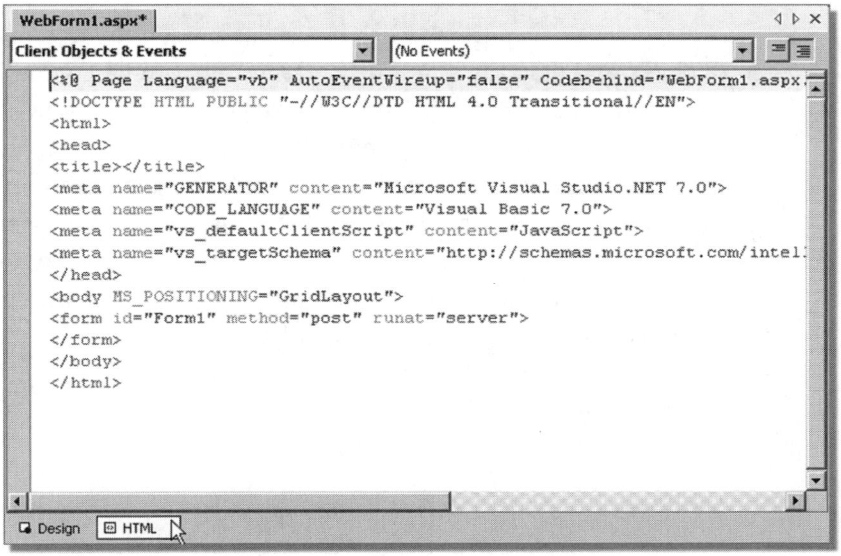

**Figure 27.2.
Selecting the HTML button to view the source code window**

```
<%@ Page Language="vb" AutoEventWireup="false" Codebehind="WebForm1.aspx.
<!DOCTYPE HTML PUBLIC "-//W3C//DTD HTML 4.0 Transitional//EN">
<html>
<head>
<title></title>
<meta name="GENERATOR" content="Microsoft Visual Studio.NET 7.0">
<meta name="CODE_LANGUAGE" content="Visual Basic 7.0">
<meta name="vs_defaultClientScript" content="JavaScript">
<meta name="vs_targetSchema" content="http://schemas.microsoft.com/intel
</head>
<body MS_POSITIONING="GridLayout">
<form id="Form1" method="post" runat="server">
</form>
</body>
</html>
```

Deleting Objects from ASP.NET Web Forms

Using visual techniques, it is simple to delete objects from your ASP.NET Web Forms. Basically, you select the object that you want to delete and then press the Delete key. For instance, you can delete the Label object that you created earlier by clicking on it and then pressing the Delete key. You can delete any object on your ASP.NET Web Forms by using this same technique rather than having to search for the appropriate code.

Adding Basic Form Elements

Some of the basic elements of forms, even in HTML forms, are text fields and buttons. You can easily add this basic form functionality to your Web Forms, as well by using the TextBox control and the Button control. You can select either of these controls by clicking on the appropriate control icon in the Toolbox panel, as shown in Figure 27.3.

**Figure 27.3.
The TextBox and
Button control
icons in the
Toolbox panel**

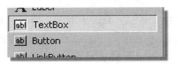

You can then simply click in the position on the Web Form that you want the object and drag the size that you want for it to be, as you learned in Lesson 25 with the Label control. Using the Properties panel, you can adjust features of both the TextBox and Button objects, such as the background color and the font.

To add functionality to a Button object when the user clicks on it, simply double-click on the Button object in the Design view and then Visual Studio.NET will create the shell of the subroutine, ready for you to insert your own code. Visual Studio.NET will also present you with the Visual Basic.NET source code window, as a result of you double-clicking on an object. The following is an example of the wrapper code created by Visual Studio.NET to handle the Click event:

```
Private Sub Button1_Click(ByVal sender As System.Object,
ByVal e As System.EventArgs) Handles Button1.Click

End Sub
```

You will recognize that this is basically the same as the shell of a subroutine in VBScript, which you have been using throughout this book and undoubtedly in your own Active Server Pages projects. Within the wrapper code you can insert any of your own code, which will execute whenever the user clicks on the Button object.

Try an example to understand this more clearly. You will create a TextBox object and two Button objects. When you click on one button you want the text in the TextBox object to change to Blue, and when you click on the other button you want the text to change to Green. It is a simple example, but it teaches some important fundamentals of working with events with ASP.NET.

In a new project, net2, add a TextBox object and two Button objects, and leave their ID property as the default. For all three of the objects, change the Font property's Size property to Medium. Change the Text property of one Button object to Blue and the other Button object to Green. Change the BackColor property to a light blue color for the Blue button and a light green color for the Green button. If you want, you can change the BackColor property of the TextBox object too.

Note: *When selecting colors for objects in Web Forms, it is a good idea to select colors from the Web tab in the color drop-down list.*

Double-click on the Green button to have Visual Studio.NET create the wrapper code for its Click event subroutine. The aim of this subroutine is to change the text in the TextBox object to Green. You can modify the Text property of an object by using standard Visual Basic dot notation between the object name and the property name and then assigning the value to it. This is identical in VBScript, Visual Basic 6, and Visual Basic.NET, so there is nothing new to learn here and you should feel quite comfortable by now with this syntax. So the code to change the Text property of the TextBox object to Green would be as follows:

```
TextBox1.Text = "Green"
```

Repeat the process to create a subroutine for the Blue button's Click event that changes the TextBox object's text to Blue. The final code for both subroutines should appear as follows:

```
Private Sub Button2_Click(ByVal sender As System.Object,
ByVal e As System.EventArgs) Handles Button2.Click
   TextBox1.Text = "Green"
End Sub

Private Sub Button1_Click(ByVal sender As System.Object,
ByVal e As System.EventArgs) Handles Button1.Click
   TextBox1.Text = "Blue"
End Sub
```

Save the project and from the Build menu select the Build option to create the ASP.NET page. Load the page in a browser. Figure 27.4 shows the TextBox object and two Button objects as seen from a browser:

Figure 27.4.
A Web Form showing a TextBox object and two Button objects

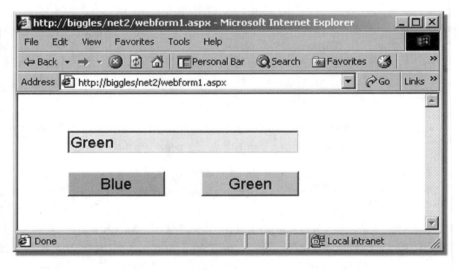

Adding Graphics to Web Forms

Rather than simply using text labels, you will at times have the need to insert graphics into your Web Forms. The main control for adding graphics to your Web Forms is the Image control, whose icon you can find in the Toolbox panel, as shown in Figure 27.5.

Figure 27.5.
The Image control
icon in the Toolbox
panel

You are going to create a simple page that displays an image as the title for an imaginary company called *PaperWorks*. You can either create a new project or use the *net* project from earlier in this lesson. Make sure you are in Design view by clicking on the Design button. Select the Image item from the Toolbox Panel. The Image item is under the Web Forms section. Drag the size of the Image object on the Web Form.

Using the Properties panel, change the ID property from *Image1 to imgTitle*. You can refer back to Lesson 25 or earlier examples in this lesson if you are unsure about changing properties in the Properties panel. Setting the ID of an object allows you to identify the object easily from code if you have multiple objects of the same type.

In the AlternateText property field, add the text *PaperWorks*. The text in the AlternateText property is shown if the user's browser cannot show images. Therefore, in the context of this example you are applying the company title in this field.

When you want to set the actual image to use in the Image object, you must use the ImageUrl property. In the Properties panel, there is a small button with an ellipsis in the ImageUrl field. Click on this button and Visual Studio.NET will open the Select Image dialog box, as shown in Figure 27.6.

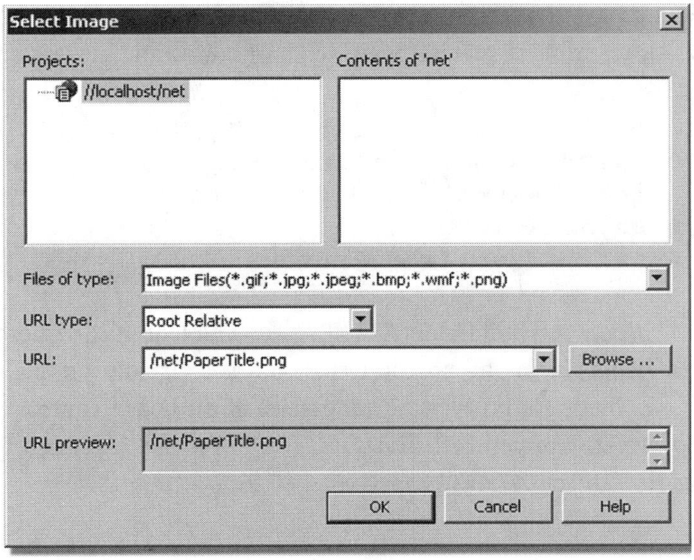

Figure 27.6.
The Select Image
dialog box

Select Image			✕
Projects:		Contents of 'net'	
//localhost/net			

Files of type:	Image Files(*.gif;*.jpg;*.jpeg;*.bmp;*.wmf;*.png) ▼
URL type:	Root Relative ▼
URL:	/net/PaperTitle.png ▼ Browse ...
URL preview:	/net/PaperTitle.png

OK   Cancel   Help

You can set the URL type field to be Absolute, Document Relative, or Root Relative. Absolute indicates an absolute path including logical drive letter. Document Relative indicates a path from the same folder that the current page is in. Root Relative indicates a path from the root Web server folder. Often, you can use any of these URL types to get the same image file. Select a URL type and type the correct path and filename of the image that you want to load or use the Browse button to select the image file from a list.

Set the BorderStyle property to Inset. You will notice an immediate update to the border of *imgTitle* in Design view if there is an image already loaded into the object.

You can position the image using a visual approach by simply pointing over the image so that the cursor changes to a four-directional arrow for positioning, as shown in Figure 27.7. When you have a four-directional arrow for a cursor, hold down the left mouse button and simply drag the image to the location that you want it to appear in the final Web page.

**Figure 27.7.
The cursor
changes to a four-
directional arrow
for positioning**

When you are adding many images to your ASP.NET pages, you can see that the visual approach is simpler and faster than coding by hand, as well as being less prone to human error. The code that Visual Basic.NET adds to the page, with the exception of precise position coordinates and measurements, is as follows:

```
<asp:ImageButton id="imgTitle" style="Z-INDEX: 101;
LEFT: 67px; POSITION: absolute; TOP: 9px" runat="server"
Height="62px" Width="446px" AlternateText="PaperWorks"
ImageUrl="/net/PaperTitle.png"
BorderStyle="Inset"></asp:ImageButton>
```

When you build your ASP.NET page and load it into a browser, the image will be displayed in the position that you set visually on the Web Form in Design view. There are many more properties of an Image control that you might want to investigate yourself. However, you have seen the main functionality, as well as how to set various properties.

Responding to Mouse Clicks with an Image

If you want to use an image in your Web Form that responds to mouse clicks from the user, you should use an ImageButton control instead of an Image control. An ImageButton control not only allows you to respond to mouse clicks, but also allows you to simulate the behavior of a command button through use of the OnCommand event handler, as well as the CommandName and CommandArgument properties.

Adding DropDownList and ListBox Controls

In this section you will learn about two very similar controls in ASP.NET: a visual method of working with the DropDownList control and a programmatic method of working with the ListBox control through code. However, both controls can be modified through either code or using visual methods, depending on your preference.

You can easily add a DropDownList object to your Web Forms in ASP.NET. A DropDownList allows you to provide a preset selection of items for the user to choose from. To add a DropDownList control to your Web Form in Design view, select the DropDownList control icon in the Toolbox panel, as shown in Figure 27.8, and then simply drag the size and position that you want for the object onto your Web Form.

**Figure 27.8.
The DropDownList
control icon in the
Toolbox panel**

To begin with, you will simply see a single row with the word *Unbound* and an icon with an arrow pointing down at the right of this. The word you have seen indicates that you do not have any data items bound to the DropDownList yet, and the small arrow icon shows where to click for the full selection of items. You can modify the contents of the Items collection at design time using the ListItem Collection Editor dialog box, as shown in Figure 27.9, which provides a visual approach for adding items.

Figure 27.9.
Adding items to the
DropDownList object
at design time

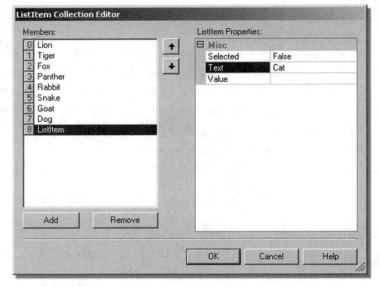

To add an item with the ListItem Collection Editor dialog box, simply click on the Add button and then enter the text that you want to display in the Text field. Unless you explicitly specify something different for the Value field, it will be set identically to the Text field. When you have finished adding items, click on the OK button to close the dialog box.

When you build and load your ASP.NET page into a browser, it appears as shown in Figure 27.10.

Figure 27.10.
The populated
DropDownList
control from the
user's perspective

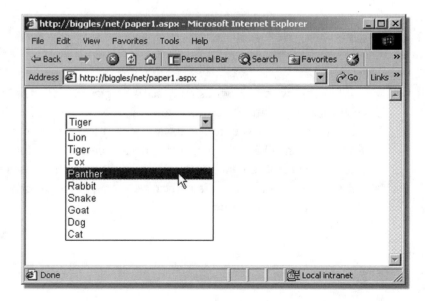

If you want to access the current selection of a DropDownList object, perhaps to display it in a Label control, use the SelectedItem property (which technically is inherited from the ListBox control), as follows:

```
Label1.Text = DropDownList1.SelectedItem.Text
```

A ListBox control is almost identical to a DropDownList control except for three things:

- It has a Rows property
- It has a SelectionMode property
- It has a preset display size

The Rows property indicates the number of rows in the ListBox object, which you might like to think of as the height of the ListBox object. The default value of the Rows property is 4, but you can set it to anything from 1 to 2000. You will get an error if you try to set the Rows property outside of this range.

The SelectionMode property allows you to choose whether you want to have single or multiple items selected in the ListBox object.

The preset display size can be set by you at design time in the Design view simply by dragging the window to the size that you want it to appear on the Web form. Visual Studio.NET will create the appropriate code for you, as follows:

```
<asp:ListBox id="ListBox1"
style="Z-INDEX: 103; LEFT: 51px; POSITION: absolute; TOP: 94px"
runat="server"
Width="172px" Height="188px">
</asp:ListBox>
```

You can add items to your ListBox object by inserting code within the <asp:ListBox> block. Each item that you want to add to your ListBox object must be placed within an <asp:ListItem> block. For example, if you want to add the item Red to your ListBox, you would write code similar to the following:

```
<asp:ListItem>Red</asp:ListItem>
```

To see a more holistic view of adding multiple items to a ListBox through a coding method, the following example shows how to add four color items to a ListBox object:

```
<asp:ListBox id="ListBox1"
style="Z-INDEX: 102; LEFT: 24px; POSITION: absolute; TOP: 18px"
runat="server" Width="173px" Height="116px">
<asp:ListItem>Red</asp:ListItem>
<asp:ListItem>Blue</asp:ListItem>
<asp:ListItem>Yellow</asp:ListItem>
<asp:ListItem>Green</asp:ListItem>
</asp:ListBox>
```

The result of compiling an ASP.NET page with this code and then loading it in a browser is shown in Figure 27.11.

**Figure 27.11.
The result of
compiling the
ListBox items code**

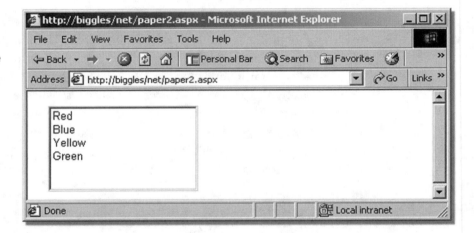

WHAT YOU MUST KNOW

Web Forms process the script in ASP.NET pages on the server, including when the user triggers events. In Lesson 28, "Advancing with Web Forms," you will learn to use some of the other controls in ASP.NET as you continue working with Web Forms. Before you continue with Lesson 28, make sure you know the following key concepts:

- Web Forms replace HTML forms and client-side scripting in Active Server Pages. Web Forms also provide a greater level of reliability and functionality than traditional Active Server Pages.

- You can view the source code for any ASP.NET page or Web Form in Visual Studio.NET by simply clicking on the HTML button at the base of the main window.

x You can use the TextBox control to add text field functionality to your Web Forms similar to that of traditional HTML forms. The addition of various properties make the TextBox control more customizable than its predecessor.

x You can easily add a button to your Web Forms by using the Button control. Double-clicking on the Button object in Design view causes Visual Studio.NET to create the wrapper code for the Button object's Click event subroutine.

x The Image control allows you to add graphics to your Web Forms. You can use the Select Image dialog box to easily set the Image object's ImageUrl property to specify the URL path of the image source.

x The ListBox control and the DropDownList control are very similar, and both present the user with a list of preset items. The items can be set either by code or the ListItem Collection Editor dialog box.

28

LESSON

myValue = 7

Integer

As String = {"Red", "Blue", "yellow", "Green"}

o UBound(colors)

Items.Add(colors(i))

cb12 Items.Add(colors(i))

Next

ADVANCING WITH WEB FORMS

*i*n Lesson 27, "Working with Web Forms," you learned how the new Web Forms in ASP.NET work, as well as some of the simple controls that you can use with your Web Forms. In this lesson, you will continue working with Web Forms, learning to use some of the other controls in ASP.NET to add further functionality to your Active Server Pages and advance the skills you already have with Web Forms. By the time that you complete this lesson, you will have covered the following key concepts:

- You can add hyperlinks to your Web Forms using a visual approach by adding a HyperLink control and setting its NavigateUrl property in the Properties panel or through code.
- Many of the controls that you use with Web Forms have a ToolTips property that lets you display some text explaining a control that the user hovers his mouse over.
- There are two check box controls in the Web Forms section of the Toolbox panel —the CheckBox control and the CheckBoxList control.
- With the focus of consistency being placed on Visual Studio.NET and all the .NET languages, you can find many occasions where learning how to use a property in one particular control means that you automatically know how to use the same property in other controls.
- You can populate the Items property of a CheckBoxList in many ways, including using the ListItem Collection Editor dialog box, the Add method, arrays, and ListBox controls.

- You can use the UBound method to discover the upper bound of an array, and use the Count method to return the number of items in a collection.
- The RadioButton control and the RadioButtonList control behave almost identically to their check box counterparts, except that you can only have a single selection within any radio button group. The GroupName property determines the group of a RadioButton object.
- The Panel control is a container that holds other controls, allowing you to create groups and work easily with them.
- By using the PlaceHolder control, you can reserve a place in a Web Form and then add controls dynamically using code.
- The Calendar control lets you easily provide a visual calendar in your Active Server Pages for users to either select dates or view them.

Adding Hyperlinks Visually

As simple as they are, you also have a method of adding hyperlinks to your Web Forms using a visual approach. To add a hyperlink you must use the Hyperlink control, whose icon you can find in the Toolbox panel, as shown in Figure 28.1.

**Figure 28.1.
The Hyperlink control icon in the Toolbox panel**

Adding the HyperLink control to your Web Form is the same as other Web controls, utilizing a visual click and drag approach. Click on the HyperLink control icon and drag the size that you want for your hyperlink on the Web Form, so that you see something similar to Figure 28.2.

**Figure 28.2.
After adding a HyperLink object to your Web Form**

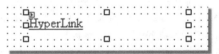

In the Properties panel you can change the Text property to set the visible text to something of your own choosing. To actually set the target URL you must set the NavigateUrl property in the Properties panel or through code. You can click on the small button with an ellipsis to bring up the Select URL dialog box, which is essentially the same as the Select Image dialog box that you learned about in the previous lesson.

Almost all of the Web controls that you use with your Web Forms have two useful properties, for hiding or disabling, and the HyperLink control is no exception. The Visible property can be set to False to hide the control or True to show it. This is useful if you have links to sensitive pages that you only want to show under certain conditions. You can place the controls and set their size at design time, even if you do not want to show them yet. The Enabled property will enable a control when set to True or disable it if set to False. This is useful when you want to show controls to the user but not activate them, perhaps because some prerequisite has not been met yet.

Telling the Difference Between a HyperLink and a LinkButton

A LinkButton control is similar to a HyperLink only in appearance. In functionality, it is like the Submit button in an HTML form—sending the data in the form to the server for processing. You can also configure the LinkButton control to behave like an ordinary command button, if you want to. The main point to be aware of when starting with Web Forms is that the LinkButton control is not to be confused with a HyperLink control, despite its initial visual similarity.

Adding ToolTips to Your Web Form Objects

Many of the controls that you can use in your Web Forms have a ToolTip property. You can enter text into the ToolTip property field or assign text to it through code. The text that you assign to the ToolTip property will be displayed in a small yellow rectangle when the user points his mouse over the object. You will have seen this functionality in many applications, and it is now available in your Active Server Pages.

For example, if you had a HyperLink object that had a caption of *weather,* you could set the ToolTip property to something more explanatory, such as *See today's weather forecast*, as shown in Figure 28.3.

**Figure 28.3.
Showing a ToolTip
associated with a
HyperLink object**

Working with Check Boxes

You have already used check boxes in traditional HTML forms, as well as made use of them through classic Active Server Pages. You can continue to use check boxes in ASP.NET and Web Forms, yet you now have much greater flexibility and customization available to you. There are two separate check box controls for use with Web Forms in ASP.NET. One is an individual check box that you can create using the CheckBox control in the Web Forms section of the Toolbox panel. The other is a list of multiple check boxes that you can create using the CheckBoxList control, which is also in the Web Forms section of the Toolbox panel. Figure 28.4 shows both check box controls and their associated icons, as they appear in the Toolbox panel.

**Figure 28.4.
The two Web Form
controls for working
with check boxes**

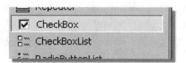

The CheckBox control has two key properties: the Text property and the Checked property. The Text property is where you can enter the caption that the user will see next to the check box. The Checked property takes a Boolean value of either True if it contains a check mark or False otherwise. You can set both of these properties from the Properties panel in Design view or using code similar to the following:

```
CheckBox1.Text = "Red"
CheckBox1.Checked = False
```

As with many other controls that you can use with Web Forms, the CheckBox control can be highly customized to fit your requirements. You can change the font, the border, the background color, the text alignment, or any combination thereof using the standard properties, such as BackColor, BorderColor, BorderStyle, BorderWidth, Font, ForeColor, and TextAlign. Once again, these properties can be set either from the Properties panel in the Design view or through code. In Figure 28.5 you can see various renditions of the CheckBox control, showing just a small part of the much greater range of flexibility that this control has in ASP.NET.

**Figure 28.5.
Various renditions
of the CheckBox
control**

You will notice that Microsoft has gone to a lot of trouble to make the interface and language consistent so that developers quickly become very comfortable, even with new controls. For instance, once you learn how to use the BackColor property in one control, you can apply that same knowledge immediately to the BackColor property of many other controls.

The CheckBoxList control is very similar to the CheckBox control except that you can work with multiple check boxes with greater ease. Unlike the CheckBox control, the CheckBoxList control does not have a Text property to set the caption, which makes sense when you consider that there are multiple check box items. Instead, the CheckBoxList control has an Items property, which is a collection and behaves in the same manner as the DropDownList control's Items property or the ListBox control's Items property. Once again, you can see the subtle improvement of consistency that shines through in Visual Studio.NET. You can add items to the Items collection using the Properties panel and the ListItem Collection Editor dialog box. You can also use code to add items to a CheckBoxList control's Items property, using the Add method and assigning the string that you want to use for the caption, as follows:

```
' cbl1 is the ID of your CheckBoxList object
cbl1.Items.Add("Red")
```

Alternatively, you can use an array to hold all the items and then add this data to multiple CheckBoxList objects in code, a method that could save considerable work. Arrays are simply a single reference point for multiple values of the same basic type. Notice that you use a pair of parentheses after the array variable name to indicate that it is an array. Also notice that you can directly declare variables and initialize them in the same line. To do this, you must separate each element of the array using a comma and place all of these elements within a set of curly braces. The final point to be aware of is the UBound method, which returns the upper bound of any given array. In the example, *cbl1* is the ID of one CheckBoxList object and *cbl2* the other. The source code for assigning items to an array once and then to multiple CheckBoxList objects is as follows:

```
Dim i As Integer
Dim colors() As String = {"Red", "Blue", "Yellow", "Green"}
For i = 0 To UBound(colors)
    cbl1.Items.Add(colors(i))
    cbl2.Items.Add(colors(i))
Next
```

Figure 28.6 shows the result of running this code to populate two CheckBoxList objects, as well as some of the diversity that is available in final appearance.

Figure 28.6.
After using code
to populate two
CheckBoxList
objects

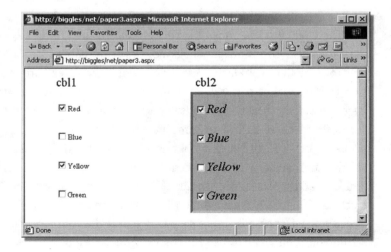

Another method that you might use to populate your CheckBoxList object's Items property collection is to use a ListBox control that you have already created and populated. You have already learned about using a ListBox control in the previous lesson. Similarly to what you did with the array, you can use a loop to populate the CheckBoxList object's Items property using the ListBox object's Items property. You cannot use the UBound method with the Items property, however, because it is a collection and the UBound method only works with arrays. Instead, you can use the Count method that every collection has. Because all collections have a base 0 in Visual Basic.NET, your loop must go from 0 through to Count-1 (not Count). The example code for showing how to populate a CheckBoxList object using a ListBox object is as follows:

```
Dim i As Integer
For i = 0 To ListBox1.Items.Count — 1
    cbl1.Items.Add(ListBox1.Items(i))
Next
```

Working with Radio Buttons

After working with the CheckBox control and the CheckBoxList control in your ASP.NET Web Forms, and also taking into consideration the consistency feature throughout the .NET framework, you will be very comfortable working with radio buttons in ASP.NET. There are two radio button controls in the Web Forms section of the Toolbox panel, as shown in Figure 28.7, namely the RadioButton control and the RadioButtonList control.

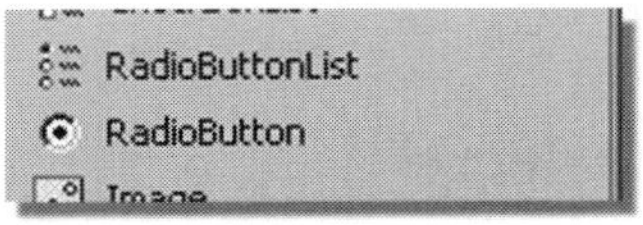

Figure 28.7. The two radio button controls in the Web Forms section

The RadioButton control and the RadioButtonList control, although almost identical to their check box counterparts, can only have a single selection within any particular group of radio buttons. Like their check box counterparts, you can determine by the value of the Checked property if a radio button is checked or not.

The RadioButtonList is automatically a group of items by nature and, as such, does not require any specific grouping. You can add items to the

RadioButtonList object using the Items property. The Items property is something that you are by now quite familiar with.

The RadioButton control is an individual control and so it requires something more to indicate the group that it belongs to. The GroupName property allows you to specify a unique string that indicates the name of a group to which the RadioButton control belongs. Any other RadioButton controls that have the same value for their GroupName property are part of the same group, and will automatically uncheck themselves if another radio button in the same group is checked by the user.

To see an example of using the RadioButton controls and attain a better understanding of how to use them, you will create an ASP.NET page that has two radio buttons—one for male and one for female.

First, add two RadioButton controls to a new Web Form, using the Toolbox panel. Make sure that you select the RadioButton control icon and not the RadioButtonList control icon. Change the Text property of one RadioButton object to *Male,* and then change the other to *Female*. Change the GroupName property of both RadioButton objects to *gender*. You might want to add a Label control that asks the user to select their gender. Build the ASP.NET page and load it into your browser. The result should appear something similar to Figure 28.8. Try alternately selecting each of the two radio buttons and see how they automatically uncheck themselves.

**Figure 28.8.
Two RadioButton
controls with the
same GroupName
property value**

Working with the Panel Control

The Panel control is basically a container for other controls. It allows you to easily group a number of objects for easy management. For instance, you could place thirty different controls within a Panel control and then by simply moving the Panel control, you move all of the thirty contained controls, maintaining their position relevant to each other. You can also make a Panel control hide or show a group of objects by making use of the Panel control's Visible property. When the Visible property is set to False, none of the controls that are contained within the Panel control are visible. This is useful because it can save many lines of code when you toggle between displaying and hiding a large group of objects.

You can add a Panel control to your ASP.NET pages by selecting the Panel control icon under the Web Forms section of the Toolbox panel, as shown in Figure 28.9.

**Figure 28.9.
The Panel control icon under the Web Forms section of the Toolbox panel**

As with many controls in ASP.NET and Web Forms, the Panel control is quite customizable, and you can change the border, background, color, font—even set a background image if you want. Figure 28.10 shows two Panel controls, giving you an example of the variety of customization that you can work with.

**Figure 28.10.
Two Panel controls showing an example of the variety of customization available**

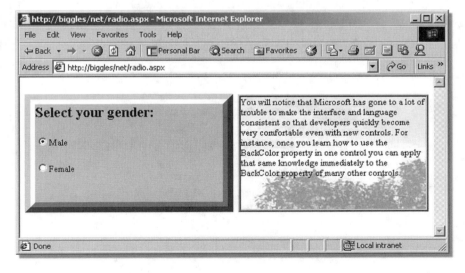

Reserving a Place in a Web Form

You can reserve a place in a Web Form and then add controls dynamically through the use of code. To reserve a place in a Web Form, you must use the PlaceHolder control. Figure 28.11 shows the PlaceHolder control icon, which is located in the Web Forms section of the Toolbox panel.

**Figure 28.11.
The PlaceHolder
control icon in
the Toolbox panel**

When you add the PlaceHolder control to your Web Form in Design view, the first thing you notice is that it does not maintain the position and size that you give it. This is because the PlaceHolder control does not actually provide any visible output in the resultant Web Form. There are only four properties for the PlaceHolder control in the Properties panel. The two most useful of these properties are the ID property to be able to identify the PlaceHolder control from code, and the Visible property to be able to hide or display the contents of the PlaceHolder control in a single line of code.

The basic idea behind the PlaceHolder control concept is that you create and prepare some other control, including setting any properties for it, and then you add the control to the PlaceHolder control using the Controls collection's Add method.

The following example code shows how to add three Button controls to a PlaceHolder control with an ID of PlaceHolder1 (the default ID). You could place this code in a subroutine for when the page starts, or perhaps in a button's Click event subroutine—that choice is up to you.

```
Dim myButton As Button

myButton = New Button()
myButton.Text = "Red"
PlaceHolder1.Controls.Add(myButton)

myButton = New Button()
myButton.Text = "Blue"
PlaceHolder1.Controls.Add(myButton)

myButton = New Button()
myButton.Text = "Yellow"
PlaceHolder1.Controls.Add(myButton)
```

You can see clearly the pattern of preparing each Button control before adding it to the Controls collection. Figure 28.12 shows the result of this code when you view it in a browser. Notice the difference in size of the buttons, due largely to the fact that the only property you set was the Text property.

Note: *Although the example is adding Button controls, you can add other controls using the same format.*

**Figure 28.12.
The result of using a
PlaceHolder control
and adding Button
controls to it**

If you want to delete the second button, *Blue*, you would need to use the Remove method of the Controls collection, as follows:

```
With PlaceHolder1.Controls
   .Remove(.Item(1))
End With
```

Even though it is the second item that you want to remove, you use the index 1 because the collection is zero based (i.e. starts from 0).

Working with the Calendar Control

You can easily add a graphical calendar to your ASP.NET pages using a Calendar control. Figure 28.13 shows the Calendar control icon from the Web Forms section of the Toolbox panel.

**Figure 28.13.
The Calendar
control icon in the
Toolbox panel**

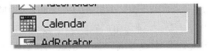

The Calendar control has a minimum value for its Height and Width properties, so when you drag the position and size for it onto your Web Form, do not be surprised if it jumps to a slightly different size at first. After you have initially placed the Calendar control, you can point your mouse over the edge of the control and the cursor changes to a multi-directional arrow. You can then hold down the left mouse button and resize the Calendar control accordingly.

The Calendar control provides a means of easily providing a graphical representation for users to either select a date or view a date. You would typically use it for sites where your clients must often select dates, such as airline agencies, or wedding venues. You might also use it to give your clients a visual way of selecting a date for small, one-time services, such as lawn-mowing, window washing, or general consultation availability. The uses for the Calendar control are varied and many, as working with dates is such an integral part of people's lives.

You can configure the Calendar control to have a specific appearance by using the many properties that it provides. Many of these are standard properties that you will already be familiar with, such as the BackColor property or BorderStyle property. Other properties are specific to the Calendar control, such as the DayNameFormat, FirstDayOfWeek, and ShowDayHeader properties, to name just a few. There are many options available to configure the look and feel of a Calendar control to suit your purposes and to change the overall look of your site. Figure 28.14 shows a Calendar control in Design view, clearly portraying how the control updates its appearance without the need to be compiled and viewed in a browser. This makes it easier and faster to achieve the effect you require for your Calendar control.

**Figure 28.14.
Setting up a
Calendar control
in Design view**

<		September 2001				>
Sun	Mon	Tue	Wed	Thu	Fri	Sat
26	27	28	29	30	31	1
2	3	4	5	6	7	8
9	10	11	12	13	14	15
16	17	18	19	20	21	22
23	24	25	26	27	28	29
30	1	2	3	4	5	6

You can determine what date the user selects by using the Calendar control's SelectedDate property. If the user has selected more than one date, this property will be the first date within the range of dates selected. The following code shows an example of placing the user's selected date into a Label control:

```
Label1.Text = "Date selected: " & _
Calendar1.SelectedDate.ToShortDateString()
```

You can respond to the user making changes to the Calendar control by utilizing the SelectionChanged event in your code. By simply double-clicking on your Calendar control in Design view, Visual Studio.NET will create the wrapper code for the SelectionChanged event subroutine for you.

You can set the current date in the Calendar control when your ASP.NET page first opens by simply assigning the TodaysDate property to the SelectedDate property within the Page_Load event subroutine, as follows:

```
Sub Page_Load(ByVal sender As Object, ByVal e As EventArgs)
    Calendar1.SelectedDate = Calendar1.TodaysDate
End Sub
```

The Calendar control is a simple and useful control; however, it has many properties that you can use to customize it, which make it appear more complex than it is. There are far more features that the Calendar control provides than can be covered in this section, including multiple selection of dates, selecting entire weeks or months, title customization, and more. If you are doing work with dates, be sure to look more closely at the Calendar control and the huge range of customization that it gives you within your Active Server Pages.

WHAT YOU MUST KNOW

In this lesson, you were exposed to a variety of Web Form controls that are able to increase the productivity of Active Server Pages developers, such as yourself, by providing specific functionality that you can easily add to your pages and work with. You now know how to work with hyperlinks, check boxes, and radio buttons in Web Forms. You also learned about adding tooltips to the controls in your Web Forms, as well as using Panel controls, and reserving places in your Web Forms. In Lesson 29, "Using Validation Controls," you will learn how to use more Web Form controls, all of which focus on providing validation functionality of one form or another. Before you begin Lesson 29, make sure that you understand the following key concepts:

- Hyperlinks can be added to your Web forms using the HyperLink control and setting its NavigateUrl property.

- You can create a small explanatory caption for many controls in your Web Forms by setting the ToolTips property for that particular control.

- You can add check boxes to your Web Forms using the CheckBox control and the CheckBoxList control. The CheckBoxList control is a group of check boxes, while the CheckBox control is a single check box.

- The ListItem Collection Editor dialog box can be used to populate the Items property of a CheckBoxList using a visual approach. You can also use the Add method, arrays, and ListBox controls.

- Due to a strong consistency focus, you will find that you become quickly comfortable with many properties because they have identical functionality and are used in many controls.

- When you are using a collection, you can use its Count method to determine the number of items that it contains. When you are using an array, you can use the UBound method to discover the upper bound, and therefore determine the number of items that it contains.

- Within a group of radio buttons, you can only ever have one particular radio button selected. Other than this main point, the RadioButton control and the RadioButtonList control behave almost identically to their check box counterparts.

- For RadioButton controls, you can use the GroupName property to determine the group that it belongs to by supplying a unique string.

- When you want to group multiple controls together easily, you can use the Panel control. The Panel control is a container for holding other controls.

- The PlaceHolder control has no visible output itself, but allows you to reserve a place in a Web Form where you can later add or remove controls dynamically by using code.

- Highly customizable, the Calendar control allows you to provide a graphical calendar in your Active Server Pages. You can use the Calendar control to select dates or display dates.

USING VALIDATION CONTROLS

n Lesson 28, "Advancing with Web Forms," you learned how the new Web Forms in ASP.NET work, as well as some of the simple controls that you can use with your Web Forms. In this lesson, you will continue working with Web Forms, learning to use the validation controls in ASP.NET to make sure data entries meet your requirements.

- Only certain controls can be used with the six validation controls supplied by Microsoft.
- The RequiredFieldValidator control makes sure that there is some data in the control to be validated. This ensures that important fields are not missed in forms.
- The ValidationSummary control lets you display further information for all validation controls that failed to be satisfied. The ErrorMessage property of each of the validation controls is used for this.
- You can make sure that the data is within a given range using the RangeValidator control.
- The CompareValidator control lets you use comparison operators against either another control or a value. You can also compare for data types.
- The RegularExpressionValidator control lets you do more complex validations on controls by using pattern matching algorithms.
- Use the CustomValidator control if you cannot attain the level of validation that you require using the standard validation controls. This allows you to write your own validation script or procedure.

Introducing the ASP.NET Validation Controls

There are six validation controls that Microsoft provides with ASP.NET. You can use them to make sure that entries made by a user conform to necessary requirements before being submitted to the server for processing. Not all controls are able to be validated in Web Forms. Within the Toolbox panel, the controls that you can validate consist of only four from the Web Forms section and four from the HTML section. Furthermore, the validation is performed against only one particular property in each case, as shown in Table 29.1.

Control	Validation Property
DropDownList	SelectedItem.Value
HtmlInputText	Value
HtmlTextArea	Value
HtmlSelect	Value
HtmlInputFile	Value
ListBox	SelectedItem.Value
RadioButtonList	SelectedItem.Value
TextBox	Text

Table 29.1. The various controls and properties that you can validate

The six validation controls, and a brief description of each, are shown in Table 29.2.

Control	Description
CompareValidator	Makes sure the entry compares correctly with a value
CustomValidator	Makes sure the entry meets custom conditions
RangeValidator	Makes sure the entry is between specified lower and upper boundaries
RegularExpressionValidator	Makes sure that the entry matches a pattern defined by a regular expression
RequiredFieldValidator	Makes sure that a control contains data
ValidationSummary	Displays a summary of all the validation errors on a page

Table 29.2. The validation controls for use in your Web Forms

Using the RequiredFieldValidator Control

The RequiredFieldValidator control ensures that a user inputs data into a required field on a form. For instance, if you had a form that contains a text field for the user's name and one for the year of birth, you may only require the name field to be filled in by the user. You can find the RequiredFieldValidator icon in the Web Forms section of the Toolbox panel, as shown in Figure 29.1.

Figure 29.1. The RequiredField Validator control icon in the Toolbox panel

To give you some practical experience while learning about the various validation controls, you will create a simple form and validate one of the controls in it. The form will indicate when validation is successful.

Create a new Web Form by selecting the Project menu Add Web Form item. Select a file name for your new Web Form, such as *form.aspx*. Using the skills that you already have, add two TextBox controls for the name and year of birth fields, with accompanying Label controls to indicate to the user what they are. Add another Label control for the result with a red font color by setting its ForeColor property. Change the ID of the result Label control to lblResult. Add a Button control and change its Text property to Submit. Finally, add a RequiredFieldValidator control to the right of the Name field TextBox control. Set the Text property of the RequiredFieldValidator control to * *Required*. Remaining in Design view, you should see something similar to Figure 29.2 for your general control layout.

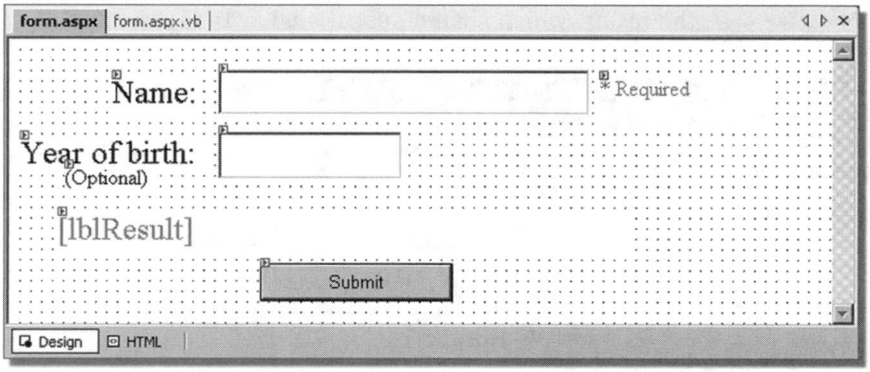

Figure 29.2. Designing a Web Form with a RequiredField-Validator control

form.aspx	form.aspx.vb

Name: [   ] * Required

Year of birth: [   ]
(Optional)

[lblResult]

[Submit]

Design | HTML

You must associate the RequiredFieldValidator control with the control that you want to validate. Remember that this must be one of the controls that you saw in Table 29.1 earlier in this lesson. To associate the RequiredFieldValidator control with another control you use the ControlToValidate property in the Properties panel. The drop-down list for this property makes it easy to select a control because it only lists those controls that qualify for validation. Set the ControlToValidate property to the TextBox control that holds the user's name.

You will use the Submit button's Click event to determine if the page is validated successfully or not. If the page is valid, then you want to display a message to the user using the lblResult Label control, as well as removing the Submit button from sight by making its Visible property False. To determine if the page is valid from code, you can use the Page.IsValid method, which will return True if the page is valid and False otherwise. Double-click on the Submit button in Design view. Visual Basic.NET will create the wrapper code for the Click event subroutine for you, in which you can place your code, as follows:

```
Private Sub Button1_Click(ByVal sender As System.Object,
ByVal e As System.EventArgs) Handles Button1.Click
    If Page.IsValid Then
        lblResult.Text = _
        "Page has been validated successfully."
        Button1.Visible = False
    End If
End Sub
```

Save your work and then compile your ASP.NET page using the Build menu Build option. Load your page into a browser and type some data into the Year of birth field, but not into the Name field. Click on the Submit button to observe the result, which should appear similar to Figure 29.3.

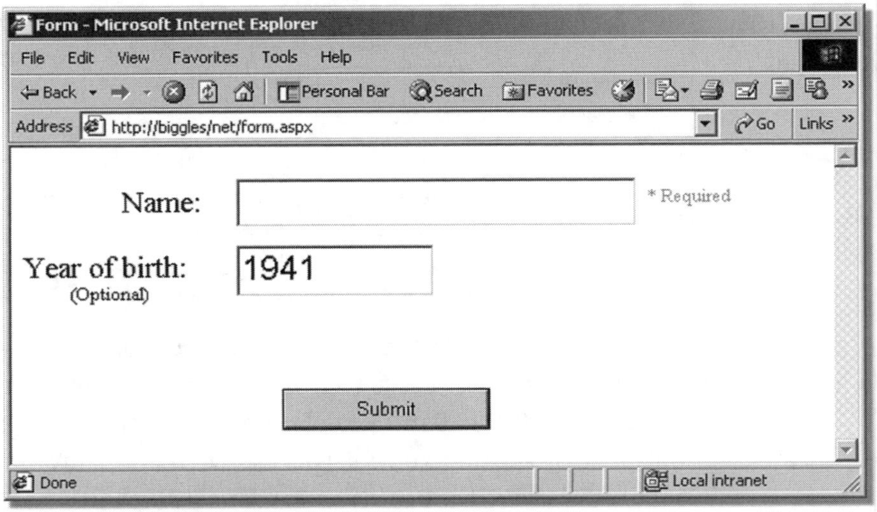

**Figure 29.3.
The result of not
entering data in
the required field**

When no data is entered into the Name TextBox control, in this case the required field, the RequiredFieldValidator control becomes visible and displays the text in its Text property, thus informing the user of what is required. Enter a name into the Name field of the form, and click on the Submit button again. This time, the RequiredFieldValidator control detects data in the Name field and the page is valid. The RequiredFieldValidator control hides its text, and the Button control's Click event causes your Label control to display a success message, as shown in Figure 29.4.

Figure 29.4.
The same form after entering data in the required field

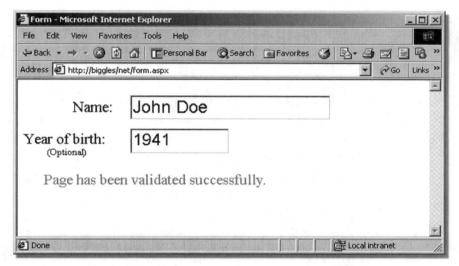

Validating an ASP.NET Page When Loading

You will notice that you used the Page.IsValid method as part of the button's Click event when attempting to find out if the data in all of the controls were valid. There are occasions when you would like to check the validation of a page when the page first loads. This might be to inform the user straightaway about which controls or fields require data, rather than leaving them to guess or find out upon submission. In these cases you would need to use the Page_Load event for the page.

You cannot use the Page.IsValid method in the Page_Load event. However, the Page.Validate method manually validates the page and can be used in this event. The exception to using the Page.IsValid method in the Page_Load event is after a Page.Validate method call has been made.

Using the ValidationSummary Control

You can give the user even more feedback by utilizing the ValidationSummary control. The ValidationSummary control will take note of the controls that failed to pass validation and present them to the user in the form of a bulleted list, a list, or a simple paragraph. To add this functionality to your Web Form, you must select the ValidationSummary control icon in the Web Forms section of the Toolbox panel, as shown in Figure 29.5.

Figure 29.5. The ValidationSummary control icon in the Toolbox panel

After selecting the ValidationSummary control icon, you can drag the size and position of the control on your Web Form. Like the RequiredFieldValidator control, the ValidationSummary control becomes visible when validation fails. You can use the ValidationSummary control's HeaderText property to set some text at the top (or header) of the control. Set the HeaderText property to *The following fields require data:*, which indicates to the user what information the control is asking for.

To make things more interesting, as well as to highlight the ValidationSummary control with greater clarity, add a RadioButtonList control and add two items—Male and Female. Add a Label control in front of the RadioButtonList control and change its text property to *Gender:*. Set the RadioButtonList control's border and font to match the other controls on the form. Copy the RequiredFieldValidator control that you have associated with the Name field by right-clicking on it and choosing Copy from the pop-up menu. Point your mouse to a blank section of the Web Form, right-click again, and select Paste from the pop-up menu. Visual Studio.NET will create a second RequiredFieldValidator control. Position this new control to the right of the RadioButtonList control.

Unlike the RadioButtonList control, you do not associate the ValidationSummary control with any particular control, as it takes into consideration all other validation controls on the page. In each of these other validation controls, you must set the ErrorMessage property to something meaningful. Set the ErrorMessage property of RequiredFieldValidator1 to *Name*. Set the ErrorMessage property of RequiredFieldValidator2 to *Gender*. If the control that these validation controls are associated with fails validation, then the ErrorMessage property will be displayed.

To add greater visibility, set the ValidationSummary control's BorderStyle to solid. The DisplayMode property indicates how you want the ValidationSummary control to display any ErrorMessage properties. Leave the default setting of BulletedList for now, but be aware that you can change this property if you want to.

Save your work and then select Build from the Build menu. Load *form.aspx* into your browser. With no data in any of the fields and then clicking on the Submit button, you should see something similar to Figure 29.6, where the two RequiredFieldValidator controls have shown up and also the ValidationSummary control.

**Figure 29.6.
The result of submitting no data and using a ValidationSummary control**

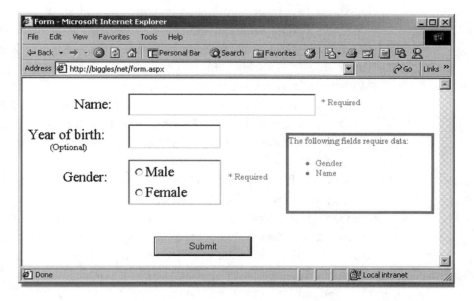

Select your gender from the radio buttons and click on the Submit button again. Notice the update to both the RequiredFieldValidator control and the ValidationSummary control. Enter your first name in the Name field, but leave the Year of birth field empty. Clicking on the Submit button now results in all the validation controls being hidden. Because the Year of birth field is not associated with any validation control, it has no impact on the validation of the form.

Using the RangeValidator Control

Rather than just checking to see that data has been entered into a field, you can make sure that the data is within a given range using the RangeValidator control. You can add a RangeValidator control to your Web Form by selecting its icon from the Toolbox panel, as shown in Figure 29.7.

**Figure 29.7.
The RangeValidator
control icon in the
Toolbox panel**

You will continue working with the example form that you have been working with so far in this lesson; place a validation that makes sure that the Year of birth field contains an integer between 1900 and 2001.

Add a RangeValidator control to your Web Form beside the Year of birth field. Using the same technique that you have already learned, prepare an appropriate caption for the ErrorMessage property so that the ValidationSummary control reports it properly. Using something that shows the valid range is always a good idea, such as *Year of birth (1900-2001)*. In the Text property assign the string * *Required,* to match the general format of your form. Set the ControlToValidate property to your TextBox control that contains the Year of birth field.

So far, all the properties of the RangeValidator control have been the same as what you saw with the RequiredFieldValidator control. There are some additional properties that require your attention though. The MaximumValue property indicates the maximum value of the valid range and the MinimumValue property indicates the minimum value of the valid range. From the Properties panel, set the MaximumValue property to *2001* and the MinimumValue property to *1900*.

You must also set the Type property of the RangeValidator control. The Type property informs the RangeValidator control of the data type that you are validating. If it is not in the appropriate data type, the RangeValidator control will attempt to convert it if possible, however, if conversion fails then the validation fails. The available data types that you can compare are as follows:

- String
- Integer
- Double
- Date
- Currency

In the example that you are working through, you require the data to be of type Integer in the Type property. This is because the value of a year is a whole number that does not require many digits. Within the Properties panel, scroll down to the Type property field and select Integer from the drop-down menu.

Save your work and compile the form by using the Build menu Build option. Load the form in the browser and leave all of the fields empty. Within the Year of birth field, type in a year that is less than 1900 and try submitting the form. Even though there is data in the TextBox control, the RangeValidator control rejects the data because it is outside of the given range that you set earlier, as shown in Figure 29.8.

Figure 29.8.
A field not accepting data as valid because it is outside of the given range

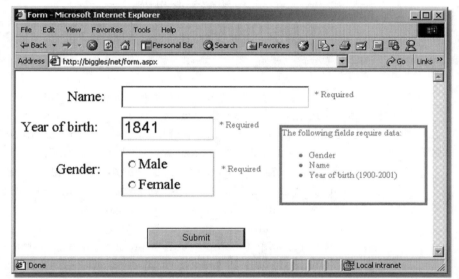

Using the CompareValidator Control

The CompareValidator control also adds a higher level of validation than simply checking if data exists in a control. You can use the CompareValidator control to make sure that the value in a control is equal to the value in another control. You are not limited to equality comparisons, but can use any of the comparisons shown in Table 29.3.

Operator	Description
DataCheck	True if the value is the right data type
Equal	True if two values are equal
GreaterThan	True if first value is greater than second value
GreaterThanEqual	True if first value is greater than or equal to second value
LessThan	True if first value is less than second value
LessThanEqual	True if first value is less than or equal to second value
NotEqual	True if two values are not equal

Table 29.3. The various operators for use with the CompareValidator control

You can find the CompareValidator control icon in the Web Forms section of the Toolbox panel, as shown in Figure 29.9.

**Figure 29.9.
The Compare-
Validator control
icon in the Toolbox
panel**

As an example of using the CompareValidator control, you will set up a simple form that accepts bulk orders of 5 or more units of quantity. This also gives you some idea of when you might require this type of control.

Create a new Web Form and add a Label control. Set its Font Size property to Large and its Text property to *Bulk Order Form*. Add another Label control and set its Text property to *Enter the quantity of the bulk order:*. Add a TextBox control to the right of the second Label control. At the bottom of the form, add a Button control and change its Text property to *Submit*.

Finally, add a CompareValidator control and set its Text property to *Quantity must be 5 or greater for a bulk order*. Change the ControlToValidate property in the Properties panel by selecting the TextBox control from the drop-down menu. The drop-down menu for the ControlToValidate property only shows those controls that are eligible for validation, so the TextBox control should be the only choice in the list. From the drop-down menu in the Operator property select GreaterThanEqual. The last task to complete with the CompareValidator control is to set its ValueToCompare property to *5*. The ValueToCompare property, in this case, indicates the value that you want to compare the data in the TextBox control against.

Save your work and compile your ASP.NET page using the Build menu Build option. Load the page into a browser and enter a value less than 5 before clicking on the Submit button. The CompareValidator control fails validation and displays the text in its Text property, as shown in Figure 29.10.

Figure 29.10.
The Compare-
Validator control
failing validation

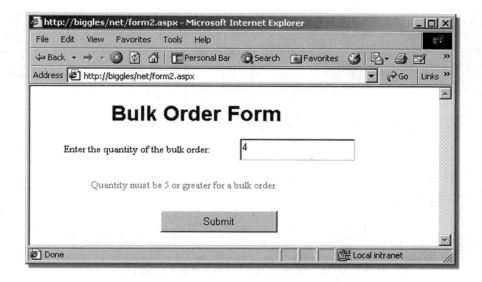

You can also use the CompareValidator control to handle validation by comparing two controls. This is an interesting and useful feature because it means the value that you are comparing against could also change without you having to change your code at all.

For example, consider the case of selecting a minimum and maximum values, where both values can be between 1 and 100 but the minimum value must always be less than or equal to the maximum value. You could set a RangeValidator control for each input field to ensure that the value is between 1 and 100, but you also need to make sure that the minimum value is less than or equal to the maximum value, which could change to anything from 1 to 100 depending on the user. You can use the CompareValidator control similarly to the way you used it earlier to set most of its properties. For instance, you could set its ControlToValidate property to the TextBox control used for the minimum value. You could also set its Operator property to LessThanEqual. The ValueToCompare property should remain completely empty, and instead, you should set the ControlToCompare property to the TextBox control used for the maximum value.

Working with this form in a browser, a user would be warned by the RangeValidator controls if either the minimum or the maximum fields were outside of the range 1-100. If both values were within the given range but the minimum value was greater than the maximum value, the CompareValidator control would give a warning to the user. Figure 29.11 shows the browser result when the minimum value is both outside of the range and greater than the maximum value.

Figure 29.11.
A validation where the data in one control is directly determined by another

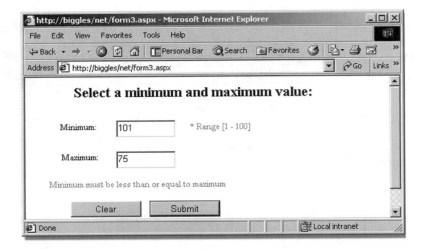

Most of the CompareValidator control's operators are fairly comprehensive at first glance, with the exception of the DataCheck operator. The DataCheck operator basically checks that the control associated with the CompareValidator control contains data of the same type that you specify in the Type property.

For example, if you have a TextBox control (text field) and ask the user to input his age in years, the data that is entered by the user should be of type Integer. By setting the Type property of the CompareValidator control to Integer, you can then use the DataCheck operator for the Operator property to validate the control actually contains an integer and not something else, such as text.

Using the RegularExpressionValidator Control

You can do more complex validation using the RegularExpressionValidator control, which is located in the Web Forms section of the Toolbox panel, as shown in Figure 29.12.

Figure 29.12.
The Regular-ExpressionValidator icon in the Toolbox panel

The RegularExpressionValidator control allows you to set a specific pattern that the data has to match to be valid. This is extremely useful when you are trying to retrieve accurate information for credit cards, e-mail addresses, phone numbers, and other similar types of data. It also prevents typing mistakes, such as when the user misses a digit in his phone number—allowing the mistake to be corrected immediately.

Working with the original example that you have been developing through-out this lesson, you will add a field to retrieve the user's e-mail address. You will use the RegularExpressionValidator control to make sure that the data entered into this new field conforms to the pattern required for an e-mail address.

Add a TextBox control and then a Label describing the new e-mail field. To the right of the e-mail Textbox control add a RegularExpressionValidator control. Set the ControlToValidate property to the new e-mail TextBox control. Assign some appropriate text in the ErrorMessage property for the ValidationSummary control. Set the Text property to * *Required,* to fit in with the general format of the form.

The only other property that you must set is the ValidationExpression property. In the properties panel, click on the ellipses button in the ValidationExpression property field. Visual Studio.NET will present you with the Regular Expression Editor dialog box from which you can select the stan-dard expression Internet E-mail Address, as shown in Figure 29.13.

Figure 29.13.
The Regular
Expression Editor
dialog box

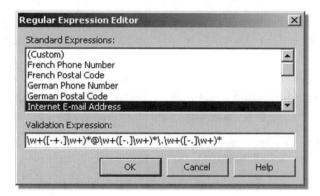

Saving your work and compiling the ASP.NET page, will give you a form where the e-mail address is matched properly to make sure that it conforms to the required pattern expected of an Internet e-mail address.

Using the CustomValidator Control

You can also create your own custom validation requirements using the CustomValidator control that you can access in the Toolbox panel, as shown in Figure 29.14.

Figure 29.14.
The CustomValidator
icon in the Toolbox
panel

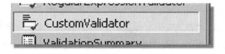

The CustomValidator control is really more of an advanced feature that you can make use of if none of the ordinary validation controls meet your requirements. You can create a client-side script using VBScript or JavaScript and indicate the name of this scripting function in the CustomValidator control's ClientValidationFunction property. Alternatively, you can create a server-side function and simply code what you want to do in the ServerValidate event subroutine. Double-clicking on the CustomValidator control in Design view will cause Visual Basic.NET to create the wrapper code for your ServerValidate event subroutine. You can then type in your specific validation code.

WHAT YOU MUST KNOW

Only some controls can be validated by the six validation controls that Microsoft provides with ASP.NET. These validation controls provide a variety of means to make sure that entries made by a user conform to necessary requirements before being submitted. You can have validation be as simple as checking that there is data in a field, to making sure that the data conforms to a certain pattern. In Lesson 30, "Working With XML in Web Forms," you will learn how to use some of the basics of XML and how you can use it within your ASP.NET pages, including using XML to configure the AdRotator control. Before you begin with Lesson 30, make sure that you know the following key concepts:

- You can ensure that users input data in specific fields when filling in a form by using the RequiredFieldValidator control.

- You can summarize all the validation controls that fail on a page by using the ValidationSummary control to list each of their ErrorMessage properties.

- When you want data to fit within a range, you can use the RangeValidator control to set a maximum and minimum value.

- You can use the CompareValidator control to compare one control against either another control or a value by setting the Operator and Type properties, you can also compare for data types.

- Through the use of pattern matching algorithms, the RegularExpressionValidator Control allows you to do more complex validations on controls. You can select these pattern matching algorithms using the Regular Expression Editor dialog box.

- You can write your own validation script or procedure using the CustomValidator control. This is an advanced technique that you would only use if you cannot attain the level of validation that you require using the standard validation controls.

WORKING WITH XML IN WEB FORMS

n Lesson 29, "Using Validation Controls," you learned how to use Microsoft's standard controls in ASP.NET for making sure that fields in Web Forms contain data or even that the data conforms to specific preset requirements before the form can be submitted. In this lesson, you will learn how to use some of the basics of XML (eXtensible Markup Language) and how you can use it within your ASP.NET pages, through use of Microsoft's Xml control and AdRotator control. By the time you complete this lesson, you will have covered the following key concepts:

- XML (eXtensible Markup Language) is a format for putting structured data into a text file.
- Rather than describing how to visually display data, XML simply specifies the data itself and lets different applications and platforms display it.
- XML has a family of related technologies, which can often confuse people. These related technologies include XSL, XSLT, CSS, XLink, DOM, XML Namespaces, and XML Schema.
- XML basically works by using opening and closing tags for various blocks. The block is the entire markup from the opening tag to its corresponding closing tag.
- Tags are closed in the opposite order that they were opened.
- You can use the Xml control in your Web Forms to display the contents of an XML file in your ASP.NET pages.
- The DocumentSource property of the Xml control indicates the XML document to use. You can either use the Select XML File dialog box or manually type in an entry to populate this property.

- The XSL transform file contains formatting instructions that you can use to display the data from an XML file with customized layout and style characteristics. This keeps the data separate from the display formatting instructions, unlike HTML.
- You can set your Xml control to use an XSL transform file by setting the TransformSource property.
- The AdRotator control uses XML to determine how to display advertisement images. The AdvertisementFile property indicates the XML file to use.

Introducing XML

XML (eXtensible Markup Language) is a format for putting structured data, such as spreadsheets, financial transactions, technical drawings, and similar data, into a text file. This format is made up of rules, conventions, and general guidelines that developers follow and conform to. By design, it is text based rather than a compiled, binary file. Developers can then easily read an XML file to fix it, regardless of platform or commercial software, as a simple text editor suffices.

XML has become hugely popular recently because of its incredible portability, and you can almost think of it in terms of the glue the binds many different platforms and technologies together. ASP.NET, in keeping with current requirements, adds support for XML through use of controls and also certain control properties. More and more, Active Server Pages developers are being asked to integrate sites with XML with many large document management systems and intranet database systems evolving with a Web interface and XML at their heart. Microsoft, too, has embraced XML, giving it considerable support throughout their new development tools in Visual Studio.NET and across the .NET framework in general.

XML appears very similar to HTML because it uses tags. However, XML is very strict to maintain consistency of form unlike HTML, where exceptions and implications are made everywhere. XML does not attempt to describe how to visually display data. Instead, it simply specifies the data itself and lets the different applications and platforms interpret it. This means that a small handheld device or cell phone might choose to show only basic text, while a desktop PC might display the data in a table. Some fancier applications might display the data in both tables and graphs. You can apply formatting to XML and pass the result to a printer. However the result is displayed and regardless of the device or application, using XML means that the same data can be accessed by a wide range of clients without the need to provide separate data sources. Figure 30.1 portrays this concept in a clear graphical manner.

**Figure 30.1.
XML specifies the
data itself and
lets different
applications
interpret it**

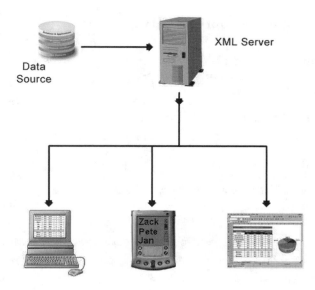

XML Server

Data
Source

XML is really a broad umbrella that is placed over a number of technologies —all of them related to XML. These technologies include XSL, which was derived from XSLT—a transform language, CSS (Cascading Style Sheets), XLink for advanced hyperlinks, DOM (Document Object Model), XML Namespaces, and XML Schema. Phew! This is why many people get confused about XML and where to start learning about it.

This lesson will focus on basic XML files, as well as XSL files—showing you how to incorporate both into your ASP.NET pages.

Learning About the XML File Format

XML basically works by using opening and closing tags for various sections or blocks. The tags do not have to be from a preapproved list (as with HTML), but can be of your own creation. This shows, in part, the extensibility of XML. The main concept to be aware of when writing XML files is that every tag that you open must be closed. The closing tag is the same as in HTML, and uses a forward slash character before the tag name. For example, the following markup shows how to open a tag and close it again immediately afterwards:

```
<FirstName></FirstName>
```

To include some data for this field, simply place the text between the two tags, as follows:

```
<FirstName>Stephen</FirstName>
```

When you use multiple tags, you must close the tag blocks in the reverse order that you opened them. The block is the entire markup from the opening tag to its corresponding closing tag. For instance, opening a tag called <Author> and then a tag called <FirstName> means that you must close the <FirstName> block first and then the <Author> block, as follows:

```
<Author>
   <FirstName>Stephen</FirstName>
</Author>
```

You can also have multiple tag blocks contained within a block, with the same basic rules for opening and closing tags applying. For instance, you might have a <FirstName> block and a <LastName> block contained within the <Author> block, as follows:

```
<Author>
   <FirstName>Stephen</FirstName>
   <LastName>Donaldson</LastName>
</Author>
```

To continue with a simple example, the following XML markup shows you how to continue with this concept of tags and blocks within blocks. You can see that the order in which tags are opened relates directly to the order in which they are closed. The example contains only three books, giving some simple information about each book. The example is very simple and in practice your XML files would be much larger than this. However, a short example is all that is necessary to show you how to create an XML file, and then later in this lesson use and manipulate the data in your ASP.NET pages. The XML example is as follows:

```
<Books>
   <Book>
      <Author>
         <FirstName>Joanne</FirstName>
         <LastName>Rowling</LastName>
      </Author>
      <Details>
         <Title>Harry Potter and the Sorcerer's Stone</Title>
```

```
            <Topic>Children's Fiction</Topic>
            <ISBN>0807281956</ISBN>
        </Details>
    </Book>

    <Book>
        <Author>
            <FirstName>Stephen</FirstName>
            <LastName>Donaldson</LastName>
        </Author>
        <Details>
            <Title>Lord Foul's Bane</Title>
            <Topic>Fiction</Topic>
            <ISBN>0345348656</ISBN>
        </Details>
    </Book>

    <Book>
        <Author>
            <FirstName>Raymond</FirstName>
            <LastName>Feist</LastName>
        </Author>
        <Details>
            <Title>Magician</Title>
            <Topic>Fiction</Topic>
            <ISBN>0385175809</ISBN>
        </Details>
    </Book>
</Books>
```

Save the XML example markup in a Notepad file, as *books.xml*.

Using the Xml Control

You know what XML is generally about now, as well as being aware of its basic format. You have also created a simple data file in XML format. To be able to utilize this XML file (or any others) in your ASP.NET pages, you must use the Xml control. The icon for the Xml control is in the Web Forms section of the Toolbox panel, as shown in Figure 30.2.

Figure 30.2.
The Xml control icon
in the Toolbox panel

You can add an Xml control to your Web Form by simply selecting its control icon and then clicking somewhere within the Web Form itself. Under some circumstances, you may receive an error when adding the first Xml control to your Web Form. When this happens, simply add a second Xml control, as shown in Figure 30.3. You can then delete the first Xml control with the error by simply clicking on it once and pressing the Delete key.

Figure 30.3.
Possibly receiving
an error when
creating the first
Xml control

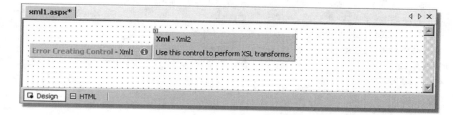

You can see from the Properties panel in Design view that the Xml control has very few properties and is therefore quite easy to use. By clicking on the button with the ellipses in the DocumentSource property field, you will be presented with the Select XML File dialog box, as shown in Figure 30.4, from which you may select your XML file.

Figure 30.4.
The Select XML
File dialog box

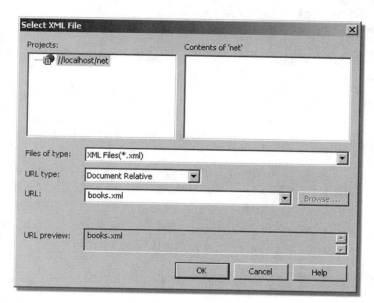

You can select a URL type of either Document Relative or Root Relative. Document Relative means that the path to the XML file is relative to the ASP.NET page that contains the Xml control you are configuring. So, if *books.xml* is in the same directory as your ASP.NET page, you can simply type in *books.xml* in the URL field. Root Relative means that the path to the XML file is relative to the root directory of your Active Server Pages server. So, if *books.xml* is in the net directory beneath your root directory, you can simply type in */net/books.xml* in the URL field.

After setting the URL field to *books.xml* in the Select XML File dialog box, click on the OK button to close the dialog box. You will notice that the DocumentSource property field in the Properties panel contains your entry for *books.xml*. Save your project and compile your ASP.NET page using the Build menu Build option. Load your ASP.NET page into your browser, where it should appear similar to Figure 30.5.

Figure 30.5. Loading the XML file without any transform

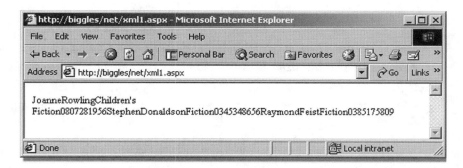

You will notice a few things about the result of loading the XML file into the browser. First, you will probably notice the complete lack of formatting. All of the data is simply strung together without any cohesive style, making it difficult to read. The only separation of the data occurs where the space character appears between the words *Children's Fiction*, which was a single data entry, as you will recall. Formatting your XML output so that it has specific style characteristics brings you to the next section—XSL transform files.

Understanding XSL Transform Files

XSL (eXtensible Stylesheet Language) transform files let you apply formatting and style characteristics to XML files. XSL goes through two distinct stages. The first stage is tree transformation, in which the data is obtained from the

XML file. The result tree from the tree transformation can be different from the source tree because of filtering, reordering, and additional content that may occur. The second stage is formatting, in which formatting semantics are placed within the result tree from the first stage. Formatting semantics allow the designer of the stylesheet to indicate desired layout, pagination, and general styling details to use when displaying the data.

You have already seen that, on its own, an XML file has no formatting instructions and results in a long string of characters. XSL transform files allow the formatting of the data to be separated from the data itself. You could therefore use different XSL transform files to display the same data in a variety of ways. These multiple XSL transform files might perhaps be for different clients—such as one transform file for displaying in browsers and another for sending to printers.

XSL transfer files have a different file extension than typical XML files. Where an XML file ends in *.xml*, an XSL transform file ends in *.xsl*. This difference in file extension is purely to make it easier to distinguish between the two types of files. Both file types are still simply text files.

Note: There is a vast amount to learn about using XSL transform files, which is beyond the scope of this lesson. You can purchase entire books on XML and using technologies associated with XML, such as XSL. You can also discover a wealth of information on these topics online if you are interested in pursuing them further and in greater detail.

To see how to construct an XSL transfer file, you will create a very simple file for formatting the *books.xml* file that you created earlier in this lesson. You will simply be separating the data so that it easy to read, with the author's name in bold and the book details below.

The main block for the XSL transform is the <xsl:stylesheet> block. Within the opening tag, you place the version and namespace of the stylesheet. Notice that the year 1999 does not refer to the version of XSL being used, but rather to the year in which the World Wide Web Consortium (W3C) allocated the URL indicated. The following code shows an example of the <xsl:stylesheet> block:

```
<xsl:stylesheet version="1.0" xmlns:
xsl="http://www.w3.org/1999/XSL/Transform">

</xsl:stylesheet>
```

Inside the <xsl:stylesheet> block you can place multiple <xsl:template> blocks for the various tags in the XML document. Within the opening tag of each <xsl:template> block, you can use the *match* attribute to which you can assign a text string. This text string must match a tag within the XML file for its contents to be processed.

The <xsl:text> block is for inserting text into the result tree. You can use any text, including spaces. The following example shows how you can use the <xsl:text> block to include a space character into the result tree:

```
<xsl:text> </xsl:text>
```

The <xsl:value-of> tag is also for inserting text into the result tree. There is no closing tag, such as </xsl:value-of>, but rather simply a forward slash just prior to the last character of the opening tag (/>). The <xsl:value-of> tag must contain a *select* attribute to which you assign the name of the tag in the XML file that contains the data that you want to display. You can traverse multiple containing blocks by using a forward slash character (/) between tag names. An example of using the <xsl:value-of> tag for inserting the data contained within the <FirstName> block, which itself is contained within the <Author> *block* would be as follows:

```
<xsl:value-of select="Author/FirstName"/>
```

The <xsl:apply-templates> tag recursively processes all of the children of the current tree node. By using the optional *select* attribute, this recursive processing applies only to the specified node. An example of processing all the children of the *Book* node recursively, would be as follows:

```
<xsl:apply-templates select="Book"/>
```

You can also mix HTML statements in with your XSL transform file for working with tables. An entire XSL transform file for displaying the XML file that you created earlier—*books.xml*—is as follows:

```
<xsl:stylesheet version="1.0" xmlns:
xsl="http://www.w3.org/1999/XSL/Transform">
   <xsl:template match="/Books">
      <xsl:apply-templates select="Book"/>
   </xsl:template>

   <xsl:template match="Book">
      <table width="100%" border="2">
         <tr>
            <td>
               <b>
                  <xsl:value-of select="Author/FirstName"/>
                  <xsl:text> </xsl:text>
                  <xsl:value-of select="Author/LastName"/>
               </b>
            </td>
         </tr>
         <tr>
            <td>
               <xsl:value-of select="Details/Title"/><br/>
               <xsl:value-of select="Details/Topic"/><br/>
               ISBN: <xsl:value-of select="Details/ISBN"/>
            </td>
         </tr>
      </table>
   </xsl:template>

</xsl:stylesheet>
```

Save the example in Notepad as *books.xsl*. From within your Visual Studio.NET environment, click on the Xml control so that its properties are shown in the Properties panel. In the same manner as you used to select the DocumentSource property, select the new XSL transform file, *books.xsl*, for the TransformSource property. You can either type *books.xsl* directly into the TransformSource property field or you can use the Select Transform File dialog box. Save your project and compile your ASP.NET page using the Build menu Build option. This time when you load the page into your browser, it should have a much more formatted appearance that is not so difficult to read, as shown in Figure 30.6.

**Figure 30.6.
Loading the XML
file with a transform**

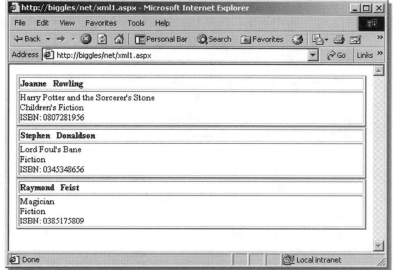

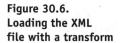

Creating Automatic Advertisement Rotation

You can create automatic advertisement rotation in your ASP.NET pages by using the AdRotator control. Basically, this means that you can have several graphic images of advertisements and show a different advertisement on subsequent loading of the same ASP.NET page. The icon for the AdRotator control is in the Web Forms section of the Toolbox panel, as shown in Figure 30.7.

**Figure 30.7.
The AdRotator
control icon in the
Toolbox panel**

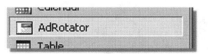

The AdRotator control uses XML to handle all of the data for the various advertisements. By now you will already be quite comfortable with XML and prepared for using this control.

The first block is the <Advertisements> block. Within this block you have multiple <Ad> blocks—one for each advertisement image. Within each <Ad> block you have several other blocks containing data. In point form, they are as follows:

- The <ImageUrl> block contains the URL for the advertisement's image file.
- The <NavigateUrl> block contains the URL to redirect to when the user clicks his mouse on the image. You can leave this block empty if you do not want the advertisement to go anywhere, but simply want to display the image.

- The <AlternateText> block contains text to display when the user's browser cannot display the image. If the browser supports the functionality, this text will also be used as the ToolTip for the image.
- The <Keyword> block contains a category for the advertisement in the form of a text string that can be used by the page for filtering purposes.
- The <Impressions> block contains an integer value indicating how much of an impression the advertisement should have relative to the other advertisements in the file. The advertisements that you want to make a bigger impression have the larger numbers.

To test your knowledge of the AdRotator control and the XML file that accompanies it, try your hand at a simple example that rotates three advertisement images. You want one of the advertisements to appear half of the time and either of the other two advertisements to appear the rest of the time. You will therefore require one advertisement to have a weighting twice that of the other two.

Open up Notepad and type in the following XML markup:

```xml
<Advertisements>
    <Ad>
        <ImageUrl>/pics/ad001.jpg</ImageUrl>
        <NavigateUrl>http://www.nowhere.com</NavigateUrl>
        <AlternateText>Welcome to Nowhere</AlternateText>
        <Keyword>Holiday</Keyword>
        <Impressions>50</Impressions>
    </Ad>

    <Ad>
        <ImageUrl>/pics/ad002.jpg</ImageUrl>
        <NavigateUrl>http://www.nowhere.com</NavigateUrl>
        <AlternateText>Stairway to Nowhere</AlternateText>
        <Keyword>Holiday</Keyword>
        <Impressions>25</Impressions>
    </Ad>

    <Ad>
        <ImageUrl>/pics/ad003.jpg</ImageUrl>
        <NavigateUrl>http://www.nowhere.com</NavigateUrl>
        <AlternateText>Sail away to Nowhere</AlternateText>
        <Keyword>Holiday</Keyword>
        <Impressions>25</Impressions>
    </Ad>
</Advertisements>
```

Save the file as *nowhere.xml*. Open up a new ASP.NET project and add an AdRotator control to your Web Form. Make sure that the size and position of the AdRotator control on the Web Form is set the way you want because it will not automatically resize according to the image that it displays, but rather it will distort the image to fit. For fine tuning, you can set the Height and Width properties of the AdRotator control to be the same size as your advertisement banners. Set the AdvertisementFile property to *nowhere.xml* either just by typing it in or using the Select XML File dialog box. Save your project and compile your ASP.NET page by selecting the Build menu Build option. Loading your ASP.NET page into your browser should show the AdRotator control in action, as shown in Figure 30.8.

**Figure 30.8.
The AdRotator
control in action**

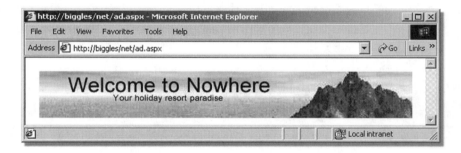

If an image was successfully shown in your browser, then you know that your AdRotator object read in your XML file properly. Refresh your browser without making any other changes. Do not recompile the ASP.NET page or modify the XML file. The result should be one of the other advertisements, as shown in Figure 30.9.

**Figure 30.9.
The AdRotator
control changing
the advertisement
when the page is
refreshed**

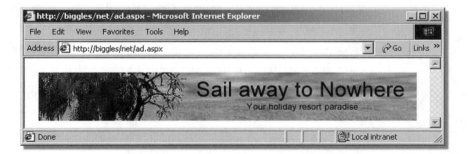

XML is a format for putting structured data into a text file, making use of tags similar to HTML format. Rather than describing how to visually display data, XML contains the data itself without display instructions. You can use an XSL transform file to format the data from an XML file so that the display is more attractive and readable. In Lesson 31, "Using ADO.NET," you will continue working with XML and see the impact that XML has in ASP.NET, as you discover the changes that ActiveX Data Objects has undergone since its last incarnation. Before you begin with Lesson 31, make sure that you know the following key concepts, as you will need some knowledge of XML basics:

- The family of related technologies, including XSL, XSLT, CSS, XLink, DOM, XML Namespaces, and XML Schema, can often confuse people when they are learning about XML.

- XML syntax appear very similar to that of HTML, using opening and closing tags for various blocks and containing data between tags. The entire markup from the opening tag to its corresponding closing tag is considered to be a block.

- The order in which tags are opened is the opposite to the order that they are closed.

- To display the contents of an XML file in your ASP.NET pages, you can use the Xml control in a Web Form and set its DocumentSource property to the XML document to use.

- You can use XSL transform files to display the data from an XML file with customized layout and style characteristics. The XSL transform file contains formatting instructions, leaving the XML file free to contain just the data, unlike HTML.

- By setting the TransformSource property, you can set your Xml control to use an XSL transform file to format the XML document that it is displaying.

- By first creating an XML file containing details about advertisement images, you can then make use of the AdRotator control to display advertisements, as well as handle redirection and automatic rotation. You can use the AdvertisementFile property to specify the XML file to use.

USING ADO.NET

*i*n Lesson 30, "Working with XML in Web Forms," you learned how to create some basic XML files and then work with them in your ASP.NET pages, using different Web Form controls. In this lesson, you will continue looking at XML as you revisit working with databases, but this time using ADO.NET—the latest version of ActiveX Data Objects. By the time you finish this lesson, you will have seen the following key concepts:

- There are considerable changes between ADO and ADO.NET. At the heart of many of these changes is the use of the dataset in ADO.NET instead of the recordset, as well as the embracing of XML.
- The use of XML in ADO.NET is fundamental to its ability to exchange datasets between components where the data must pass through a firewall—a huge advantage for enterprise sites.
- XML plays an important role in ADO.NET, with both components sharing the same architecture.
- Datasets can hold multiple tables and relations for the data of those tables through its DataTable and DataRelation collections.
- Datasets store their data using a disconnected data architecture, which means that they do not maintain a connection to any data source.
- Datasets are passive containers of data that are not aware of the source of their data and therefore make excellent objects for storing data from varying sources.
- Data providers are the link between the database and the DataSet objects.
- Data providers are made up of multiple components including the Connection, Command, DataAdapter, and DataReader objects.

Understanding the Differences Between ADO and ADO.NET

There are quite significant differences between ADO and ADO.NET. The methods that you use to interact with databases has undergone some necessary changes to improve areas of performance, especially in the enterprise environment. Although you do not need to know all aspects of both ADO and ADO.NET to use ADO.NET, a brief understanding of the differences will let you appreciate ADO.NET more. The key differences in point form are as follows:

- Representations of data in memory
- Number of tables
- Data navigation
- Disconnected access to data
- Sharing data between applications
- Richer data types
- Performance
- Penetrating firewalls

A brief description of each of these points follows to help clarify them further.

Representations of Data in Memory

As you saw earlier, ADO makes use of the recordset to store data in memory. However, in ADO.NET the data is stored in what is known as a dataset. As you will learn later in this lesson, the dataset has a different representation from that of a recordset, as well as supporting added functionality and a greater range of features.

Number of Tables

The recordset that you use with ADO appears as a single table, even if you merge multiple tables from your data source. You will recall using the JOIN query to accomplish this earlier in the book.

The dataset works in a different way. Rather than storing the result as a single table, it stores a collection of tables. There are times when there may be only a single table in the collection, but the underlying concept of the dataset is very different from the recordset. Tables within a dataset are known as data tables or DataTable objects. Due to being a collection of tables rather than just a single table, you will learn later in this lesson about more detailed aspects of datasets, such as relationships and self-relating tables, that are not possible in recordsets.

Data Navigation

When navigating through a table in a recordset, you use the ADO methods MoveNext, MovePrevious, MoveFirst, and MoveLast. Using the MoveNext and MovePrevious methods gives you sequential navigation through the table. In ADO.NET you can navigate through the table as a collection because the rows are in fact represented as collections. This basically means that you are no longer limited to sequential navigation, but can loop through rows of a table or access individual rows using index values or keys—the same freedom and simplicity that other collections offer.

You can also access records that are related to an individual record by using the DataRelation object. This feature comes from the dataset's ability to maintain multiple tables and relationships between them. An example of related records would be having a table of authors and then for one particular author, being able to retrieve records from another table for all the books that he or she wrote.

Disconnected Access to Data

When you are using recordsets in ADO, they are designed with connected access in mind. You can have disconnected recordsets, but this is not the normal mode of operation. In ADO, when you are using disconnected recordsets, you communicate with the database by means of an OLE DB provider to which you make calls.

In contrast, when you are using datasets in ADO.NET, you do not remain connected to the database. Due to being able to store multiple tables, the dataset can mimic the database to some extent without having to maintain a constant connection. Rather than communicating with the database directly through calls to an OLE DB provider, ADO.NET communicates through either an OleDbDataAdapter or SqlDataAdapter object. These data adapters handle the calls to the OLE DB provider, giving you additional benefits, such as data validation and performance optimization.

Sharing Data Between Applications

When trying to share data between applications using ADO, you must send an ADO recordset using what is known as COM marshaling. However, in ADO.NET you do not use COM marshaling to share data between applications. Instead you use an XML stream to send an ADO.NET dataset to another application. Because sending XML files is generally easier than COM marshaling, sharing data between applications is easier in ADO.NET than it is in ADO. Sending XML files instead of using COM marshaling gives you richer data types, added performance benefits, and the ability to penetrate firewalls.

Richer Data Types

ADO uses COM marshaling to send recordsets, which provides only those data types defined by the COM standard. This may be satisfactory for some conditions, but extremely limiting for others.

ADO.NET, on the other hand, uses XML to send datasets. XML is extensible, as you are already aware, and does not have a fixed set of tags or elements to work with. This basically means that there is no limitation on the data types that the components sharing the dataset can use.

Performance

Although not something that small applications will notice, small improvements in handling data can have huge performance boosts in larger enterprise situations. You have already seen that using data adapters in ADO.NET provides performance optimization benefits with disconnected data access, but there are other performance considerations.

Both ADO and ADO.NET let you keep the amount of data sent over the network to a minimum, which not only increases performance but reduces bandwidth demands, especially when the amount of data gets large.

ADO uses COM marshaling, as you already know, which requires ADO data types to be converted to COM data types before they can be sent over the network. ADO.NET does not require data-type conversions when sending a dataset over the network, which results in time and processor savings and therefore performance improvements.

Penetrating Firewalls

A firewall is software, hardware, or a combination of both that basically protects a site from being open to attack from malicious people. A site without a firewall is very vulnerable indeed. Firewalls work by blocking access to ports, particular IP addresses, or particular programs by using a set of rules. Firewalls can differentiate between incoming and outgoing traffic. A firewall sits between all of the computers at a site and the Internet (or any other computers), as shown in Figure 31.1, which means that in order to get information to a computer on the Internet somewhere, you have to get through a firewall.

Figure 31.1.
A firewall sits
between the
computers at a site
and the Internet

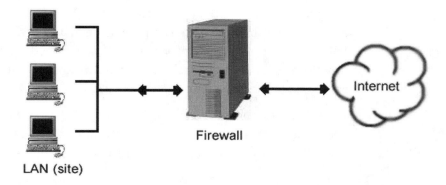

With ADO, firewalls would often prevent two components from sending ADO recordsets. This is because COM marshaling is often considered a risk through a firewall and is blocked. Figure 31.2 shows the scenario of a computer at site A trying to use ADO to share a disconnected recordset with a computer at site B.

Figure 31.2.
The firewall prevents
the ADO recordset
from reaching
computers at site B

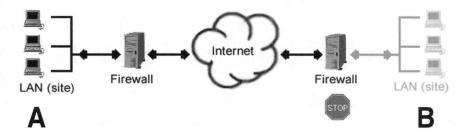

However, firewalls usually allow HTML through, depending upon their configuration, of course. XML is very similar to HTML in appearance, as you are already aware. Therefore, because ADO.NET is based upon XML to a large extent, it is able to penetrate firewalls when components exchange datasets. Figure 31.3 shows the scenario of a computer at site A trying to use ADO.NET to share a dataset with a computer at site B.

Figure 31.3.
The firewall allows
the ADO.NET dataset
to reach computers
at site B

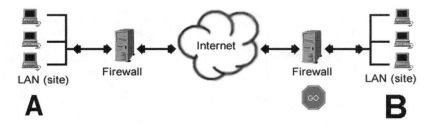

Understanding the Relation Between XML and ADO.NET

XML and ADO.NET are more closely linked together than you might realize. The whole design of ADO.NET is with the .NET XML framework in mind—to such an extent that both of these components belong to the same architecture. The following brief extract from the Microsoft documentation for ADO.NET explains clearly not only the relationship between XML and ADO.NET, but also succinctly states the company's strong support of XML.

Note: XML and data access are intimately tied—XML is all about encoding data, and data access is increasingly becoming all about XML. The .NET framework does not just support Web standards—it is built entirely on top of them.

XML support is built into ADO.NET at a very fundamental level. The XML class framework in .NET and ADO.NET are part of the same architecture—they integrate at many different levels. You no longer have to choose between the data access set of services and their XML counterparts; the ability to cross over from one to the other is inherent in the design of both.

You have seen already the usefulness of XML with firewalls. To briefly recap, using XML in ADO.NET allows data to pass through a firewall because firewalls are typically configured to allow HTML to pass through. For enterprise situations, this is a big point in favor of ADO.NET, and the credit goes to its XML ties.

In the next section you will learn more about DataSet objects and the disconnected data architecture. It is through XML that ADO.NET provides disconnected access to data. DataSet objects can contain not only data from a SQL server or other such database, but also data directly from an XML source, such as an XML file or XML stream. Regardless of the source of the data that populated a DataSet object, all DataSet objects can write out their contents as XML—including XML Schema. Another advantage of DataSet objects is that by synchronizing a DataSet object with an XmlDataDocument, you can provide real time access to data rather than disconnected data access. You will learn about the DataSet object in the next section.

Understanding Datasets

You saw earlier, when looking at the differences between ADO and ADO.NET, the introduction of the dataset and some of the functionality that it provides. Datasets are a disconnected source of data in the form of a collection of tables.

Datasets are also unaware of where their data has come from—be that an SQL server, an OLE DB provider, or an XML data file. In the overall scheme of ADO.NET, the DataSet object plays a very important role. The database is accessed by a data provider, which in turn populates the DataSet object. The DataSet object provides the data for the various display controls available in ADO.NET. Figure 31.4 shows the general place of the DataSet object in ADO.NET and gives you a clearer understanding from an overview perspective.

Figure 31.4.
The general place of
the DataSet object
in ADO.NET

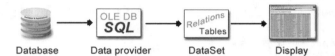

Database   Data provider   DataSet   Display

When you are learning to understand datasets in ADO.NET you must understand the concept of disconnected data. Rather than maintaining a constant connection to a database, which can affect performance when many clients are trying to access the same database, disconnected data basically copies the data (or a portion of it) from the database and closes the connection. This frees up the database for others to access.

Also on the topic of disconnected versus connected data, you might like to ponder upon the issues involved in maintaining a database connection in a Web application. Knowing when the user has finished or has no further use for a database connection can be difficult. You can apply a timeout period for a database connection, but working out the optimal time requires trial and error, not to mention adjustments due to changes with the passage of time and evolvement of the site. Another pitfall of connected data is where multiple components, possibly from different applications, all require access to the same database.

Using a disconnected data architecture, datasets address many of these issues for large sites because they are (indirectly) only connected to the database for the period of extracting data or updating it, which means there are far more opportunities for other components to access the database.

It is worth mentioning that working with datasets and their disconnected data architecture does not mean that you cannot make changes to the underlying database. It is true that the datasets are unaware of the source of their data; however, the data providers are not. Basically, if you make a change to a dataset, the data provider will handle updating the actual database. You will learn more about data providers in the next section.

Datasets are not stored in a flat table, such as a Recordset object is in ADO. Datasets hold a collection of DataTable objects, of which there is at least one but potentially more. Each of these DataTable objects contains information about the rows and columns in collections, as well as a collection of constraints. Constraints are basically rules that maintain data integrity by determining if a value is unique or how to react to foreign keys being deleted. The DataSet object also contains a DataRelation collection that you might like to think of as holding all the primary and foreign key information associated with relations between tables. Figure 31.5 shows the basic DataSet object model.

**Figure 31.5.
The DataSet
object model**

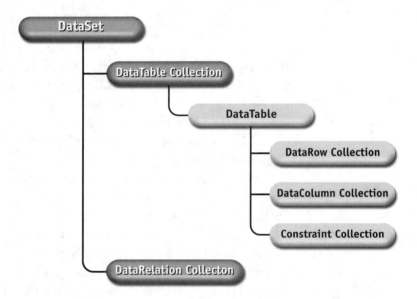

Datasets are not aware of the source of their data and are not, generally speaking, continually connected, as you are now aware. This means they are a passive container for data and provide the ideal place for multiple data sources to contribute data. Also, you could populate a dataset partially from an Access database, partially from an SQL server, partially from an XML document, and partially from another application that sends you data in a stream. You might even generate random data from code. The options are tremendous. After a dataset contains data, it does not matter at all where the data came from—you can safely work with it all in the same consistent manner. The consistency factor in Visual Studio.NET shines through once again. The value of datasets as a meeting place for various data sources is obvious and can be put to good use in many situations where data is stored in different formats.

Understanding Data Providers

You saw earlier that data providers basically sit between a database and a dataset in the general scheme of things in ADO.NET. The data providers are not a single object, but rather a set of related components that work together to achieve the necessary functionality. The components can be broken down to the following objects:

- Connection
- Command
- DataAdapter
- DataReader

The data provider gives you high-performance access to the data source because it reduces many options. For instance, it mostly provides read-only access with the exception of writing updates to datasets through the DataAdapter object. Another way that it is able to provide faster access is by only using forward navigation, eliminating the need for alternatives.

The first component is the Connection object, which you use for connecting to a database.

The second component is the Command object, which you use for sending database commands, such as SQL queries, to a database that is connected through the Connection object. You can use the Command object to run any typical database commands, such as returning data from queries or modifying data.

The third component is the DataAdapter object, which is basically the link between a data source and a dataset. The DataAdapter has methods for working with both the database and the DataSet object, utilizing the Command object for database commands. An important method of the DataAdapter object is the Fill method, which you must use to update the data in a dataset to reflect accurately what is in a database.

The fourth component is the DataReader object, which provides a high-performance stream of data from the data source.

Implementing DataSets in ADO.NET

You now know quite a bit about datasets in theory; however, you also need to take a practical approach and look at implementing datasets in your ASP.NET pages. To implement a dataset, you must first declare a variable of type DataSet, as follows:

```
Dim myDS As DataSet
```

You then must assign an instance of the DataSet object to your variable, as follows:

```
myDS = New DataSet("Books")
```

In the previous example, the term *Books* is merely a created name. The name is optional and if you do not include one, the name NewDataSet is assigned by Visual Studio.NET instead.

Note: *If you are familiar with C++ or C#, it might make it easier for you to know that you are simply creating an instance of a DataSet by calling the DataSet constructor and making use of the DataSet class.*

You can also construct an instance of a DataSet and declare a variable in a single line of code, as shown:

```
Dim myDS As DataSet = New DataSet("Books")
```

The methods for creating a DataSet that you have seen in this section should not be confused with connecting to an existing database. These methods are for implementing a new DataSet object from code for holding or storing data in memory temporarily.

Working with DataTables in ADO.NET

You are already aware that a DataSet object contains a collection of DataTable objects. When you create a DataSet object and want to create a DataTable object for it, you can create a variable of type DataTable and then assign it to your DataSet object using the Tables collection's Add method. The Add method takes the name of your DataTable object as an argument. A simple example of creating a DataSet object and then assigning a DataTable object to it, would be as follows:

```
Dim myDS As DataSet = New DataSet("Books")
Dim myTable As DataTable = myDS.Tables.Add("Authors")
```

You also know that each DataTable object contains three collections—one each for rows, columns, and constraints. To store data in a DataSet object, you need to make use of these collections.

Rather than dealing with all three collections at once, you will concentrate first on working with the collection for columns. The column headers are stored in the DataColumn collection. The first step in adding a new DataColumn object is by declaring a variable of type DataColumn and then assigning a new instance of a DataColumn object to it, as follows:

```
Dim myColumn As DataColumn = New DataColumn
```

Each DataColumn object has multiple properties. One of these properties is the ColumnName property, which you use for referencing the DataColumn object from code. The Caption property is the text that is displayed at the top of the column. If you do not set a Caption property explicitly, then it will default to using the value in the ColumnName property. You can set them differently though, as shown in the following example:

```
myColumn.ColumnName = "Auth_Full"
myColumn.Caption = "Author"
```

One of the most important properties of the DataColumn object is the DataType property. The DataType property indicates the type of data that is stored in the column, which is crucial for database integrity. You can set the DataType property to any base system type, such as String, Boolean, Int32, Int64, Decimal, and others. The recommended method is to use the System object's Type.GetType method to return the required type. You pass the required data type as a string argument to the Type.GetType method. The following shows examples of setting the column DataType property with various data types:

```
myColumn.DataType = System.Type.GetType("System.String")
myColumn.DataType = System.Type.GetType("System.Int32")
myColumn.DataType = System.Type.GetType("System.Boolean")
```

If you are using a text data type, such as String, for your column, you may also require limiting the text to a certain number of characters. You can achieve this by setting the MaxLength property to the maximum number of characters allowable, as follows:

```
myColumn.MaxLength = 20
```

You can set a default value for a column that is applied to each new row. Simply setting the DefaultValue property of a DataColumn object accomplishes this, as shown in the following example where each new row will have a value of 19:

```
myColumn.DefaultValue = 19
```

After setting the various properties of your DataColumn object, you still have to actually add your column to a DataTable object. You do this by using the Add method of the Columns

```
myTable.Columns.Add(myColumn)
```

Working with Additional Properties and Methods

The DataColumn and DataRow objects both have far more properties and methods than can be covered in this brief overview of the essentials of working with ADO.NET. You can do a lot of neat things with these objects besides basic functionality, such as having automatically incrementing values by a step of your own choosing. You can determine the DataTable object that a column or row comes from. There is even functionality for dealing with errors and error handling. If you are working seriously with the DataRow and DataColumn objects, you should refer to the Microsoft documentation that comes with ADO.NET when you install it.

The row data is stored in the DataRow collection. Using the DataRow object, you are able to insert data, retrieve data, update data, and ultimately delete data. To insert a row in your DataTable object, you must first declare a variable of type DataRow and assign a new instance of the DataRow object to it by using the DataTable object's NewRow method, as follows:

```
Dim myRow As DataRow
myRow = myTable.NewRow()
```

As with the DataColumn object, the DataRow object has many methods and properties that you can work with. You can add data to a row by following in parentheses the ColumnName property of the particular column that the data is intended for and then assigning the actual value of the data. For instance, the following example assigns the full name of an author to a column with a ColumnName property value of Auth_Full:

```
myRow("Auth_Full") = "Stephen Donaldson"
```

You can also add data to your row by using the index value of the column if you know it. For instance, if the Auth_Full column is actually index 3, you could use the following example to achieve the same result:

```
myRow(3) = "Stephen Donaldson"
```

When you have added all your data to various columns, you must add your new row to the DataRow collection using the Add method. The following example shows you how to add a new DataRow object to the DataRow collection of a DataTable object:

```
myTable.Rows.Add(myRow)
```

Finally, you must actually confirm all of these changes to the row. The simplest way to confirm all of the changes is by using the DataTable object's AcceptChanges method. This implicitly calls the EndEdit method and is also simple to implement, as you can see in the following example:

```
myTable.AcceptChanges
```

Working with DataRelation Objects

When you have a dataset with multiple tables, you can make use of DataRelation objects to assist you in forming relationships between the data in one table and the data in another. You can therefore retrieve data that is related somehow even though it resides in different tables. For instance, so far you have been looking at a simple book database. If you found a book that you liked very much and wanted to find all other books by the same author, you could do that even though the data may not reside in the original table.

Note: Although working with constraints is a more advanced topic, it is worth knowing that by creating a DataRelation object and adding it to a DataSet object, you are automatically creating a UniqueConstraint on the parent table and a ForeignKeyConstraint on the child table.

To actually create a DataRelation object, you can use the DataSet object's Relations.Add method. The Relations.Add method takes three arguments. The first argument is the name of the DataRelation that you are creating. The second argument is the parent DataColumn, and the third argument is the child DataColumn. The parent and child DataColumn objects should contain common data. In the following example, both parent and child DataColumn objects contain the full name of authors, even though the two tables have a different ColumnName property value:

```
myDS.Relations.Add("BooksByAuthor", _
myDS.Tables("Authors").Columns("Full_Name"), _
myDS.Tables("Books").Columns("Auth_Full"))
```

Later, after you have established a DataRelation object, you can use the DataRow object's GetChildRows and GetParent methods to retrieve relational data when working with the tables.

WHAT YOU MUST KNOW

ADO.NET has introduced some new features to working with databases, and there are some significant differences between it and ADO. The dataset in ADO.NET replaces the recordset in ADO and works with disconnected data rather than maintaining a connection to the database. Data providers are the link between databases and datasets. In Lesson 32, "Working with DataGrids," you will look at using ADO.NET to actually display data in your ASP.NET pages, utilizing controls for many of the objects that you learned about in theory throughout this lesson. Before you rush off to Lesson 32, make sure that you feel comfortable with the following key concepts, as they will form the basis of the next lesson:

- Datasets and the use of XML are the two single biggest factors that represent the changes from ADO to ADO.NET.

- You can penetrate firewalls when sharing datasets between components with ADO.NET because of the use of XML in ADO.NET. This feature of ADO.NET is significant for enterprise sites.

- The .NET XML framework shares the same architecture as ADO.NET, which shows how deeply tied these two technologies are.

- A DataSet object contains a DataTable collection that can hold multiple DataTable objects. Each DataTable object itself contains collections for rows, columns, and constraints.

- A DataSet object can also hold relations for the data of those tables, through its DataRelation collection.

- A disconnected data architecture is when a connection to the database is not maintained. The DataSet object uses a disconnected architecture and basically stores a separate cache of data for working with.

- Because datasets are not aware of the source of their data, you can use them for storing data from varying sources and still be able to work with all of the data in a consistent manner.

- Data providers form a connection to a database and then pass on data to DataSet objects.

- A data provider is not a single object, but actually a combination of components. The four main objects that make up a data provider are the Connection, Command, DataAdapter, and DataReader objects.

myValue = 7

<AlternatingItemStyle
  BackColor="Cornsilk">
</AlternatingItemStyle>
<HeaderStyle
  BackColor="Wheat"
  Font-Bold=True>
</HeaderStyle>

WORKING WITH DATAGRIDS

*i*n Lesson 31, "Using ADO.NET," you learned about working with databases through theory and code using XML and the latest version of ActiveX Data Objects—ADO.NET. In this lesson you will take a look at databases using a very useful Web Form control—the DataGrid control—which you can use for displaying data in your ASP.NET pages. By the time you finish this lesson, you will have learned about the following key points concerning this control:

- The DataGrid control can save you the effort of having to create HTML tables manually, giving you a simple means of displaying data in your ASP.NET pages.
- The DataGrid control icon is in the Web Forms section of the Toolbox panel and although visible at run time, will automatically resize if necessary to display data and therefore does not require sizing in Design view.
- The DataSet control icon is actually in the data section of the Toolbox panel. It is not visible at run time.
- The Add Dataset dialog box is where you determine whether you want to use a typed or untyped dataset.
- Use the DataGrid control's DataSource property to associate it with your DataSet control.
- The OleDbConnection control icon can be found in the Data section of the Toolbox panel. The OleDbConnection control actually makes the connection to a database.
- Using the OleDbConnection control's ConnectionString property will bring up the Data Link Properties dialog box, where you can set a provider type, connection settings, and even passwords and permissions.

- The OleDbDataAdapter control connects the database connection with the dataset using the Data Adapter Configuration wizard, along with the Query Builder wizard, to determine what portion of the data to pass.
- You can configure style properties for a DataGrid control within the tags of other specific properties, such as the HeaderStyle, FooterStyle, and AlternatingItemStyle properties.

Introducing the DataGrid Control

You have already seen how to use ADO.NET to connect to databases and other sources of data. You can use the DataGrid control to make it easy to display that data in your ASP.NET pages. The DataGrid control saves you from the hassle of creating static HTML tables that require considerable time to rework when changes occur. Basically, you indicate a data source and make use of the many customization options to display the data in a variety of forms.

The DataGrid works with tables of data. You have already seen earlier how to create a table in a database, such as Microsoft Access. Rather than repeating that exercise, this lesson is focusing on using controls to display data from these tables in ASP.NET, and therefore will simply make use of the Microsoft example Northwind database. You can just as easily apply what you learn to other databases, which you will see later in this lesson.

Adding a DataGrid Control

Adding a DataGrid control is very straightforward due to the fact that you have already added plenty of controls to your Web Forms, as well as Microsoft's consistency efforts with Visual Studio.NET. In a similar manner to that which you used with previous controls, select the DataGrid control icon, as shown in Figure 32.1, from the Web Forms section of the Toolbox panel.

**Figure 32.1.
The DataGrid
control icon in
the Toolbox panel**

After selecting the DataGrid control icon, drag the size of the control on your form. You do not have to concern yourself too much over the size of the DataGrid control because it dynamically increases in size to accommodate the data that it has to display at run time.

Now that the DataGrid control has been added, you can add other components that you require before looking more closely at the many properties and options available with the DataGrid control.

Adding a DataSet Control

You already know a lot about DataSet objects from the previous lesson, including their purpose and place in the overall ADO.NET scheme. In this lesson you will look at creating your DataSet by simply using the DataSet control in Design view. The only trick to adding the DataSet control to your Web Form in design view is locating the control icon itself. Unlike the other controls that you have been using so far in the ASP.NET section of this book, the DataSet control is not located in the Web Forms section of the Toolbox panel but rather under the Data section, as is shown more clearly in Figure 32.2.

Figure 32.2.
The DataSet control icon is in the Data section of the Toolbox panel

To show the Data section of the Toolbox panel, simply click on the Data tab, and Visual Studio.NET will then display all of the icons for the controls in the Data section. Accompanying the DataSet control in the Data section of the Toolbox panel are other controls related to working with data—some of which you will be using shortly.

Select the DataSet control icon by clicking on the icon. The DataSet control does not have any visible element and therefore you do not have to drag a size for the control on the Web Form, as you do with most of the other controls. Clicking anywhere on the Web Form will suffice to begin the Add Dataset dialog box, as shown in Figure 32.3.

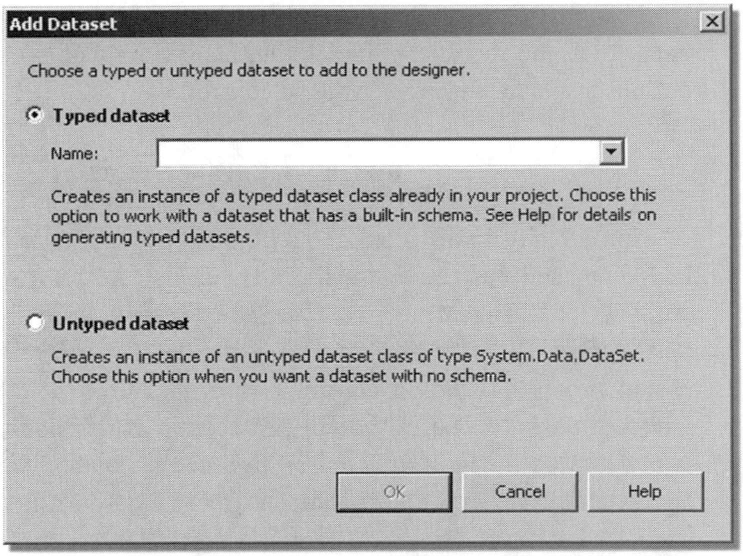

Figure 32.3.
The Add Dataset
dialog box

Add Dataset   ☒

Choose a typed or untyped dataset to add to the designer.

⦿ **Typed dataset**

Name: [▼]

Creates an instance of a typed dataset class already in your project. Choose this
option to work with a dataset that has a built-in schema. See Help for details on
generating typed datasets.

○ **Untyped dataset**

Creates an instance of an untyped dataset class of type System.Data.DataSet.
Choose this option when you want a dataset with no schema.

                              [OK]   [Cancel]   [Help]

The options in the Add Dataset dialog box are for either a typed dataset or
an untyped dataset. A typed dataset is a dataset that already exists in your
project or that will be available to your project at run time. It might also be an
XML schema (*.xsd* file) that you already have available. The untyped dataset is
basically an empty dataset that you populate through code. As this is exactly
what you want to do in this example, select the untyped dataset option and
click on the OK button.

After adding the DataSet control to your Web Form, you have to associate it
with your DataGrid control. You can do this by clicking on your DataGrid control
and then selecting your DataSet control from the drop-down menu for the
DataSource property in the Properties panel, as shown in Figure 32.4.

Figure 32.4.
Associating the
DataSource property
of your DataGrid
control with your
DataSet control

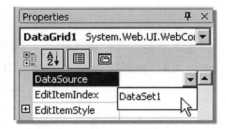

Adding an OleDbConnection Control

To actually make a connection to a database file, such as the Microsoft Access example database *Northwind.mdb*, you must add an OleDbConnection control to your Web Form. Like the DataSet control, the icon for the OleDbConnection control, as shown in Figure 32.5, can be found in the Data section of the Toolbox panel.

Figure 32.5.
The OleDbConnection control icon in the Data section of the Toolbox panel

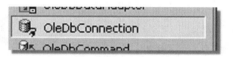

Click on the OleDbConnection control icon to select it and then click anywhere in the Web Form to add the control. Like the DataSet control, the OleDbConnection control has no visible element at run time, so you do not have to worry about dragging any size for the control, and a single click of your mouse will suffice to add it.

Associating a Database with the OleDbConnection Control

After you add the OleDbConnection control to your Web Form, you can associate it with a database to provide a connection between your ASP.NET page and the database.

If it is not already selected, select the OleDbConnection control in your Web Form by clicking on it once. From the Properties panel, select New Connection from the ConnectionString property's drop-down menu. Visual Studio.NET will display the Data Link Properties dialog box with the Connection tab preselected. The default provider in the Data Link Properties dialog box is Microsoft SQL Server and therefore all the options in the Connection tab refer to this type of connection.

Recall that you are trying to connect to the Microsoft Access example database file—*Northwind.mdb*. To change the options in the Connection tab to refer to a Microsoft Access database, you must first correctly set the database provider. Select the Provider tab and from the list of OLE database providers, select the Microsoft Jet 4.0 OLE DB Provider option, as shown in Figure 32.6.

**Figure 32.6.
Selecting the
Microsoft Jet 4.0
OLE DB Provider
option**

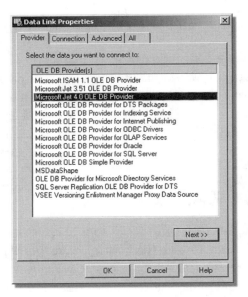

After selecting the Microsoft Jet 4.0 OLE DB Provider option, click on the Next button to go to the Connection tab. You will notice that the options in this tab have changed to reflect options suitable for selecting a Microsoft Access database, rather than a Microsoft SQL Server database.

In the Connection tab, click on the small button with the ellipses located to the right of the database name field. This will allow you to browse to the location of the *Northwind.mdb* database file or any other Microsoft Access database file that you may want to open in other ASP.NET projects. After you have navigated to the *Northwind.mdb* file in the Select Access Database dialog box, click on the Open button. This will add the path to the *Northwind.mdb* database file to your database name field.

The second option available to you in the Connection tab allows you to specify the username and password to use with the database. As you learned earlier in the book, protecting your database files by locking them with a password is a good security action to take. The OleDbConnection control makes it easy to make use of this feature without leaving the password open to prying eyes. You can also select the check box for allowing the password to be saved or not.

After setting all the options in the Connection tab appropriately, click on the Test Connection button to make sure that a connection can be made, as shown in Figure 32.7.

Figure 32.7.
The Connection tab
after filling in the
options and testing
the connection

Data Link Properties

| Provider | Connection | Advanced | All |

Specify the following to connect to Access data:

1. Select or enter a database name:

C:\Inetpub\wwwroot\net\Northwind.mdb

2. Enter information to log on to the database:

User name: Admin

Password:

☑ Blank password ☐ Allow saving password

Microsoft Data Link

ⓘ Test connection succeeded.

OK

Test Connection

OK   Cancel   Help

Setting User Access Permissions

If you want to specify any particular permissions for the type of access to the database, you can make use of the Advanced tab. A list of check boxes next to each of the available permission types allows you to check or uncheck permissions to suit your requirements. For example, if you require that a database be purely read-only when accessed by your ASP.NET application, you could check the Read check box.

Adding an OleDbDataAdapter Control

As you will recall from the previous lesson, to connect a DataSet control with a data connection, you use the OleDbDataAdapter control. The OleDbDataAdapter control icon, as shown in Figure 32.8, can be found in the Data section of the Toolbox panel.

Figure 32.8.
The OleDbDataAdapter control icon in the Data section of the Toolbox panel

The OleDbDataAdapter control is also not visible at run time so simply click once on its icon to select it, and then once on your Web Form to add it to your ASP.NET page. When you add an OleDbDataAdapter control to your Web Form, Visual Studio.NET will display the Data Adapter Configuration Wizard. At the initial welcome screen, simply click on the Next button and the Data Adapter Configuration Wizard will move on to the Choose Your Data Connection dialog box. The data connection will already contain the settings for the Northwind Microsoft Access database because you have set up the connection through your OleDbConnection control earlier, as shown in Figure 32.9.

Figure 32.9.
The Choose Your Data Connection dialog box with the settings for the Northwind database

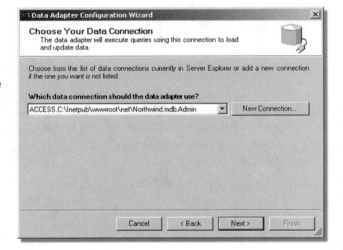

Click on the Next button to continue. The Data Adapter Configuration Wizard will move on to the Choose a Query Type dialog box. You can specify in this dialog box the type of method for querying the database. Select the Use SQL statements option because you are already knowledgeable in constructing SQL queries, and then click on the Next button.

The Data Adapter Configuration Wizard will move on to the Generate the SQL statements dialog box. Within this dialog box, you can manually type in your SQL statement. However, for purposes of learning, you will use the Query Builder wizard to generate your SQL statement instead. Click on the Query Builder . . . button. The Query Builder dialog box will be displayed with the Add Table dialog box appearing directly in front of it. From the Add Table dialog box, under the Tables tab, select the *Customers* table from the list of tables available in the Northwind database, as shown in Figure 32.10.

Figure 32.10. Selecting the *Customers* table from the Add Table dialog box

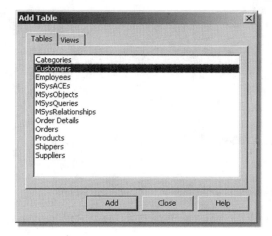

Click on the Add button, and the Customers dialog box will be shown for use with the Query Builder wizard. Click on the Add button, and the Add Table dialog box will close. Make sure that the *Customers* dialog box has a check mark next to the * (All Columns) item, as shown in Figure 32.11.

Figure 32.11. Making sure the * (All Columns) check box has a check mark

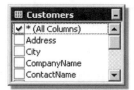

The Query Builder dialog box will now contain the necessary data to generate the SQL query, as shown in Figure 32.12.

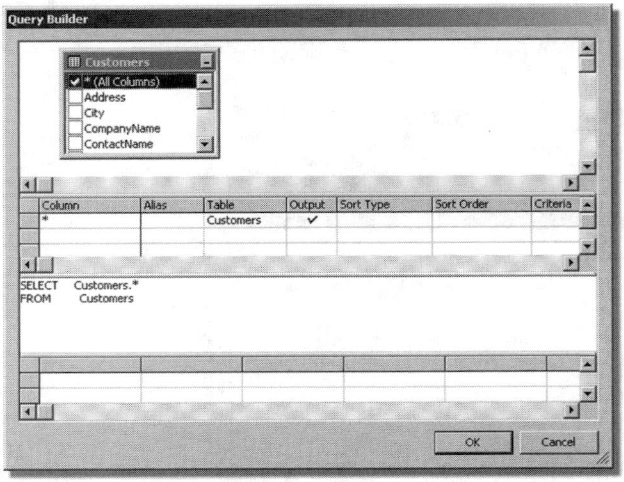

**Figure 32.12.
The Query Builder
dialog box contain-
ing the necessary
data for generating
a SQL query**

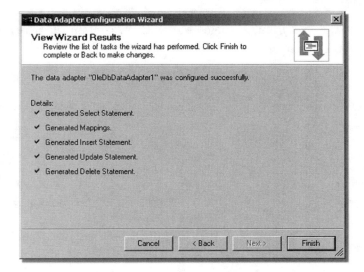

Click on the OK button to close the Query Builder dialog box. You will now
see the generated SQL query displayed in the Generate the SQL statements dialog box. The SQL query should appear as follows:

```
SELECT
     Customers.*
FROM
     Customers
```

Click on the Next button and the Data Adapter Configuration Wizard will
move on to the View Wizard Results dialog box, as shown in Figure 32.13.

**Figure 32.13.
The View Wizard
Results dialog box**

Make sure that your OleDbDataAdapter control was successfully configured, and that there are tick marks beside every point in the Details section, otherwise you will most likely experience problems with your ASP.NET page. Click on the Finish button to close the Data Adapter Configuration Wizard. You have now configured your OleDbDataAdapter control to query your OleDbConnection control for all the columns in the *Customers* table using a SQL statement.

Putting It All Together

To populate the DataSet control, you must return to code and use the Fill method and the DataBind method, as you saw in the previous lesson. The following code shows an example of what to place in your Page_Load event subroutine:

```
OleDbDataAdapter1.Fill(DataSet1)
Page.DataBind()
```

Save your project and then build your new ASP.NET page by using the Build menu Build option. Open the page in your browser, and it should appear similar to Figure 32.14, with a very simple look to it.

**Figure 32.14.
Plain table using
the DataGrid control**

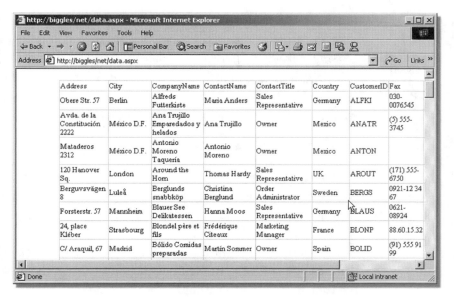

Spicing Up Your DataGrid Control

You can customize the look of your DataGrid control by using some of the many properties it contains that set specific styles. You can specify font settings, background colors, foreground colors, header settings, footer settings, item settings, and others.

To see an example of setting styles to customize the look of your DataGrid control, you will set properties so that the header is in a darker shade background color with a bold font. You will also set every other row to be displayed with a different background color but a lighter color than that used for the header.

Changing Styles Using the Properties Panel

You can change all the style settings for your DataGrid control using the Properties panel in Design view. There are many properties that confusingly appear to be identical, but you will soon discover that they are merely another instance of consistency seen so thoroughly in Visual Studio.NET. For instance, the BackColor property under the AlternatingStyle property is not the same as the top level BackColor property, even though these two distinct and separate properties have the same name.

Note: *If you are using Visual Studio.NET Beta 2, the changes you make to the style properties will only affect the DataGrid control in Design view, and will not actually make any difference when viewed in a browser. This is because the style properties do not actually create corresponding code, as they undoubtedly will in the release version. You can still use the Properties panel for such tasks as discovering the names of colors that you want to assign.*

To change the background color of the header, you must change the BackColor property within the HeaderStyle property. A color, such as Wheat, should work well here because it is strong yet does not make the column headings difficult to read. To make the font bold in the header, you must set the Font-Bold property to True within the HeaderStyle property. You will need an opening and closing tag for the HeaderStyle property. You set the BackColor property and the Font-Bold property within the opening tag of the HeaderStyle property.

To change the background color of every other row, you must change the BackColor property within the AlternatingItemStyle property. A light color, such as Cornsilk, should work well. You will need an opening and closing tag for the

AlternatingItemStyle property. You set the BackColor property within the opening tag of the AlternatingItemStyle property.

Switch to the HTML view of your ASP.NET page and place the following code just prior to the </asp:datagrid> tag:

```
<AlternatingItemStyle
  BackColor="Cornsilk">
</AlternatingItemStyle>
<HeaderStyle
  BackColor="Wheat"
  Font-Bold=True>
</HeaderStyle>
```

Save your work and compile your ASP.NET page again using the Build menu Build option. Loading the page into your browser should show clearly the improved appearance of the DataGrid control, as shown in Figure 32.15, thanks to your customization efforts.

Figure 32.15.
The DataGrid control
with a customized
appearance

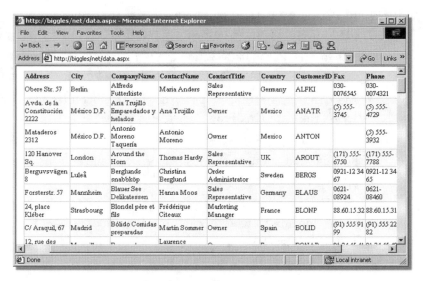

WHAT YOU MUST KNOW

Rather than going to all the effort of having to create HTML tables manually, the DataGrid control can make it easy to display data in your ASP.NET pages. You can use a populated DataSet control to provide a data source for the DataGrid control. A huge amount of customization options make the DataGrid control very attractive to use for both developers and clients. In Lesson 33, "Working with Web Services," you will learn about Web Services in ASP.NET both fundamentally and practically, as you learn how to create your own. Web Services are basically a means of offering an application to clients as a service. Before you move on to Lesson 33, make sure you know the following key concepts concerning working with the DataGrid control and associated controls:

- Located in the Web Forms section of the Toolbox panel, the DataGrid control is visible at run time, displaying data in a grid or table format.

- The DataGrid control does not require sizing in Design view because it automatically resizes itself if necessary to display the given data.

- You can find the DataSet control icon in the Data section of the Toolbox panel to add to your Web Form in Design view, although it is not visible at run time.

- The typed or untyped dataset basically refers to using an existing dataset or creating a new dataset. You use the Add Dataset dialog box to determine which you want to use with your DataSet control.

- To associate a DataSet control with your DataGrid control, you must assign it to the DataGrid control's DataSource property.

- Also located in the Data section of the Toolbox panel, the OleDbConnection control can be used to actually connect to a variety of databases from your ASP.NET page.

- The ConnectionString property of the OleDbConnection control allows you to access the Data Link Properties dialog box.

- The two most significant tabs of the Data Link Properties dialog box are the Provider and Connection tabs, where you can set a provider type and appropriate connection settings, respectively.

- By adding an OleDbDataAdapter control, which is also found in the Data section of the Toolbox panel, you can provide a link between the database connection with the dataset.

- The OleDbDataAdapter control allows you to use the Data Adapter Configuration wizard and the Query Builder wizard to formulate queries to direct to the data source and then pass the results to the DataSet control.

- The DataGrid control gives developers a lot of flexibility to customize the look of the resultant table. Specific properties, such as the HeaderStyle and AlternatingItemStyle properties, contain subproperties for setting the style for that section.

WORKING WITH WEB SERVICES

n Lesson 32, "Working with DataGrids," you saw how to present data to the user by means of the DataGrid control, together with some of the other data controls involved in ADO.NET. In this lesson, you will learn to present additional functionality in the form of a Web Service, as well as learning how you can access that Web Service. By the time you complete this lesson, you will have covered the following key points:

- A Web Service is a text file that allows you to provide functionality to clients, which consume or access the service.
- The file extension of a Web Service file is always *.asmx,* which makes them easy to identify.
- The two key sections to understanding Web Services are creating them and accessing them.
- To create a Web Service, select the ASP.NET Web Service icon when creating a new project, and then go to the code window. You declare the function that you want to make available with <WebMethod()> in front of it.
- You can use the Add Web Reference dialog box to easily add a reference to your Web Service from within your ASP.NET client applications.
- Simply treating your Web Service as a data type and creating a variable of this, lets you make use of any of the exposed functionality provided by the Web Service.
- The Universal Description Discovery and Integration (UDDI) is an initiative for business commerce, which includes a description for publicly searching and sharing Web Services on the Internet.

Introducing Web Services

Web Services in ASP.NET are basically a means of supplying functionality in the form of a service to clients. Although a Web Service file is a text file similar to an ordinary ASP.NET page, *(.aspx)* the file extension is *.asmx* rather than *.aspx,* making the two file types easy to distinguish.

There are two main parts to understanding or working with Web Services. One part is knowing how to create a Web Service and the other part is knowing how to access a Web Service. When you create a Web Service, you are providing functionality to various clients in need of that particular functionality. When you access a Web Service, you are simply making use of functionality that is provided without having to reinvent the wheel by coding everything from scratch.

When you consider the client that will make use of a Web Service, you may automatically think this term only applies to Web applications. However, you can actually utilize the functionality within Web Services from many client types, including other Web Services, standard Windows applications, and console applications. Any client that is going to access a Web Service must be able to communicate properly with the Web Service, by receiving, sending, and processing messages.

Web Services are basically a new standard for working with distributed applications. Rather than being purely Windows based, Web Services have their roots in XML and XSD (XML Schema Definition) to allow them to be accessible from basically any platform.

As their name suggests, Web Services are best utilized in a Web environment. The applications for Web Services are practically limitless. To give you an idea, you could create a Web Service for supplying weather information based on zip code. Other example scenarios are providing a quote of the day or joke of the day, giving out courier or shipping quotes based on zip code or country, and providing thumbnails for graphics. After you are confident of creating and accessing Web Services, you will come up with a myriad of ideas.

Situations where using Web Services is not really a good choice are usually anything that is running on a single, standalone computer. Also, when working on applications that are purely for a LAN, it would be better, performance-wise, to utilize alternatives, such as DCOM (Distributed Component Object Model), TCP (Transmission Control Protocol), and RPC (Remote Procedure Calls). Using Web Services does not benefit you when there are faster and more efficient alternatives.

Creating a New Web Service in Visual Basic.NET

In this section, you will go through all the steps of creating a Web Service. To keep things simple, as there are many points to take into consideration, you will create the example Web Service that Microsoft provides, which does nothing more than return the text *Hello World*. Later, you will create your own Web Service to concentrate more on the programming involved, as well as consolidating more firmly all the pieces that go into creating a Web Service.

From within Visual Studio.NET, select the File menu New submenu Project item to display the New Project dialog box. From the Visual Basic Projects options, make sure that you select the ASP.NET Web Service icon within the Templates window, as shown in Figure 33.1.

Figure 33.1. Select the ASP.NET Web Service icon from the New Project dialog box

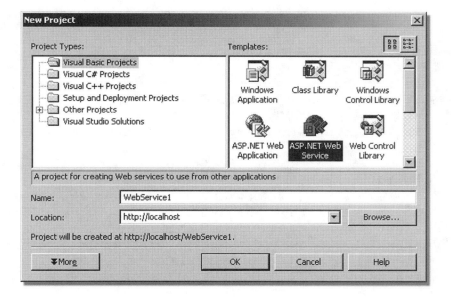

Click on the OK button and the New Project dialog box will close, and Visual Studio.NET will display the initial Web Service window in Design view, as shown in Figure 33.2.

Figure 33.2.
The initial Web
Service window
in Design view

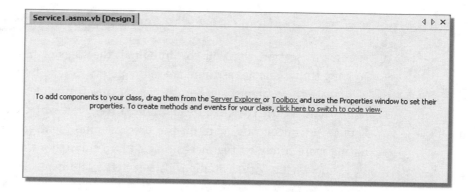

You can click on the link that says *click here to switch to code view.*
Alternatively, you can select the View menu Code item to achieve the same
result. From Code view, you will see the following default code in the window:

```
Imports System.Web.Services
Public Class Service1
    Inherits System.Web.Services.WebService
Web Services Designer Generated Code
    ' WEB SERVICE EXAMPLE
    ' The HelloWorld() example service returns the string
    ' Hello World.
    ' To build, uncomment the following lines then save and
    ' build the project.
    ' To test this web service, ensure that the .asmx file
    ' is the start page and press F5.
    '
    '<WebMethod()> Public Function HelloWorld() As String
    '      HelloWorld = "Hello World"
    ' End Function
End Class
```

As mentioned earlier, to begin with you will simply make use of the
Microsoft default Web Service example. Uncomment the last three lines of the
Service1 class by deleting the apostrophe character. You will note that the color
of the function *HelloWorld* and its code changes from the default comment color.
Press the F5 key to test the Web Service, which simply saves the page and loads
it in a browser for you. The initial browser window should appear similar to
Figure 33.3.

**Figure 33.3.
The initial browser
window after
loading the
Service1.asmx file**

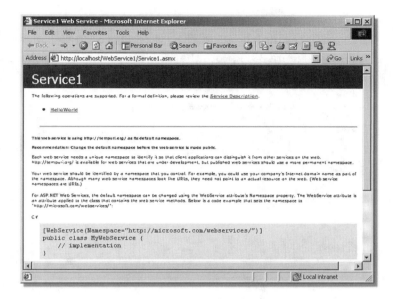

To test the *HelloWorld* function of your Web Service, click on the HelloWorld link at the top of the page. This action will open the HelloWorld test page, within which you must click on the Invoke button, as shown in Figure 33.4.

**Figure 33.4.
Clicking on the
Invoke button in
the HelloWorld
test page**

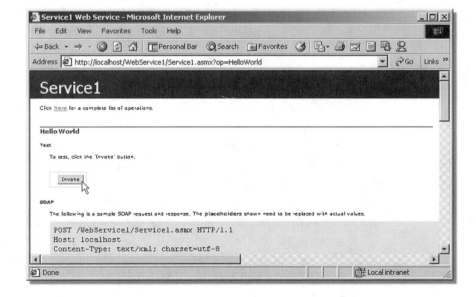

When you click on the Invoke button, another browser window is opened, showing the code that is returned by the *HelloWorld* function, as shown in Figure 33.5. Note that the code is in XML format, which allows it to be highly portable.

In this case, the result of invoking the *HelloWorld* function simply returns the string, *Hello World*.

Understanding Web Service Description Language (WSDL)

When you have created a wonderful Web Service that you want everyone to make use of, you must have a means of describing what functionality it has, including the specific calls. This includes the names of procedures and the arguments that those procedures take. You can write up a simple document that tells others in plain text all of this information. Unfortunately, this will not aid a computer application trying to make use of your Web Service without significant human interaction and assistance.

The solution is to use WSDL (Web Service Description Language) to make the list of functions available to any client that wants to make use of your Web Service automatically. WSDL is basically an XML document. You do not have to create the WSDL file for your Web Services manually, although you can. Visual Studio.NET automatically creates this document for your Web Service. There is also a command line tool that comes with the ASP.NET SDK called *wsdl.exe* that you can make use of if you require.

In the case of the *HelloWorld* Web Service example, you can view the WSDL document by clicking on the Service Description link at the top of the initial browser window. When you click on the Service Description link, the browser displays the WSDL code for the Web Service, as shown in Figure 33.6.

Figure 33.6.
The WSDL code
for the *HelloWorld*
Web Service

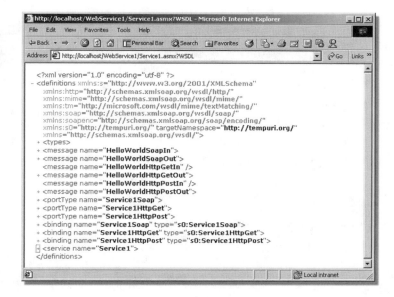

Consuming Web Services

When a client application accesses a Web Service, it is said to *consume* the Web Service. This does not mean that the Web Service is gone completely. Rather, it is more an indication that the client application has taken in the functionality offered by the Web Service for internal use.

To consume a Web Service from a client application, you will start off with a new ASP.NET project. Add a Label control to your Web Form and then double-click anywhere on the background of the form. Visual Studio.NET will present the Page_Load event wrapper code in the Code view window.

You must create a Web reference to the Web Service from within your ASP.NET project before you can reference your Web Service from code. Select the Project menu Add Web Reference item to open up the Add Web Reference dialog box. In the Address field, type in the URL to your Web Service. For example, *http://localhost/webservice1/service1.asmx* should be your URL if you did not change any of the defaults. After you type in the URL, hit the Enter key or click on the green arrow to the right of the Address field. The Web Service information will show up in the left panel and the available references will show in the right, as shown in Figure 33.7.

**Figure 33.7.
The Add Web
Reference dialog box
showing available
references in the
right panel**

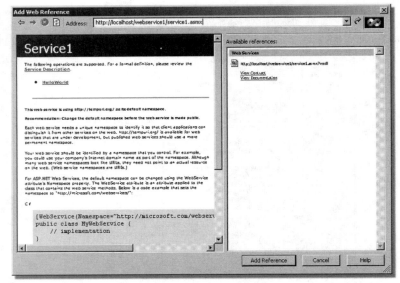

After you add the reference to your Web Service, you can access it from your code. You simply treat the Web Service as if it were a new data type and assign it to a variable, as follows:

```
Dim myWS As New localhost.Service1()
```

The variable of this new Web Service data type can now be utilized by accessing its functions. In this case, the *HelloWorld* Web Service only has the function *HelloWorld,* so you could assign it to your Label control's Text property with the following line of code:

```
Label1.Text = myWS.HelloWorld()
```

Save your project and compile your ASP.NET page by selecting the Build menu Build option. Load your ASP.NET page into a browser, where you will see the text *HelloWorld* shown.

Creating Your Own Web Service

Now that you have seen the overall aspects involved with creating a Web Service, it is time to examine the code involved with creating a Web Service more closely. In this section you will create your own Web Service that contains two functions dealing with strings. One function will simply take two string arguments, concatenate them, and then return the result. The other function will also take two string arguments and will then return a mix of the two arguments, which will be exactly twice the length of the shortest argument string.

Create a new Web Service project and name it *StringService*. Click on the link to take you to the code window, as you did in the *HelloWorld* example. You will now look more closely at the code involved with creating your own Web Service. The first part of your code imports functionality through the System.Web.Services, as shown:

```
Imports System.Web.Services
```

You can then assign the name of your class. If you have not done any work with classes before, simply think of them as a custom type for now. You use the Public keyword to indicate that the scope of the class is open to everyone and follow this by the Class keyword. Following the Class keyword, you can place the name of your class. For instance, in this example you may want to name your class as *StringSvc,* in which case you would use the following line of code:

```
Public Class StringSvc
```

The next part of your code is where you declare the functions that you want to make available. Each function will begin with <WebMethod()> in front of it, after which it is the same as any other function. You can also have arguments with the function, although as you saw in the *HelloWorld* example, this is optional. The basic parts of a function that you want to expose in your Web Service would appear as follows:

```
<WebMethod()> Public Function myFunctionName() As String
    'Your statements go here
End Function
```

For this example, you can delete all of the comment code for *HelloWorld*. Add a function called *Concatenate,* which takes two string arguments and returns the concatenation of the two. Add another function called *Mix,* which also takes two arguments and returns a mixture of the two. The *Len* function returns the length of a string and the *Mid* function returns part of a string. You might like to use these functions to create your own *Mix* function rather than using the one shown here.

The entire code (with the exception of the generated code region) should appear as follows:

```
Imports System.Web.Services
Public Class StringSvc
    Inherits System.Web.Services.WebService
Web Services Designer Generated Code
    <WebMethod()> Public Function Concatenate(_
    ByVal a As String, ByVal b As String) As String
        Return (a & b)
    End Function

    <WebMethod()> Public Function Mix(_
    ByVal a As String, ByVal b As String) As String
        Dim result As String = ""
        Dim i As Integer = 1
        While (i <= Len(a)) And (i <= Len(b))
            result = result & Mid(a, i, 1) & Mid(b, i, 1)
            i = i + 1
        End While
        Return (result)
    End Function
End Class
```

Save your Web Service and create all the necessary files by selecting the Build menu Build item. This completes the creation side of your new *StringService* Web Service.

Create a new ASP.NET project and add two Button controls and three TextBox controls. You can add other optional Label controls if you want to. Change the Text property of the Button controls to *Concatenate* and *Mix*. Add a reference to your Web Service using the URL *http://localhost/stringservice/ stringsvc.asmx*.

Double-click on the Concatenate button and within its Click event wrapper place the following code:

```
Private Sub Button1_Click(ByVal sender As System.Object,
ByVal e As System.EventArgs) Handles Button1.Click
    Dim ss1 As localhost.StringSvc
    TextBox3.Text = _
    ss1.Concatenate(TextBox1.Text, TextBox2.Text)
End Sub
```

Repeat this process for the Mix button, adding the following similar code:

```
Private Sub Button2_Click(ByVal sender As System.Object,
ByVal e As System.EventArgs) Handles Button2.Click
    Dim ss1 As localhost.StringSvc
    TextBox3.Text = _
    ss1.Mix(TextBox1.Text, TextBox2.Text)
End Sub
```

Save your client application as *sstest.aspx* and then compile it using the Build menu Build item. Open your new ASP.NET page in a browser, type some text into the first two text fields, and then try both of the buttons to see the result in the third text field. An example of testing your StringService Web Service from an ASP.NET page in a browser is shown in Figure 33.8.

Figure 33.8.
Testing the
StringService Web
Service from the
***sstest.aspx* page**

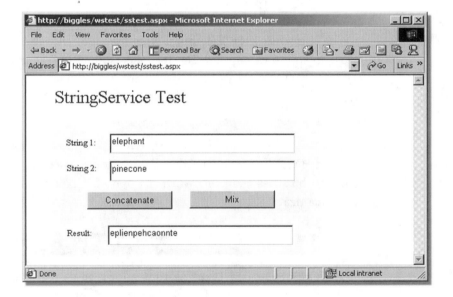

Working with Microsoft UDDI

Microsoft UDDI (Universal Description Discovery and Integration) is an initiative for working with Web Services on a global scale. It is basically a set of standards that describe how to share and discover Web Services. It is the first cross-industry initiative that is primarily aimed at business to business (B2B) commerce and is driven by platform providers, software providers, marketplace operators, and e-Commerce leaders.

When you are using Visual Studio.NET and you open the Add Web Reference dialog box, the initial window portrays links to both the Microsoft UDDI Directory and the Test Microsoft UDDI Directory, as shown in Figure 33.9.

**Figure 33.9.
Microsoft UDDI links
from the initial
Add Web Reference
dialog box**

Web Service Directories:

Microsoft UDDI Directory
Query the UDDI Business Registry to find companies and production Web Services.

Test Microsoft UDDI Directory
Locate test Web Services to use during development.

Web References on Local Web Server

The Microsoft UDDI Directory is where production Web Services can be found for use in your own projects. The Test Microsoft UDDI Directory is where you can test your Web Service while it is still in development. You can find out more about Microsoft UDDI at the following Microsoft Web site:

http://uddi.microsoft.com/help/default.aspx

One final point about UDDI is that it is not purely a Microsoft technology, although Microsoft was involved in its creation, but rather it is run by industry leaders in business commerce with the aim of it being handed over to a standards organization.

WHAT YOU MUST KNOW

A Web Service is a text file that allows you to provide functionality to clients over the Internet. The client applications consume or access the Web Service for internal use. There is much more that you can learn about Web Services, including WSDL, SOAP, and UDDI. However, the aim of this lesson was merely to introduce you to the concept of Web Services in ASP.NET and hopefully whet your appetite for looking into this area further. In Lesson 34, "Developing Custom Controls," you will learn how to create reusable controls of your own that you can then make use of in many of your own ASP.NET applications, or perhaps even offer them to the wider ASP.NET developer world. Before you move on to Lesson 34, make sure you know the following key concepts:

- Web Service files are easily identifiable by their *.asmx* file extension.

- To understand Web Services in general, you only need to know two main points—creating them and accessing them.

- Select the ASP.NET Web Service icon from the New Project dialog box to create a Web Service.

- From the code window, you can create functions that you want to expose through your Web Service by placing them within the main class, with a <WebMethod()> in front of them.

- Before you can access a Web Service from your ASP.NET client applications, you must add a Web reference. The Project menu Add Web Reference item presents a dialog box to simplify this for you.

- You can consume a Web Service by declaring a variable and creating an instance of the Web Service. You can then use simple dot notation to access any of the exposed functionality of the Web Service.

- Primarily an initiative for business commerce, UDDI (Universal Description Discovery and Integration) also includes a description for publicly searching and sharing Web Services on the Internet.

DEVELOPING CUSTOM CONTROLS

*i*n Lesson 33, "Working with Web Services," you learned how to create reusable pages of functionality that can be accessed by client applications over the Web. In this lesson, you will look at a similar concept of creating reusable components, known as controls, which can save considerable time through the functionality that they offer. By the time you complete this lesson, you will have covered the following key concepts:

- Custom controls let you customize the interface or functionality of existing controls or create your own. They are also timesaving devices, since code that is required often is only written once.

- There are different types of custom controls in ASP.NET. User controls and custom Web controls are the two distinct types, with composite controls being a special type of custom Web control.

- User controls are similar to ASP.NET pages, but they have a file extension of *.ascx*.

- A user control also requires a control directive at the top line to indicate to browsers and runtime components that this is a user control and not a standard ASP.NET page.

- When using a user control, an ASP.NET page requires the register directive at the top line to indicate where the containing Web control library is.

- One of the key steps to creating a custom Web control is creating a class that derives from System.Web.UI.WebControls. The other key step is overriding the Render method.

- Use the Add Reference dialog box to add a reference to your custom Web control library.
- Use the Customize Toolbox dialog box to add your custom Web control to the Toolbox panel.

Introducing Custom Controls

Within ASP.NET, you can see that Microsoft has supplied plenty of controls that you can easily drag and drop onto your Web Forms for instant functionality. At first you may feel that this is an overwhelming amount of controls to learn and use, and that there could not possibly be anything else required. However, with time you realize that there are particular types of interfaces that require a lot of setup to get just right, or perhaps just a style that you repeatedly find yourself adding to a Web Form, which may include a number of controls. Perhaps there are controls that lack functionality that you think you could improve on.

By creating your own custom controls, you become the master of your environment and give yourself the freedom to realize your ideas. The practical side of creating a custom control is that it can save a great deal of time for you and perhaps others in your company, after the initial time investment of creating the control. The use of custom controls within a company is often a good means of conforming to styles, as well as adding functionality. For instance, you may have a particular height requirement for all of your progress bars throughout the company's site, which custom controls can help.

There are different types of custom controls in ASP.NET, each of which have similarities. The three main types of custom controls are shown in Table 34.1.

Control Type	Description
User Control	This is basically an ASP.NET page except with an *.ascx* file extension
Custom Web Control	This is a compiled server control for ASP.NET pages
Composite Control	This is a compiled custom Web control that is designed using Web Forms, similar to user controls

Table 34.1. The three main types of custom controls in ASP.NET

Creating a Custom User Control

You can easily create reusable components by utilizing the same techniques that you have already learned concerning Web Forms, except that instead of saving the page as an ASP.NET page *(.aspx)* you will save it as a custom user control

(*.ascx*). The file extension of a user control (*.ascx*) is important not only to differentiate it from normal ASP.NET pages for developers, but also for the ASP.NET runtime environment to realize that it is dealing with a user control. The *.ascx* extension also prevents the user control from accidentally being loaded in a browser on its own, rather than through an ASP.NET page.

User controls are excellent for quickly creating a reusable component, especially one that deals with interface concerns.

To get a better idea of how easy creating custom user controls is, you will create a new project in Visual Studio.NET that simply displays some text in two different Label controls. You will later enhance this user control and add further functionality to it.

Start by selecting the File menu New submenu Project option from within Visual Studio.NET. This will open the New Project dialog box from which you can select the ASP.NET Web Application icon. Change the Name field to *wuc,* which stands for Web user control, before clicking on the OK button.

Add two Label controls to your Web Form—one above the other. Change the Text property in the top Label control to *Login Control*, and change its font to a larger size. Change the Text property of the lower Label control to *A specially designed custom user control*, and change its font to a smaller size. Even though it is very simple, this is all that you want to do in your user control at this stage, but be aware that you could design a more complex interface if you want to.

From the File menu, select Save WebForm1.aspx As . . . and as a result Visual Studio.NET will present the Save File As dialog box. In the Save as type: field, select ASCX Files (*.ascx) from the drop-down menu. Change the name of the file in the File name: field to *wuc.ascx* and then click on the Save button, as shown in Figure 34.1.

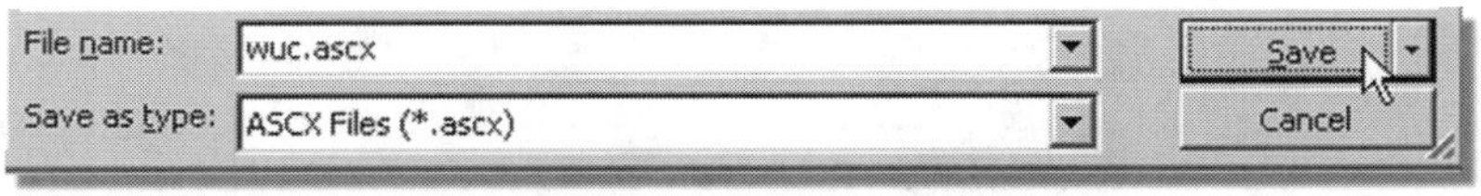

Figure 34.1. Clicking on the Save button after setting the fields correctly

After you have changed your ASP.NET page into a user control, you must also modify some of the HTML code to indicate that this is a user control. Click on the View menu HTML Source item to go from Design view to Code view. At the top of your HTML source, you should have a line that reads something similar to the following:

```
<%@ Page Language="vb" AutoEventWireup="false"
Codebehind="WebForm1.aspx.vb" Inherits="wuc.WebForm1"%>
```

This line is redundant and you can replace it with the following line of code, which is a control directive:

```
<%@control description="ASP.NET custom user control"%>
```

This is an important step to creating a user control. The directive at the top line informs browsers and runtime components that this is a user control, not a standard ASP.NET page.

Testing Your User Control

You can test this simple user control without having to create a new project. Select the project menu Add Web Form item to create a new ASP.NET page. From the File menu select the Save WebForm1.aspx As option. Save the file as *test.aspx*. You now have your new ASP.NET page with nothing in it. The next step is to add your user control.

Select the View menu Solution Explorer item to show the Solution Explorer tab. From within the Solution Explorer tab, scroll down the list of files and components until you come to your user control—*wuc.ascx*, as shown in Figure 34.2.

Figure 34.2. Locating your user control in the Solution Explorer window

After you locate your user control, drag and drop it onto your ASP.NET form in the same manner that you have been using with other controls in previous lessons. Do not worry about sizing the control, because the size and appearance only become apparent at run time. When the user control is added to your ASP.NET page, it will appear as a small gray rectangle, as shown in Figure 34.3.

**Figure 34.3.
After dragging and
dropping your user
control to an
ASP.NET page**

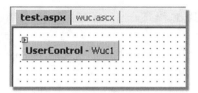

Click on the View menu HTML Source item to view the HTML source. Notice the a register directive has been added for the first line, which will appear as follows:

```
<%@ Register TagPrefix="uc1" TagName="wuc" Src="wuc.ascx"%>
```

This first line is very important and basically registers your user control in the ASP.NET page. Before you can use a user control in any ASP.NET page, it must be registered. If you are manually writing all of the code for your ASP.NET pages and want to use your own user control, be sure to include a register directive. If you are using Visual Studio.NET, the register directive is added for you.

The Visual Studio.NET environment prepares a lot of code for you, but in this particular situation parts of the code must be removed. To help understand this, notice that the code for your test page has a <form> block inside which it inserts your user control. Recall that your user control was using a <form> block in its code; this means you now have two <form> blocks, one within the other, which causes a problem. You must remove the empty <form> block from the test page by deleting both the <form> tag and the </form> tag. If you compile and run the test page without removing the <form> tags, you will get a runtime error.

You have now completed adding your user control to your ASP.NET page. Save your project and compile your ASP.NET page, *test.aspx,* by selecting the Build menu Build option. Load *test.aspx* into a browser and you will see your user control interface, as shown in Figure 34.4.

**Figure 34.4.
The user control
interface as it
appears from an
ASP.NET page in
a browser**

Modifying an Existing User Control

You can easily make changes to an existing user control that you have created. To demonstrate how easy this is, you will modify your existing user control that you created earlier in this lesson and add additional functionality to it. Be as creative as you wish in this section.

Open up your *wuc* solution in Visual Studio.NET and click on the *wuc.ascx* tab. Select the View menu Design (not Designer) item. Add some other interface controls that will give you visual feedback, such as a Panel control, two TextBox controls, and two Button controls. You can use one of the TextBox controls for a username field and one for a password field.

Using a TextBox Control as a Password Field

When using a TextBox control for a password field, you will most likely not want the characters to actually appear when the user types them. By setting the TextBox control's TextMode property to Password, all characters will appear on screen as asterisk characters.

The two Button controls can be used for both a login button and a clear button. You could write code for the Click event of both of these buttons. If you want to seriously create a login user control, you should also consider making use of the various validation controls that you learned about earlier in this book.

After you have completed making your changes to the user control, you can simply save the project and compile the page using the Build menu Build option. Loading *test.aspx* in a browser now, should show the new and improved, modified user control, with its interface changes, as shown in Figure 34.5.

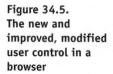

**Figure 34.5.
The new and
improved, modified
user control in a
browser**

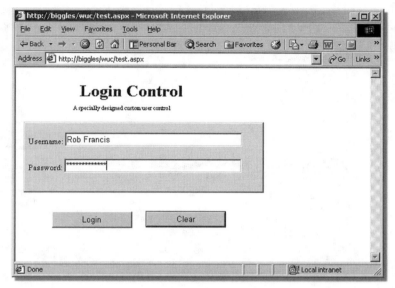

Learning the Fundamentals of Custom Web Control Creation

There are two main steps that you must learn to be able to create your own custom Web controls. Those two steps are creating a class that derives from System.Web.UI.WebControls and overriding the Render method. In this lesson you will only cover the basics of creating a custom Web class, whereas in the next lesson you will look much more closely at the whole concept of class, inheritance, and more advanced custom Web control creation.

The example for this section will create a custom Web control that is based on a Label control. This means that it will have all of the functionality of a Label control plus any additional functionality that you might add.

From Visual Studio.NET, select the File menu New submenu Project item to open the New Project dialog box. From within the Templates section of the New Project dialog box, select the Web Control Library icon so that it is highlighted, as shown in Figure 34.6.

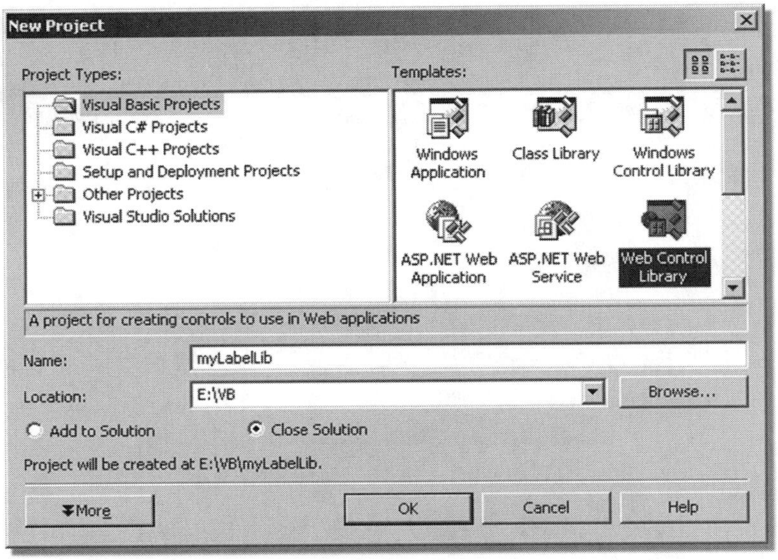

Figure 34.6.
The New Project
dialog box with the
Web Control Library
icon selected

In the Name field type in *myLabelLib*. This is the name of the Web control library that will contain your custom Web control. Click on the OK button to create the new project.

You will see that Visual Studio.NET automatically creates some code for your new Web control library. For now, delete all of this code so that you can create just the necessary code that you require in small pieces. This will help you understand what is necessary.

First, you must import any necessary libraries.

```
Imports System
Imports System.Web
Imports System.Web.UI
```

You then use what is know as a Namespace for your Web control library. This is a unique identifier that you create, such as *LabelNS*. Using the name of the Web control library with the Namespace avoids ambiguity. You can think of the Namespace as a necessary container for holding custom controls. You simply use the keyword Namespace followed by your identifier, as well as a closing line that is always End Namespace, as shown in the following example:

```
Namespace LabelNS

End Namespace
```

Within the Namespace block, you must include your actual class structure. You will want to declare this with the Public keyword to indicate that it has the widest possible scope, allowing it to be accessible from any ASP.NET page. You will also require the Class keyword followed by the name that you want to give your class. This name is basically what the name of your control will be. In many ways that makes it the most important, so do not use a cryptic or meaningless name. For this example, use the name *MyLabel* because you are creating a custom Label control. Finally, to end the class structure, you would follow the End keyword with the Class keyword. The following shows an example code of an empty class structure:

```
Public Class MyLabel

End Class
```

Within your class structure, you must indicate inheritance if there is any. In this example you want to inherit the Label class, so that your custom Web control has all of the features of the Label control. You will learn more about inheritance in the next lesson. For now, simply use the keyword Inherits followed by the Label class, as follows:

```
Inherits System.Web.UI.WebControls.Label
```

Finally, you must override the Render method. Unlike a normal Label control, your custom Web control is always going to output the same text regardless of what value the Text property has. You can accomplish this by setting a fixed string as the argument to the Output.Write method, as follows:

```
Protected Overrides Sub Render _
(ByVal Output As HtmlTextWriter)
    Output.Write ("<h2>Welcome to Total Control!</h2>")
End Sub
```

The entire code for your Web control library will appear as follows:

```
Imports System
Imports System.Web
Imports System.Web.UI

Namespace LabelNS
    Public Class MyLabel
    Inherits System.Web.UI.WebControls.Label
        Protected Overrides Sub Render _
        (ByVal Output As HtmlTextWriter)
            Output.Write _
            ("<h2>Welcome to Total Control!</h2>")
        End Sub
    End Class
End Namespace
```

Save your project and compile your Web control library by selecting the Build menu Build option. Congratulations, you have just created your first custom Web control! Now you will go about the task of learning to incorporate it into your ASP.NET pages.

Adding Your Custom Control to an ASP.NET Page

From the Project menu, select the Add Reference item to open the Add Reference dialog box. Click on the Projects tab. To locate your custom Web control, click on the Browse button to open the Select Component dialog box. Within the Select Component dialog box, navigate to the folder that you saved your custom Web control in and within the bin folder, select the .dll file with the name of your custom Web control. The bin folder's name is short for binary and it is used to store binary files. In this example, the name of the file would be *myLabelLib.dll*. Click on the Open button to return to the Add Reference dialog box. The Add Reference dialog box should now be displaying your custom Web control within its Selected Components: section, similar to Figure 34.7.

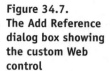

Figure 34.7.
The Add Reference
dialog box showing
the custom Web
control

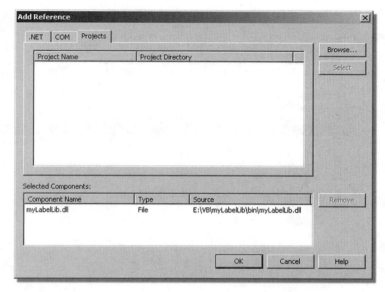

Click on the OK button to close the Add Reference dialog box and actually add the reference to your project. You will now be able to see the name of your custom Web control under the References section of the Solution Explorer panel.

To add your custom Web control to the Toolbox panel, select the Tool menu Customize Toolbox item. As a result of this, Visual Studio.NET will display the Customize Toolbox dialog box. Within the Customize Toolbox dialog box, select the .NET Framework Components tab. Once again, click on the Browse button and navigate to your custom Web control. In this example, the name of the file to select would be *myLabelLib.dll* inside your bin folder. Click on the Open button to return to the Customize Toolbox dialog box. You should now see your custom Web control in the list of components. Here you must make sure that there is a check mark in the check box beside it, as shown in Figure 34.8.

Figure 34.8.
The Customize Toolbox dialog box after browsing to your custom Web control

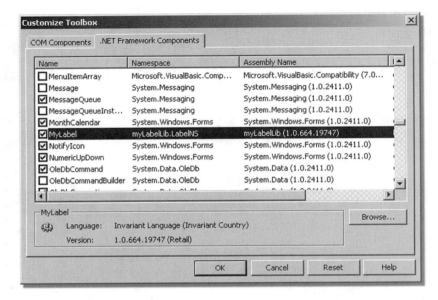

Click on the OK button to close the Customize Toolbox dialog box and add your custom Web control to the Toolbox panel. You can now drag and drop your custom Web control onto the Web Form, which should appear similar to Figure 34.9.

Figure 34.9.
After adding your custom Web control to your Web Form

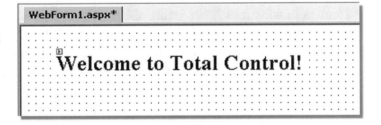

You can test your custom Web control by changing its Text property value. Unlike an ordinary Label control, nothing happens to the control to reflect the change. You can compile your ASP.NET page and view it in a browser, where you will see the same text displayed.

WHAT YOU MUST KNOW

Custom controls allow you to go beyond the limits of what is provided, and let you enhance existing controls or create your own. You can save time by creating reusable custom user controls with particular interface or style settings. In Lesson 35, "Working with Class," you will look much more closely at the whole concept of class, inheritance, and more advanced custom Web control creation. Before you begin Lesson 35, make sure that you understand fully the following key concepts, as the examples will continue from where you left off at the end of this lesson.

- There are three main types of custom controls in ASP.NET. Two of these types are distinct—user controls and custom Web controls. The other type is a composite control, which is a type of custom Web control.

- Similar to ASP.NET pages, user controls are created in the same way, but have a file extension of *.ascx* instead of *.aspx* to indicate that they are a control.

- So as not to be confused with a standard ASP.NET page by browsers and runtime components, a user control also requires a control directive at the top line.

- A register directive at the top line of an ASP.NET page is necessary to indicate where the containing Web control library is. This allows you to use a user control from the library within the page itself.

- You must create a class that derives from System.Web.UI.WebControls when creating a custom Web control.

- To allow your custom Web control any form of output, you must override the Render method.

- To add a reference to your custom Web control library, use the Add Reference dialog box.

- To add your custom Web control to the Toolbox panel, use the Customize Toolbox dialog box.

35 LESSON

Public Class myClass
Private sub mySubroutine()
'statements
End Sub

Public function myFunction() As Integer
'statements
End Function
End Class

WORKING WITH CLASS

n Lesson 34, "Developing Custom Controls," you learned about the basics of creating custom Web controls, and even created your own control based on the Label control. In this lesson, you will carry on with that example, looking at the concept of class and inheritance from the perspective of Visual Basic.NET, and gaining further insights into custom Web controls. By the time you complete this lesson, you will have covered the following key points:

- Classes are basically a special data type that can store a variety of members including events, variables, properties, and methods.
- The Class statement uses the Class keyword and the name of the new class at the start and the End Class keywords at the end. Any code within the class block is considered part of the class.
- You can have multiple classes within a single namespace and within a single library file.
- The Scope type keywords, including Public, Private, Friend, Protected, and Protected Friend, allow you to have greater control over what code is accessible within your class and outside of it.
- The Property statement contains the means for both setting a property's value and retrieving its value.
- The ReadOnly and WriteOnly keywords allow you to place additional limits on your class properties.
- You can add a method to a class by using the standard Sub and Function keywords to write a subroutine or function, respectively.

- You can save considerable time and effort using inheritance with your class construction. Inheritance allows you to reuse the existing functionality of a class and customize or expand its capabilities in the form of a new class.
- Inheritance is transitive, which means that a new class inherits not only the members created in the class that it inherits from, but also any members that its parent inherits, too.
- You cannot remove properties or other members of a class that is inherited. You can expand the list by adding additional members, but you can never diminish the list.
- For any given class that you create, you cannot inherit from more than one class.
- You can prevent a class that you create from being inherited by any other class by using the NotInheritable keyword when you are declaring your class.

Introducing the Concept of Class

As you already have learned, a class is basically a special kind of data type. However, unlike the simple data types, such as String and Integer, a class is capable of containing multiple properties, methods, events, and variables. When you create a class you use the Class statement, which in its simplest form appears as follows:

```
Class myClassName

End Class
```

This example class is empty and does nothing, but is still valid. The Class keyword is followed by the name of your class, which in the example is *myClassName*. You can create your own class names, although it is best to make them meaningful and not too long. Because you can have a single library file that contains multiple classes, this class name will need to be unique from other class names in the same file. To indicate the end of the class block, you simply use the End Class keywords. You can also precede the Class keyword with the Public keyword to indicate that the class is available to external code, as follows:

```
Public Class myClassName

End Class
```

As you know, a class can have different types of members, such as properties and methods. These members also can be preceded by the Public keyword to indicate that the class is available to external code. Table 35.1 shows the variations possible for members of a class, as well as a brief description of each.

Scope Type	Description
Public	Visible internally and externally (default in most cases)
Private	Only visible within the class
Friend	Only visible to others in the same program
Protected	Visible to its own or a derived class only
Protected Friend	A union of the protected and friend visibility levels

Table 35.1. The various scope types within a class

Note: If you declare a variable with the Public keyword, then this will be treated as a property of the class, which may not be the intention. If you want to have working variables that remain hidden within a class, be sure to explicitly use the Private keyword to declare them.

Usually you would be adding properties to a class for use by the client application or ASP.NET page. Within a class block, you can add a property by using the Property statement. A Property statement is Public by default, so you do not have to explicitly state that it is Public. If you ever worked with classes in Visual Basic, you will notice that a class in Visual Basic.NET is quite different. In Visual Basic, a class property had two distinct routines—PropertyLet and PropertyGet. In Visual Basic.NET, a single Property statement contains the means for both setting a property value and retrieving it.

To retrieve a property value, you use a Get block and place any code inside it that you require to retrieve this value. A Get block begins with the Get keyword and ends with End Get. You assign the value to the name of the Property.

To set the value of a property, you use a Set block and place any code inside it that you require to set this value. The value that the client wants to set this property to is given in the argument Value. If you do not explicitly include Value as an argument, Visual Basic.NET will implicitly include Value as an argument.

The Property block is not so complicated when you actually look at it. Also, when you are creating properties in Visual Studio.NET, the structure is created for you automatically. The following example gives you an idea of how to create a Property statement:

```
Public Class myClass
    Private mMaxVal as Integer

    Property Maximum() As Integer
        Get
            Maximum = mMaxVal
        End Get
        Set(ByVal Value)
            mMaxVal = Value
        End Set
    End Property
End Class
```

You may often have properties that you want to be read-only. In these cases you must use the ReadOnly keyword before your Property keyword. When you create a read-only property in your class, you do not use the Set and End Set block of code. An example of creating a read-only property that returns a string is as follows:

```
ReadOnly Property GetBlue()
    Get
        GetBlue = "blue"
    End Get
End Property
```

You can also have write-only properties that the client does not read from. This means that you would require the WriteOnly keyword before the Property keyword. You would also not require the Get and End Get block of code. The following is an example of a write-only property that sets the countdown seconds for a timer:

```
WriteOnly Property SetTimer()
    Set(ByVal Value)
        mSecondsLeft = Value
    End Set
End Property
```

You can add methods to a class using the Sub and Function keywords the same as you do when creating subroutines and functions usually. The only point you might want to be aware of here is that you can use the scope keywords in front of the Sub and Function keywords. For instance, in the following example the subroutine is only for use within the class, but the function is accessible to code external to the class:

```
Public Class myClass
    Private Sub mySubroutine()
        'Statements
    End Sub

    Public Function myFunction() As Integer
        'Statements
    End Function
End Class
```

Although there is quite a lot more to using classes, some of the practices are really only for quite advanced developers. What you have covered in this section should be enough to allow you to feel comfortable working with and creating your own classes for the purpose of using them in ASP.NET. This brings you to the next section of working with classes—inheritance.

Introducing the Concept of Inheritance

To understand inheritance, it is probably easiest for you to look at a class example. Class A has three available properties—Text, Bold, and Italic. You can specify text to display in the Text property and both the Bold and Italic properties take a Boolean value to indicate their state. Although you have seen how to declare properties in a class, to keep things simple throughout this section on inheritance, you will simply see pseudo-code that states the name of the actual property. For instance, the following code would represent Class A and its three properties:

```
Class A
    Text
    Bold
    Italic
End Class
```

Now suppose that you really liked Class A except that you must additionally specify color for the text as well. There may be many reasons why you cannot modify Class A directly. For instance, you may not have the source code. Or perhaps it is required in its current state for some other program. Whatever the reason may be, for now you cannot modify Class A directly.

This is where inheritance comes in. Rather than rewriting the entire code and functionality that already exists in Class A, you just want to add to it. You can do this by using the Inherits keyword with the name of the class whose functionality and properties you want to inherit.

```
Class B
    Inherits A
    Color
End Class
```

When Class B is actually used in your ASP.NET pages, it would be as if all of the properties belonged to it. Class B would appear to the client as follows:

```
Class B
    Text
    Bold
    Italic
    Color
End Class
```

Therefore, by using inheritance together with classes, you can see how easily you can add functionality to an existing class with very little work on your part. This is fundamentally the whole design plan of inheritance, and the same idea exists in other languages, such as C++ and C#, so as a developer it is important to understand this concept.

To take this concept of inheritance one step further, you should also understand that inheritance is transitive. This means that a new class that inherits an existing class, also inherits all the properties and functionality of the class that the existing class was inheriting. For example, earlier you looked at Class B inheriting Class A. If you were to create a new class, Class C, that inherits Class B, it would contain all of the additional properties in Class B, as well as any properties inherited from Class A. If Class C inherits Class B and adds a new property, Size, then it would appear (in simple format) as follows:

```
Class C
    Inherits B
    Size
End Class
```

The structure of Class C would then appear with the following properties, even though Class B only added the Color property:

```
Class C
    Text
    Bold
    Italic
    Color
    Size
End Class
```

The next important point to be aware of with inheritance is that you can add to the properties that a class inherits, but you cannot remove inherited properties. For instance, if Class D inherits Class C but wants to be a simpler class without the Color property, it cannot just remove it. So the following two pieces of code are not equal:

```
Class D
    Inherits C
End Class

Class D
    Text
    Bold
    Italic
    Size
End Class
```

The Color property that Class D inherits from Class C remains. Remember that you can only extend the properties available when you inherit a class and that you can never remove them. With that very important point being made, there are workarounds to this problem (if indeed you perceive this as a problem).

One method is to inherit from an earlier class and add only the desired properties that were added thereafter. For instance, carrying the example forward, if

Class D were to inherit from Class A instead of Class C and then simply add the Size property, the desired result would be reached. So the following would be valid:

```
Class D
    Inherits A
    Size
End Class

Class D
    Text
    Bold
    Italic
    Size
End Class
```

Another method, although time consuming, would be to write Class D without any explicit inheritance. Although an unattractive option, there are times when this is the only feasible way to obtain the desired result.

When creating any new class, you cannot inherit from more than one class. Therefore, you could not create Class D and have it inherit from Class A and Class C. This is a simple point but an important one—any class can only ever inherit from one other class. So the following would not be valid:

```
Class D
    Inherits A
    Inherits C
End Class
```

There is a special class that is known as Object. When a class does not use inheritance explicitly, then Visual Basic.NET will implicitly make its direct base class Object. This means that every single class in Visual Basic.NET can be traced back to the Object class. It also means that every single class in Visual Basic.NET can be converted to the class that it inherits, and therefore eventually to the Object class. An example of a class that implicitly uses the Object class is Class A from this section's examples.

You may at times have a special class that you do not want anyone to be able to simply subclass. You can prevent a class that you create from being inherited by any other class by using the NotInheritable keyword when you are declaring your class, as shown in the following example:

```
NotInheritable Class A
     Text
     Bold
     Italic
End Class
```

One final point on working with inheritance in Visual Basic.NET is that you cannot inherit from containing classes. In theory, this would cause an infinite loop, although in practice it would cause an error. An example of this scenario would be if Class A was to inherit from Class B, when Class B was already inheriting from Class A. So you could never have the following:

```
Class A
     Inherits B
     Text
     Bold
     Italic
End Class

Class B
     Inherits A
     Color
End Class
```

Enhancing Your Custom Web Control

You now know a lot more about working with classes and inheritance. You can use this new knowledge to accomplish more advanced tasks when working with custom Web controls for use in your ASP.NET pages. Returning to your custom Web control from the previous lesson, it currently is not a very practical control. Instead, you will maintain the inheritance of the Label class and will add an additional property, TextRev, which will return the reverse of the text in the Text property.

The entire code for your Web control library should currently appear as follows:

```
Imports System
Imports System.Web
Imports System.Web.UI

Namespace LabelNS
    Public Class MyLabel
    Inherits System.Web.UI.WebControls.Label
        Protected Overrides Sub Render _
    (ByVal Output As HtmlTextWriter)
            Output.Write _
        ("<h2>Welcome to Total Control!</h2>")
        End Sub
    End Class
End Namespace
```

You no longer want to actually change the output and behavior of your custom Label control, so you do not require the override of the Render method. Remove this whole section so that the class section appears as follows:

```
Public Class MyLabel
    Inherits System.Web.UI.WebControls.Label

End Class
```

At this point, your custom Web control behaves exactly like a Label control, with all of the same properties and methods, due to inheritance. You want to extend this functionality by adding a read-only property, TextRev, which will return the reverse of the text in the Text property. You can accomplish this using the ReadOnly keyword before your Property keyword, as you will recall from earlier in this lesson. An easy way to reverse a string is to use the StrReverse method and pass it the Text property as an argument. Any properties that a class inherits can be accessed through code by placing brackets [] around the property name. The final code for your class will now appear as follows:

```
Public Class MyLabel
    Inherits System.Web.UI.WebControls.Label

    ReadOnly Property TextRev()
        Get
            TextRev = StrReverse([Text])
```

```
        End Get
     End Property
End Class
```

Save your project and compile your control from the Build menu Build option. Open your test solution that you created in the previous lesson to test your custom Web control. The following two lines of code should be added to your Page_Load event subroutine:

```
MyLabel1.Text = "Hello World"
MyLabel1.Text = MyLabel1.TextRev
```

When you are adding the second line, you will notice that your read-only custom property, TextRev, is available from code in the dropdown menu, as shown in Figure 35.1.

Figure 35.1.
Your read-only property, TextRev, is available from code in the drop-down menu

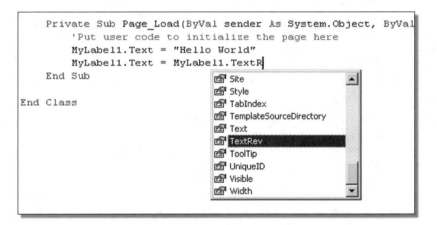

Try changing the Size sub property under the Font property. Set it to Large. This will give you some indication that the inherited properties of the Label class are still functioning. Save your test page and compile it using the Build menu Build option. When you load your ASP.NET test page into a browser, you will see that your custom Web control is working because it will be displaying the text in reverse, as shown in Figure 35.2.

**Figure 35.2.
The result of
loading the test
page in a browser**

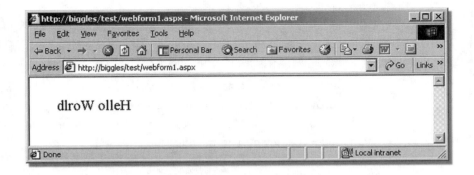

Moving Forward with Active Server Pages

You have now attained a solid level of understanding of both classic Active Server Pages and the newer ASP.NET. This book has hopefully served its purpose in presenting a good overview of Active Server Pages as a whole, showing you many of the important parts of Active Server Pages.

If you find yourself fascinated with any particular areas that you covered in the book, then by all means go and research them more closely. Work with these areas and set yourself goals to achieve. If you do not know how to accomplish a given task, stop and write down methodically on paper all the steps involved to add clarity. It is highly recommended that you purchase books that go into specific detail, such as books that deal solely with ADO.NET or XML. Of course, this depends on where your interests lie.

You can always refer to the Microsoft site for additional documentation, although it tends to be aimed at experienced developers rather than beginners, and there is not always a lot in the way of tutorials, especially for newer technologies. In addition to the Microsoft site, you can make use of the search engines to locate tutorials, references, and much more Active Server Pages information from experienced developers and Active Server Pages enthusiasts.

There are also numerous Active Server Pages forums where you can ask questions. If you are not in a development environment already where you are surrounded by other IT professionals that you can fire questions at when you get stuck, a forum or interactive list is the next best thing because you can ask a specific question and people will answer it for you. After you become more proficient, you may find yourself answering the questions of others!

Look for more VBScript commands that you might find useful at:

http://msdn.microsoft.com/library/en-us/script56/html/vbscripttoc.asp

Obviously this book cannot contain everything about both types of Active Server Pages, and this basically means that there is a lot more out there for you to learn or, better yet, to create. Remember that the skills you are learning are like building blocks—you can use them to create sites and pages using your imagination.

Good luck with your Active Server Pages development!

WHAT YOU MUST KNOW

Classes are basically a special data type that can store a variety of members including variables, properties, methods, and functions. Using inheritance, you can save considerable time and effort by reusing the existing functionality of a class when you want to customize or expand the capabilities of a class. Combining your knowledge of classes and inheritance, you can create custom Web controls for your ASP.NET pages that conform exactly to your requirements. You have now completed the book and will be looking to move on, but before you do, make sure you know the following important concepts that you covered in this lesson:

- The Class statement begins with the Class keyword and ends with the End Class keywords. Within the class block, you place any code for adding or changing members of the class.

- Within a single library file or within a single namespace, you can have multiple classes.

- When dealing with classes, the Scope type keywords include Public, Private, Friend, Protected, and Protected Friend. These Scope type keywords give you better control over the accessibility of code within your class.

- The Property statement uses the Get block to retrieve a property's value and the Set block to set a property's value. The End Property keywords mark the end of the Property block.

- You can use the ReadOnly keyword directly before the Property keyword when you want a property in your class to be read-only. You can do the same with the WriteOnly keyword when you want a property in your class to be write-only.

- By using the standard Sub and Function keywords to write a subroutine or function, you can easily add a method to a class. Using the Private keyword in front will make them only visible within the class.

- When a class inherits another class, it makes use of the existing members of that parent class. This allows you to save considerable time and effort in your class construction, as well as giving you the opportunity to expand the capabilities of the parent class in the form of a new class.

- Inheritance is transitive in classes. This means that a new class inherits all members created in its parent class, as well as any members that its parent class inherits too.

- When you inherit from an existing class, you cannot remove inherited members. You can, however, add additional members.

- A class cannot inherit from multiple classes. Each class can only inherit from a single other class, so long as that class is not a containing class.

- When you have a class that you do not want to allow others to inherit, you can use the NotInheritable keyword prior to the Class keyword.

INDEX